Along the Maysville Road

Along the Maysville Road

The Early American Republic in the
Trans-Appalachian West

Craig Thompson Friend

The University of Tennessee Press • Knoxville

Copyright © 2005 by The University of Tennessee Press / Knoxville.
All Rights Reserved.

Cloth: 1st printing, 2005.
Paper: 1st printing, 2017.

Library of Congress Cataloging-in-Publication Data
Friend, Craig Thompson.
Along the Maysville Road : the early American republic in the
trans-Appalachian West / Craig Thompson Friend.— 1st ed.
    p. cm.
Includes bibliographical references and index.
ISBN 978-1-62190-307-9
1. Roads—Kentucky—Maysville Region—History.
2. Maysville Region (Ky.)—History.
I. Title.
F459.M47F75 2005
976.4'545—dc22 2004010581

To Betty Clagg and Elaine Jenkins

Contents

Acknowledgments	xi
Chronology of Events Related to the Maysville Road	xv
Introduction: The Road and the Republic	1
1. Origins of Community: Pioneers along the Old Buffalo Trace	9
2. Great Settlers: The Gentry and the Rage for Republican Order	59
3. Upsetting the Balance: Crises of Economy and Ecology	103
4. New Americans? Revising Identities and Communities	157
5. Changing Landscapes: The Triumph of the Middle Class	217
Appendix: Tables	283
Notes	291
Index	363

Illustrations

FIGURES

Abraham Drake's House	2
Maysville Turnpike Marker	3
The Old Buffalo Trace	7
The Buffalo Hunt	11
Daniel Boone	14
Simon Kenton	24
Arriving at the Point	47
The Station at Lexington	62
John Breckinridge	73
Robert Patterson's Cabin	75
George Nicholas	78
Charles Julian's Farm Plan	106
John Corlis's House Plan	113
John Wesley Hunt	133
Cane Ridge Meeting House	160
Adam Rankin's House	163
1806 Seal of Transylvania University	177
The Marquis de LaFayette in Maysville	194
Henry Clay's Law Office	195
George Trotter	202
James McCalla's House	212

Transylvania University Main Building	213
Woodcut of John Robert Shaw's Injury	215
Joseph Duncan's Tavern	224
James McKee's House	225
John Johnston's House	226
Thomas Marshall's House	227
James Ellis's Tavern	250
Henry Clay	255
Mayor Charles Erb Wolf's Gravemarker	273
William "King" Solomon	281

MAPS

The World along Alanantowamiowee	16
Group Migrations along the Road	28
The Village West	44
Lexington's Town Grid, 1780	54
Elihu Barker's Map, 1795	57
Victor Collot's Map, 1795	61
The Mercantile Empire of George and Samuel Trotter, 1805	132
Lexington Neighborhoods, 1810s	209
Lexington's Middle-Class Neighborhood	211
Ellis and Darnaby's Survey, 1827	243

Acknowledgments

Historical research and writing is a solitary process. Inspiration, motivation, and creativity have to come from within. But it is not a lonely process. Hours spent in diaries, census records, and collections of letters introduced me to a new crowd of friends—all of whom have been dead for a century and a half! Although they spoke to me, they could not answer me. Other voices filled the silence, encouraging me, teaching me. I must thank in particular Theda Perdue, who welcomed me into the profession as both mentor and friend. While I had many excellent teachers in my life, Theda tirelessly shaped my writing, my research skills, and my analytical thinking. I began to write that she is the most important person in my life beyond my immediate family, but in actuality she and Mike Green are my family as well.

I have benefited greatly from sharing ideas with two other people I highly respect. Matthew Schoenbachler met me as a fellow graduate student. More often than not, we disagreed about history and its lessons, but I think Matt will be surprised at the degree to which he influenced my thinking. I hope that I have enriched his scholarship as much. In 1995, I traveled with Stephen Aron to the Cumberland Gap. He was finishing his book on early Kentucky; I was starting mine. On that journey, Steve shared his perspectives on frontier history and how to produce a manuscript. He has continued to be generous with his time and ideas since that day when we glanced, much as I imagine Daniel Boone once did, through the gap into Kentucky. I thank him for his challenges and suggestions.

I am also fortunate that fellow scholars shared their thoughts on my work and helped to answer the many questions that my subjects could not. I have long admired the works of Joyce Appleby, Andrew R. L. Cayton, and Warren Hofstra, and I am most appreciative to them for critiquing the entire manuscript. Sean Adams, Rose Beiler, Eric Christianson, Mary Wilma Hargraeves, James C. Klotter, Michael Morrison, Tim Silver, and Daniel Blake Smith commented on various parts and drafts. Via the internet, Timothy L. Bratton and Thomas Rightmyer assisted me as I struggled with Julian calendars; and Jon Roland and Steve Gimber enlightened me on the Militia Act of 1792. I thank all of these individuals, and Nancy Rauscher and Christine Persons who cheerfully helped in retyping text and endnotes.

I am indebted to the librarians and archivists who directed me to the sources: Trace Kirkwood, Becky Rice, and James Holmberg at the Filson Historical Society; Mary Margaret Kendall and Molly Kendall at the Mason County Historical Museum; Bill Marshall at the University of Kentucky Special Collections; B. J. Gooch at Transylvania University Special Collections; and the staff at the Kentucky Department for Libraries and Archives. I also thank Mark Wetherington and the Filson Historical Society who have been very supportive of my scholarly pursuits, conferring upon me the first Filson Fellowship in 1997 and allowing me to hone my thoughts on trans-Appalachia through the "Eden of the West" program, cosponsored by the Tennessee Historical Society, the Kentucky Humanities Council, and the National Endowment for the Humanities.

I have always loved American history and was blessed to have parents who planned family trips around Saratoga and Kennedy Space Center, Abraham Lincoln's birthplace and the Tomb of the Unknown Soldier. My sisters, Becky and Dixie, suffered through many of these trips, sometimes very dramatically, but I am thankful for their tolerance of my enthusiasm. I remember my mother and grandmother quizzing me about history after dinners. They can no longer question me, but they still speak to me. My father and stepmother, Gaylord and Billie, continue to support my passion for the past. I love and thank them.

Between 1974 and 1976, I had two extraordinary teachers who deserve recognition. Elaine Jenkins and Betty Clagg taught my social studies classes at William C. Friday Junior High School in Dallas, North Carolina. They encouraged my passion for American history and sparked my imagination in ways that continue to shape my conceptualization of the past. Mrs. Jenkins loved the subject, which made a tremendous impression on me. During my eighth-grade year, our school sponsored a regional history fair, and with her encouragement, I won with a scale replica of historic Williamsburg. The following year, Mrs. Clagg's ninth-grade curriculum was largely American government, and in the atmosphere of the American Revolution bicentennial, she further inspired my interests in and appreciation for history. At the close of ninth grade, when I received the Woodsman of the World award for outstanding history student, a historian was born because two teachers had taken a sincere interest in a student who loved American history. Reflecting on it, more than the many years of graduate preparation, of collegiate teaching, and of research and writing, those years in junior high enkindled my love for early American history and my decision to pursue it as a career. For that reason, it is to Elaine Jenkins and Betty Clagg that this book is dedicated.

Finally, I want to thank my partner, Rod G. Turner. My journey with him began simultaneous to my research for this study. His interest in my work manifested in the photographs found throughout this book. He is my best friend, most sincere critic, and biggest fan. The road I travel with him has been the most rewarding of my life.

Sections of this book were published as earlier drafts in "'Fond Illusions' and Environmental Transformations Along the Maysville–Lexington Road," *Register of the Kentucky Historical Society* 94 (1996): 1–23; "Merchants and Markethouses: Reflections on Moral Economy in Early Kentucky," *Journal of the Early Republic* 17 (1997): 553–74; and "'Work & Be Rich': Economy and Culture on the Bluegrass Farm," in *The Buzzel about Kentuck: Settling the Promised Land,* ed. Craig Thompson Friend (Lexington: University Press of Kentucky, 1999), 125–51. I thank these journals and the press for permission to recycle the materials.

Chronology of Events
Related to the Maysville Road

1775	Pennsylvanians, exploring Kentucky wilderness, name campsite Lexington upon news of violence between Minutemen and British troops.
1776	The Continental Congress adopts the Declaration of Independence. Virginia situates its trans-Appalachian settlements in newly formed Kentucky County.
1777	British encourage Indian allies to wage war on Kentucky settlements.
1778	Kentucky militiamen join George Rogers Clark in attacking northwestern Indian villages and British forts. Pennsylvania emigrants establish Millersburg.
1779	Robert Patterson leads construction of Fort Lexington at southern terminus of road.
1779–80	"Hard Winter" strikes central Kentucky settlements.
1780	Creation of Fayette County.
1782	Shawnees and allies defeat 170 white pioneers at Battle of Blue Licks. George Rogers Clark leads retaliatory forces against northwestern Indian villages.
1783	Treaty of Paris ends American Revolutionary War.
1784	Simon Kenton builds blockhouse at juncture of road and Ohio River to protect new arrivals.
1785	Creation of Bourbon County. William Wood and Arthur Fox lay out the village of Washington. Lawrence Protzman lays out the village of Hopewell. Virginia legislature requires three days of road maintenance per year from males sixteen years and older.
1788	Constitution of the United States ratified. Creation of Mason County. New Jersey migrants settle Mayslick.
1789	George Washington inaugurated as president. John Bradford prints first issue of *Kentucky Gazette*.

1790	Virginia legislature renames Hopewell as Paris. Census indicates 24,055 whites, 32 free blacks, and 4,889 slaves living in counties along the Maysville Road.
1792	Kentuckians ratify first constitution and gain statehood. John Bradford publishes first issue of *Kentucky Almanac*.
1793	Transylvania Academy permanently situated in Lexington. Lexington Democratic Society forms to petition for Mississippi River trading rights.
1793–94	Lexington experiences smallpox epidemic.
1794	Anthony Wayne's army defeats Shawnees at Fallen Timbers.
1795	Treaty of Greenville opens Ohio up to white settlement, eliminates Indian attacks on central Kentucky settlements. Lexingtonians construct first markethouse.
1796	John Adams elected president. Ebenezer Zane clears Zane's Trace through Ohio Territory to connect with old buffalo trace at Limestone. Thomas Paine's *The Age of Reason* becomes a rage in Lexington.
1797	Kentucky legislature permits slaveowners to employ slaves as replacements in road maintenance.
1798	Creation of Fleming County. Kentucky slave code categorizes slaves, free blacks, mulattos, and Indians.
1799	Creation of Nicholas County. Name of Limestone officially changed to Maysville. Second Kentucky constitution written and ratified. Transylvania Academy becomes Transylvania University.
1800	Thomas Jefferson elected president. Census indicates 33,286 whites, 244 free blacks, and 8,430 slaves living in counties along the Maysville Road.
1801	Great Revival held at Cane Ridge.
1802	James Blythe becomes president of Transylvania University.
1803	United States purchases Louisiana from France; New Orleans opened to Mississippi River trade. Kentucky legislature charters Kentucky Insurance Company.
1806	Washington markethouse constructed. Bank of Kentucky incorporated.
1808	James Madison elected president.
1810	Census indicates 41,508 whites, 464 free blacks, and 14,764 slaves living in counties along the Maysville Road.
1811	Humphrey Marshall publishes first *History of Kentucky*.

1812	War erupts between the United States and Great Britain.
1813	Kentucky troops participate in American victory at Battle of the Thames.
1815	Kentucky troops participate in American victory at Battle of New Orleans. *Enterprise,* the first steamboat to travel from New Orleans to Pittsburgh, passes Maysville.
1816	James Monroe elected president.
1817	Kentucky assembly charters Lexington and Maysville Turnpike Road Company, but no action taken to improve road.
1818	Horace Holley becomes president of Transylvania University, recruits outstanding faculty to boost reputation. Kentucky Insurance Company fails. Kentucky legislature charters forty-six independent banks.
1819	Economic panic strikes United States. President James Monroe and General Andrew Jackson visit Lexington.
1819–23	Debt and bank foreclosures lead to Relief Crisis throughout Kentucky.
1820	Census indicates 43,278 whites, 473 free blacks, and 18,503 slaves living in counties along the Maysville Road.
1822	Kentucky legislature revokes Bank of Kentucky charter.
1823–26	Attempts to resolve Relief Crisis lead to Old Court–New Court Controversy.
1824	John Quincy Adams elected president.
1825	The Marquis de LaFayette visits Lexington and Maysville.
1827	Kentucky legislature charters Maysville, Washington, Paris, and Lexington Turnpike Company. William Ellis and James Darnaby survey new turnpike route. New markethouses built in Lexington.
1828	Andrew Jackson elected president.
1829	Work begins on macadamizing new turnpike in Mason and Bourbon Counties.
1830	Jackson vetoes Henry Clay's bill to fund improvement of Maysville Road with federal monies. Census indicates 44,128 whites, 1,010 free blacks, and 23,429 slaves living in counties along the Maysville Road.
1831	Lexingtonians lay cornerstone for railroad from Lexington to Cincinnati.
1833	Cholera sweeps southward along road.
1835	Lexington and Maysville Turnpike completed. Bourbon County exports no tobacco or wheat.

Introduction

The Road and the Republic

Abraham Drake built his farmhouse along an old buffalo trace in Mayslick, Kentucky. Adjacent to every thoroughfare in the Early American Republic were houses like his, primarily private residences that opened as public taverns to paying guests. Their owners lived comfortably, confident in the opportunities created by steady traffic. Today, they are restored homes, handsome buildings painted in crisp whites or colonial blues or reds with neatly trimmed yards. But full restoration is compromised by modern sensibilities. Gone are aesthetically displeasing sights, sounds, and smells of the past: the mud and dust, the faded clapboards, the distorted hand-blown glass windows, the horse posts, the chamber pots and outhouses, the wandering livestock. Also gone are the old dirt roads that ushered guests to their front doors.

The only hint that a historic road passed alongside Drake's farmhouse is a restored cast-iron sign attached to the cleanly painted white rail fence. One of many that workers erected at crossroads and forks as they transformed this old buffalo trace into a turnpike in the 1830s, the marker is today only one of two that remain along the road. It directed travelers to Maysville at the northern end of the trail or to Lexington at the southern end. It also connected them farther: northeastward to Zanesville in Ohio and toward the old colonies; and southwestward to Nashville, on to Florence, Alabama, and toward the American future. Today, the marker is more ornamental than functional, adorning the farmhouse grounds in this tranquil setting. Mayslick is sleepy, and few tourists see the cast-iron road marker. Modern traffic bypasses along U.S. Highway 68.

Abraham Drake House (c. 1792), Mayslick. Photograph by Rod G. Turner.

There was a busier time, however, when pedestrians moved from house to house and village to village, when wagons laden with commercial goods lumbered along the old dirt trace, when carriages and coaches sped people northward or southward. Between 1775 and 1835, tens of thousands traveled this sixty-five-mile thoroughfare through north-central Kentucky. As the first whites wandered into the region in the mid-1770s, the old buffalo trace directed them from the Ohio River into the center of the Bluegrass region. By the mid-1830s, the route became known throughout the nation as the Maysville Road, the centerpiece of Henry Clay's 1830 internal improvements bill that President Andrew Jackson vetoed. In the intermittent six decades, this beaten path was one of the most exceptional roads of the Early American Republic, serving as the spine of the most dynamic region of the American West.

As part of Clay's American System and as the object of Jackson's veto, the Maysville Road is a staple of early American history. Sadly, the most extensive studies of the road to date are Cecil Harp and J. Winston Coleman's "The Old Lexington and Maysville Turnpike" in 1939, R. S. Cotterill's "The Old Maysville Road" in 1916, and Samuel M. Wilson's "The Old Maysville Road" in 1909: all brief articles highlighting efforts to improve the road before and after the veto. Such meager historical inquiry is surprising. After all, we have little doubt of the role that the nineteenth-century's Transportation Revolution—with its thousands of miles of new canals, railroads, and macadamized thoroughfares—played in remaking the physical and cultural landscapes of the United States. Before the National Road, the Erie Canal, and other artificial routes, however, the old buffalo trace and other natural routes were the "avenues to progress."[1]

As individuals traveled natural routes, so too did goods and ideas, spreading American cultures to the new nation's frontiers. The intimate relationship between transportation and Americanization was understood, indeed expected. Until 1803, Kentuckians complained bitterly about their inability to trade freely along the Mississippi River and through New Orleans, demanding federal intervention to open the river and casting the debate in terms of republicanism and national expansion more often than not. Between 1817 and 1830, Kentuckians also yelled loudly about the poor condition of the Maysville Road, demanding federal funds for internal improvement and again portraying the issue in terms of political economy and national destiny. The opening of the Mississippi River and the improvement of the Maysville Road were primarily about economic opportunity, but the commerce and migrations facilitated by better transportation would bring cultural change.

Kentuckians spoke with a far less unified voice when they considered the forms and directions of that cultural development. The Early Republic was diverse with communities of interests that, in their contact with others, sought to protect, negotiate, or contest racial, ethnic, gendered, religious, social, and economic boundaries.

Maysville Turnpike Marker (c. 1830), Mayslick. Photograph by Rod. G. Turner.

Each community was idiosyncratic in that its members shared a worldview and promoted a moral economy to preserve appropriate values. The premise of this book is that, in an era when American culture and society were evolving and in a place where expectations about the extension of the American experiment ran high, the individuals and communities that lived along the Maysville Road were particularly aggressive in contesting each other for communal autonomy and cultural dominance. Only one community—the men of commerce—survived the chaos, harnessed the transformative forces unleashed by the American Revolution, consolidated or marginalized other communities, and gained control over the form and direction of cultural development in this early American West.

More than a narrative of regional internal improvements and national political personalities, then, *Along the Maysville Road* employs the biography of a road—the life of the Maysville Road from its beginnings as a buffalo trace to its role in peopling and transforming an early American West to its decline in regional and national culture—as a microhistory of social and cultural change in the Early American Republic. Integral to this story are the groups and people who traveled and settled along the road. In the first thirty years, three distinct albeit overlapping phases of settlement occurred along the Maysville Road. Pioneers arrived in the 1770s and 1780s. Most were of lower and middling economic circumstances, driven by individual goals of securing good land and ensuring more comfortable futures for themselves and their children. Gentry of more substantial wealth and status showed up in the 1790s, determined to recreate eastern lives of grace and refinement within an established pioneer society. In the same decade, heavy waves of migration brought thousands of poorer and middling farmers into Kentucky, alongside a commercially oriented cohort drawn westward by maturing village economies where artisans' crafts could flourish and merchants could profit by assuaging the gentry's demand for luxuries and farmers' cries for exportation. And, of course, women and slaves arrived with all four groups, their expectations largely uncommunicated, but their presence undisputedly significant to the opening of an American West along the Maysville Road.

These were Americans of great ethnic, regional, and economic diversity. They wanted to create good lives for themselves and their families, and as do all parents, they hoped their children would inherit their mores and worldviews. But they had to create a society in which those values meant something. Along the Maysville Road, they lived in a place distant from the small worlds in which they had originally forged identities. Decisions to relocate to the new West took settlers off the beaten path of American society, forcing them to construct it anew on what they considered to be the blank slate of the western wilds.

None of these people knew, as we do now, that only the men of commerce would survive as a viable community of common interests, although even they had to adjust to changing times and circumstances. Between 1805 and 1835, their commitment to commercial development, material acquisition, and individual achievement prophesied the triumph of liberal economic order throughout nineteenth-century America. None of these people realized either that the democratic experiment initiated by the Revolutionary War would accelerate as commercialism transformed into capitalism, inspiring personal and familial gain through expanded economic opportunities. Finally, none of these people fathomed that alongside that same individualistic impulse would arise increasing pressure to abandon older identities based on regional origins and ethnic backgrounds, to accept a collective historical memory, and to become "Americans." The irony of the story, however, is that even as they constructed a dynamic society along the Maysville Road, these early Kentuckians sought a more monolithic culture, one that discouraged diversity and celebrated the universal qualities of their peculiar brand of "Americanism."

Many people living and working along the Maysville Road looked askant at commercialism, individualism, nationalism, and the emerging cultural matrix. Oddly enough, even though they celebrated the American rift with Great Britain, many remembered fondly the hierarchical, deferential, and localistic patterns of colonial society. And they were not only of the older generations. Having migrated westward, many younger settlers nostalgically replicated their parents' world as closely as possible. Work ethics, gender constructions, familial and communal relations, religious doctrines: so much had been defined in the context of colonial America, and while the Revolution had incited new ways of viewing the world, it eradicated few old ones. Consequently, as they migrated and settled along the Maysville Road, more traditional Americans attempted to construct a society based upon their values, even as those with more modern perspectives worked to create America as they imagined it.[2]

The Maysville Road offers a focal point for the cultural reconfiguration under way in the Early American Republic. Some of the era's most important phenomena reverberated along its length: republicanism, democracy, urban development, refinement, an awakening middle class, evangelical Christianity, racial slavery, and nationalism.[3] Historians have heavily studied these topics, but the story of the road offers examination into the large interwoven patterns of cultural transformation as they manifested in small places. Specifically, how did these forces revise communities and cultural landscapes along the Maysville Road? How did they affect the values, beliefs, and aspirations of these early Americans? How did cultural change manifest in the road itself? And how did this particular road advance and shape an evolving American culture and an emerging American identity?

In seeking answers to these questions, I hope to resituate the Maysville Road in the young nation's history. Certainly, it is not remembered as the most important. The National Road of the late 1810s became, as historic geographer Karl Raitz explained, "a means to consolidate national power and to take control of resource flows, gained through penetration into the country's rich interior," by opening the Old Northwest. And beyond its expansion of market networks, the Erie Canal of the mid-1820s contributed to a more mobile population and "more fluid society of competing impulses, interests, and classes," as historian Carol Sheriff claimed.[4] Yet, the impacts of the artificial routes of the Transportation Revolution were late. Beginning in the 1770s, the Maysville Road served as a stage upon which people wrestled with issues of power, resources, and worldviews as they moved into a new American West.

Yet, there is more to the road and the republic than settlement patterns, economic developments, and cultural transformations. In "Song of the Open Road," Walt Whitman poetized another possibility, more powerful still, for conceptualizing early American roads:

> A foot and light-hearted I take to the open road
> > Healthy and free, the world before me,
> > The long brown path before me leading wherever I choose . . .
> > . . . You road I enter upon and look around, I believe you are not all that is here,
> > I believe that much unseen is also here . . .
> > . . . O public road, I say back I am not afraid to leave you, yet I love you,
> > You express me better than I can express myself,
> > You shall be more to me than my poem.[5]

The road was an adventure in itself. For the individual, it was a search for economic, social, and political stability and advancement. For the community, it was a metaphoric trail connecting the colonial past to the unknown

*The Old Buffalo Trace, Blue Licks Battlefield State Park.
Photograph by Rod G. Turner.*

future, a way through the wilderness where many routes were plausible, a quest for American identity before any identity had been fixed.

Today, a vestige of the old buffalo trace serves as a small length of hiking trail in Kentucky's Blue Licks State Battlefield Park, but the rest is buried beneath paved stretches of U.S. Highway 68 or has been lost to farm plows and horse ranches. There was a time, however, when the Maysville Road was alive with the energy of a young nation on the move. Between 1775 and 1835, the trail and the events along its corridor reflected, indeed were integral to, the American narrative. This is a story of that beaten path and the peoples, worldviews, communities, and cultural forces that shaped an old buffalo trace into one of the Early American Republic's most important and controversial roads.

# 1 • ORIGINS OF COMMUNITY

Pioneers along the Old Buffalo Trace

Situated along an old buffalo trace about five miles north of Fort Lexington, Bryan's Station was a neighborhood of log cabins. Although residents slipped beyond the protective walls to gather food, water, and wood, they lived most of their days inside the stockade. As early land promoter Gilbert Imlay explained, "[I]t was not only prudent to keep close in their forts at times, but it was also necessary to keep their horses and cows up, otherwise the Indians would carry off the horses, and shoot and destroy the cattle."[1]

On 16 August 1782, Indians turned their vengeance on Bryan's Station. During their hunting trips over the previous decade, Shawnees and other northern Indians discovered increasing numbers of animal carcasses in the hunting lands known as "Kentucke." More than the actual killing of the animals, the despoliation of wildlife angered Native Americans. In most cases, white hunters sought only skins, leaving tasty meat and useable sinew and bones to rot. As more whites crossed the Appalachian Mountains, bison, elk, bear, and deer were wasted. In one of their many attempts to purge Kentucke of its white invaders, over three hundred Indians laid siege to Bryan's Station. But over the course of forty-eight hours, the fort withstood attack. Unable to defeat their nemeses, the Indians retreated northward.

An alarm had gone out to militias at other forts and stations, however, and 182 Euramerican men gathered at Bryan's Station to chase down the assailants. The makeshift army marched northward along the old buffalo trace, arriving at Blue Licks on the following day. Located in a horseshoe bend of the Licking River, Blue Licks had long been a major salt spring for bison and deer. The natives knew it well and stationed themselves on the north side of the river, on the ridge of a hill that overlooked the lick. As the pioneers divided into three detachments and advanced up the hill, Shawnees and Wyandots attacked. The Euramericans' right detachment came under heaviest fire, and when their leader fell, they scattered back toward the river. As the center collapsed, the left flank under the command of Daniel Boone fled the battlefield, many swimming to safety across the river. Survivors rapidly returned to Bryan's Station to report the tragedy. Nearly seventy whites died. The Indians stripped and scalped the corpses of those who had fallen.[2]

A ROAD THROUGH THE WILDERNESS

The siege of Bryan's Station and the Battle of Blue Licks were bound to each other by the dirt trail that connected the Indians' world to the encroaching white world. Alanantowamiowee, the Indians' name for the primordial buffalo trace that stretched southward from the Ohio River to the center of Kentucke, originated in the million-years-old collision of land masses that produced wrinkles and cracks in the North American continent. Minor geologic faults materialized beneath the soils of trans-Appalachia, including one marked by shale from the lower strata piercing the upper strata of limestone and topsoils. Because the shale fault line was less fertile and absent of vegetation, it became an accessible path to bison and other animals in search of salt licks and canebrakes. In time, the narrow fault line became one braid in a web of interwoven trails that accommodated herds of great mammals and throngs of prehistoric peoples who relied on bison for sustenance.[3]

By 1000 BC, the road ushered throngs of Native American settlers into the region. Pioneers in a primeval wilderness, they depended upon hunting for sustenance. To stimulate growth of berry-producing shrubs and canebrakes, by which they enticed bison and other game, peoples of the Woodland period (circa 1000 BC to AD 1000) burned out underbrush and engendered a game preserve. In the southern reaches of Kentucky, this practice created expansive barrens. In the northern and central sections, including along the old buffalo trace, it cultivated lush meadows and thick canebrakes that the descendants of Woodland peoples would discover to be prime hunting grounds.[4]

The Hunt. From a mural on the Maysville Flood Wall, Maysville. Photograph by *Craig Thompson Friend*.

Alanantowamiowee was just one of many dirt trails that crisscrossed the trans-Appalachian wilderness, joining native cultures and inspiring environmental transformations in order to produce goods for trade, sustenance, and diplomacy.[5] These connections became increasingly important as the Woodland period faded around AD 1000. Along the Mississippi River, a new culture arose based upon a primarily maize-oriented economy. Even as they continued to hunt, Mississippian peoples abandoned ancestral reliance on large mammal hunting and targeted much smaller game—turtles, fish, wild turkeys, and whitetail deer—to supplement their corn-based diets. Peoples in Kentucky and throughout the Ohio River valley reluctantly abandoned old ways, however. The residual ash of Woodland-era burnings had replenished topsoils and allowed continued dependence on large mammal hunting. Still, as populations grew, hunting was not enough to support the sustenance needs of this Fort Ancients culture. These Indians and their progeny employed shifting-field agriculture, frequently relocating farms and families from exhausted lands to more fertile fields. Their agriculture balanced maize, bean, and squash cultivation; their hunting remained focused on bear, elk, whitetail deer, and bison. While Mississippians created more sedentary societies, the Fort Ancients of the Ohio River valley depended on long-distance hunting parties and temporary fields.[6]

Fort Ancients' reliance on large mammal hunting required maintenance of hunting grounds, and the corridor along Alanantowamiowee provided an ideal location. By the mid-seventeenth century, canebrakes and salt licks attracted bison, deer, and elk in densities perhaps as great as twenty per square kilometer (in contrast to forest lands east of the Appalachians, which supported densities of one per square kilometer). The weight of the herds compacted the trail, and where the bison trampled small trees and underbrush, they created barrens occasionally more than a mile wide.[7]

In the 1640s and 1650s, the powerful Iroquois of present-day New York squeezed native peoples from the Ohio River country in efforts to expand their own arena of economic control. To the east of the old buffalo trace, the Shawnee town of Eskippakithiki lingered, but other villages along Alanantowamiowee rapidly disintegrated. Under the protection of the Iroquois League, the displaced migrated eastward into the Pennsylvania mountains. Kentucke became largely uninhabited for nearly a century until, in the 1730s and 1740s, colonial Pennsylvania's booming Euramerican populations pressured Shawnees and other Indians back into the western country. They recolonized the region, creating defensive alliances, constructing multinational villages, and establishing a new relationship with the environment.[8]

Upon their reclamation of the Ohio River valley, Native Americans rediscovered lands teeming with larger mammals common to their hunting needs. Shawnees, Wyandots, Mingos, Hurons, Obijways, Ottawas, and other eighteenth-century northwestern Indians erected villages north of the Ohio River to access Kentucke's hunting grounds more easily. They followed Alanantowamiowee southward where they found food and clothing, and established trade with southern Indians. The Cherokees also eyed the valuable hunting grounds of Kentucke. In the 1750s and 1760s, conflicts escalated in frequency and violence as the Cherokees followed the Great Warrior Path from their Overhill towns to its intersection with Alanantowamiowee at the Ohio River. Distracted by the Cherokees and convinced of the abundance of Kentucke's wildlife, the Shawnees and their allies showed little concern over the handful of white long hunters who crested the Appalachian Mountains in the early 1770s. As long as whites remained few and attended to rules of the hunt, they were the least of the northwestern Indians' worries.[9]

Yet, Indians learned very quickly that, in stark contrast to their own cultural patterns that inspired efficient use of animals, just a few whites hunting for sport were far too wasteful as they left skinned carcasses decaying near the salt licks. By the mid-1770s, enough colonists of European origin discovered the advantages of trans-Appalachian hunting to alarm native populations. Bison pelts and deerskins became trophies over which both Indians and Euramericans willingly fought, and the timeless wanderings of the great mammals along the ancient trail took on new purpose: to move beyond the range of the white man's gun.[10]

The conflict between Euramericans and Indians, then, was foremost one of ecological differences. Long hunters like Daniel Boone found ways to coexist with natives socially, but Euramericans' cultural conditioning prevented them from fully cooperating environmentally. While Native Americans thanked animals for the sacrifice of life and revered the earth for the productions of the fields, Euramericans assumed an Adamic jurisdiction over wildlife and wilderness through which they could rationalize exploiting nature as divinely appointed custodianship of the creation. When white farmers began arriving in the late 1770s and early 1780s, their propensities to survey tracts, build cabins, and clear fields for farming made the cultural differences even more apparent to Native Americans.[11]

Daniel Boone. Courtesy of the Filson Historical Society, Louisville.

The environmental challenge posed by Euramericans coincided with a plethora of social problems that plagued Indian villages throughout the Ohio River country. Native settlements, especially those with multicultural populations, were fragile entities that required vigilance and compromise to survive. By the 1750s and 1760s, village leaders like those in lower Shawnee Town and Chillicothe actively worked to thwart white merchant-traders and missionaries, denouncing alcohol as destructive to the collective identity of native peoples and condemning Christianity as pernicious to their spiritual values. But other Native Americans enjoyed too much the pleasures of white trade and the afterlife assurances of white religion. Combined with the intrusion upon their hunting grounds, the disharmony of alcoholic revelries and Christian conversions subverted Indian cultures, communities, and security.[12]

Most central to the pressures that Indians faced, however, was what happened beyond their villages, along Alanantowamiowee and throughout the hunting grounds of Kentucke. Between 1774 and 1782, Euramericans erected over one hundred stations and forts which—as bastions of agriculturally minded populations, commercially oriented merchants, conversion-minded Christians, determined hunters, and champions of private property—harbingered a human landscape foreign to Native Americans. Their response was largely offensive and violent. Euramericans and African Americans remembered Indian attacks as an encroachment on their small worlds in the western wilds, not imagining that building a fort or homestead, herding cattle or pigs, and starting a farm were part of a larger collective invasion of the Indians' world. Defending their fields and forests from waves of new settlers who seemed determined to destroy the foundations of Indian economy and culture, Shawnees and their allies took every occasion to kill or capture invaders.[13]

Consequently, by the early 1780s, the ancient trail took on specific purpose as a warpath, directing Indian warriors to American settlements in Kentucke. With their way of life under attack, northwestern Indians allied with British who were engaged in their own war with American colonists. By mid-1780, an Indian-British force captured Ruddle's Station and Martin's Station, both situated east of the old buffalo trace. In 1781, settlers on Beargrass Creek, near the Falls of the Ohio, came under attack. And in March 1782, a band of Wyandots killed residents of Strode's Station and kidnapped several slaves. Antagonisms ran high when the Indians struck Bryan's Station.[14]

The siege of Bryan's Station and the resulting Battle of Blue Licks signaled an important moment in the struggle between Native Americans and Euramerican settlers for control of Alanantowamiowee and the territory it traversed. The Indians, despite their success at Blue Licks, were discouraged by their failure to take Bryan's Station. With the close of the Revolutionary War and the waning of British support, Shawnee and Wyandot leaders less and less set their sights on protecting hunting lands south of the Ohio River, dedicating their resources to preserving villages and territory north

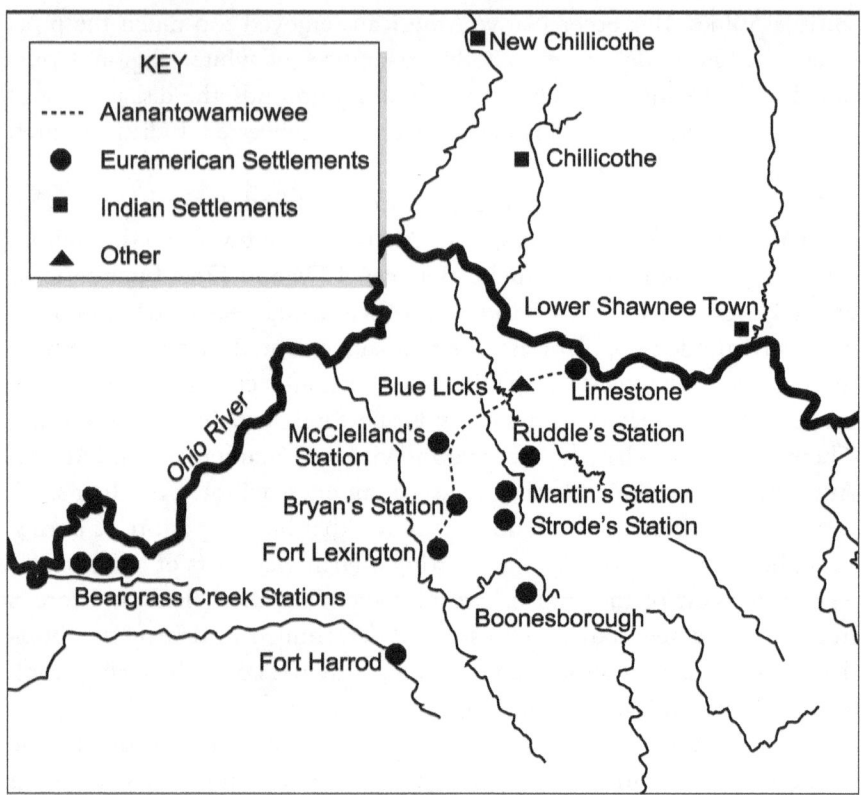

The World along Alanantowamiowee. Map by Mapsmith etc.

of the river. Over the next dozen years, Indians continued to raid white settlements in Kentucke, carried out by small bands in the most remote areas. They largely abandoned Alanantowamiowee as a warpath, just as whites appropriated it as such. The rout at Blue Licks shocked American settlers who called for larger armies to invade Indian lands in the Ohio and Indiana Territories. George Rogers Clark led an immediate retaliatory force against the Shawnee town of New Chillicothe.[15]

Even with the Indians retreating, the psychological toll that a decade of Indian attacks had taken on Euramerican and African American settlers remained. In the early 1790s, settlers Isaac Drake and Richard Ayres rested near Blue Licks as they traveled the old buffalo trace, but "soon after dark were alarmed by the yell of Indians! Unable or unprepared for any effective resistance, they escaped with their blankets into the bushes, and crouched on the ground, leaving their wagons to be pillaged and their horses to be stolen." Few Indians traversed the road in those years, and the two men discovered that they had been spooked by a pack of wolves. But the terror of the fictive Indian was powerful. Early Kentucky historian Mann Butler remembered how, in the 1790s, he "often witnessed the consternation of a whole neighborhood in consequence of a few screeches of owls." Those who had arrived in Kentucke in the 1770s and 1780s, like Drake and Butler, never forgot what the road had been. Imagined gangs of Indians roamed Alanantowamiowee, sparking a widespread rumor in 1792 that the village of Lexington would soon be torched. It was a burden that horrified both first- and second-generation settlers. Isaac Drake's son, Daniel, never forgot his mother's admonition at bedtimes to "lie still and go to sleep, or the Shawnees will catch you." Despite the many names applied to the old buffalo trace by future settlers, it was "the warpath" that remained in the recesses of their psyches, a permanent reminder of the consequences of settling someone else's land.[16]

As the Indians abandoned Alanantowamiowee, Euramericans claimed the region as their own. They anglicized the territory's name to make it more familiar and comfortable; and they began to refer to Alanantowamiowee as the Limestone Road (even though "the old buffalo trace" and "the warpath" remained colloquial favorites for decades). Migrating westward in the 1780s, these new pioneers were certain in their rights to a family size farm, hurriedly building cabins, cultivating fields to mark their lands, submitting their surveys, and putting their names to their piece of Kentucky as symbolic of their ownership.[17]

Stations provide excellent examples of the emerging relationship between property ownership and personal identity. Of the 122 that dotted central Kentucky by the mid-1780s, names of 118 denoted the property holder. McConnell's Station was the first blockhouse at the southern end of the old buffalo trace, and Kenton's Station was the first at the northern end. Some names suggested collective ethnic ownership: Dutch Station, Germany Station, Irish Station, and Scotch Station. Original owners of lands became memorialized in the names of neighborhoods: Mayslick for William May; Miller's Station (later Millersburg) for John Miller; Ellis's Station, eventually Ellisville, for James Ellis. In the 1790s, Kentucky's newly formed assembly renamed the small village of Limestone as Maysville, to honor land speculator John May. As creeks and streams became property boundaries, they too became known by the surnames of land claimants: Lee's Creek, Johnston's Fork, Hinkston Creek, Houston Creek, Lawrence Creek, and Stoner Creek. Even Town Creek, the small rivulet that ran through the village of Lexington, related a communal identity.[18]

It was a predictable yet noteworthy shift in the way humans related to the environment along the road. The earliest Euramericans in trans-Appalachia—long hunters, surveyors, and mapmakers—had depended on Native Americans for directions through the West and consequently preserved Indian words. "Ohio" and "Kentucke" marked the two grandest features of the region: the former possibly derived from the Wyandot word "Oheeza" meaning "beautiful river"; the latter may have been a Wyandot term signifying "land of tomorrow" or an Iroquoian word for "place of meadows." Regardless of their origins and literal translations, both "Ohio" and "Kentucke" represented a cultural perspective intimate with the natural environment. Even in their native English, the earliest Euramerican pioneers derived from a similar natural consciousness the names for Blue Licks, Limestone Creek, Licking River, Stoney Creek, Flat Run, Brushy Creek, Cane Run, Elkhorn Creek, and Cane Ridge. Natural names acknowledged the importance of the wilderness setting to their own identities. Pioneers anticipated savoring a "primaeval liberty" in the West, as land promoter Gilbert Imlay put it.[19]

Primeval liberty was an awkward amalgamation of Jeffersonian idealism and assumptions about wilderness regions. In *The Discovery, Settlement, and Present State of Kentucke* (1784), western promoter John Filson depicted Daniel Boone not only as the prototypical long hunter but as a republican Stoic: a natural man who interacted with the wilderness on its own terms, becoming virtuous as a result of, or more accurately despite, his immersion in an untamed wilderness. Primeval liberty held the promise that even as a rustic, uncivilized, and basically archaic pioneer culture took root, its citizens would embrace the virtuous republicanism at the heart of the American experiment. If all western hunters became republican Stoics, wilderness life offered "an opportunity of contrasting the simple manners and rational life of the Americans . . . with the distorted and unnatural habits of the Europeans," in Imlay's opinion. The ideal of the republican pioneer became common to early American literary interpretations of the West and, through its promotion of virtue and benevolence as checks to self-interest, addressed the concerns of many political theorists. But therein lay the rub: most pioneers would never be republican Stoics; even Boone's reputation tarnished with failed land speculations and questionable leadership at the Battle of Blue Licks.[20]

For many Americans of the 1770s and 1780s who observed the earliest stages of Kentucky's settlement, republican Stoicism was the last label they would have used to describe the emerging pioneer society. An untamed western wilderness symbolized evil and immorality, and the grander process of enlightening and elevating humankind required its conquest. Pioneer life became suspect. Preemption—the act of camping on a piece of land, making some improvements, and then trying to legitimately claim it—was contrary to organized land claiming, and Virginians made it illegal in 1779, the same year the first land office opened in Kentucky County. Subsistence livelihoods resembled too closely the independence of Native Americans: political theorists denounced the isolation and "savagery" of western Americans. Even hunting lost much of its appeal in the East, where most Americans had turned from wild animals to domestic livestock for meat. Hunting's association with Indians branded it a rude pursuit, a savage sport more appropriate for heathens than civilized Americans. Hence, throughout the 1780s, political rhetoric (particularly from the halls of Tidewater and Piedmont Virginia) increasingly emphasized not the retention of Kentucky's natural setting and the celebration of the republican pioneer, but the replacement of the woodlands with an agrarian civility that would save western Americans from the wilderness and transform them into virtuous republican farmers. The difficulty remained in balancing the "dignity nature bestowed upon us at the creation" (a quality that a pioneer hunting society seemingly well embodied) with civility (something it clearly lacked). In many ways, theorists had great expectations of these early pioneers to preserve the best of a previous way of life while simultaneously opening the American future.[21]

Among settlers along the old buffalo trace, there was little handwringing over the political and ideological schemata for westward expansion. By the late 1780s, they quit naming places based on natural attributes not because they suddenly realized that to create a Jeffersonian yeoman republic they had to appropriate and control this new West but because they had removed so much brush, stones, and cane that the names no longer made sense. They turned from living off the bounty of the wilderness to possessing and cultivating the land; again, a practical rather than ideological decision since overemphasis on hunting had greatly depleted the bison, deer, turkey, and elk that sustained a hunting economy. Republican

Stoicism was lost on these settlers who surveyed, mapped, and put their personal stamps on the landscape in order to eradicate the foreboding, malevolent wilderness in which Boone had thrived. Still, their actions gave hope to political theorists back east who thrilled that Americans would harness the western wilds, claim and label the land, plow and harvest it, and tap into its beneficent and regenerative qualities. By the 1790s, such ideological expectations became institutionalized in regional literature. In contrast to almanacs elsewhere in the new nation, Jeffersonian John Bradford's *The Kentucky Almanac* was filled with the term "garden," using it loosely and regularly to inspire farmers to create an agrarian republic.[22]

With the promise of agricultural prosperity drawing thousands westward, the land rush was on. Pioneer homesteads arose in a dispersed and erratic pattern, symbolizing the frenetic grab for property under way. Eighteenth-century travelers failed to find mile after mile of farmsteads lining the Limestone Road and attributed underdevelopment to the overpowering wilderness. Few realized that the forests were a facade hiding thousands of isolated cabins. Even though claims were in proximity to and even overlapped other claims, settlers built homesteads far apart, separated from others by fields and forests, emphasizing family over community.[23]

Like many Americans, Kentucky's early pioneers perceived their world through a traditional worldview rooted in the Christian intellectual atmosphere of Europe's Middle Ages. Situating human life between heaven and hell, they took for granted that the earthly and unearthly worlds abounded in meaning and that balance between the two was central to human happiness.[24]

Believing that an animate nature illuminated the will of God, they connected with the power and majesty of the natural and supernatural by relating to the earth, particularly as farmers. Folklore preserved astrological, medicinal, and magical formulas that joined humans to nature and the divine. Among these traditionalists, nature offered healing and security, a perception rooted in their Judeo-Christian heritage as the contrast between the stable cosmic and moral order of the Edenic garden and the deterioration of order that accompanied human eviction from the garden. In that manner, nature was hallowed, especially where it appeared primeval and pristine. The imagery of a virgin land enticed these Americans because it suggested greater agricultural yields *and* it represented a sacred space where supernatural met natural, where humans could be enlivened by God or, if not cautious, tormented by Satan.[25]

Daniel Drake recalled sitting through an awesome thunderstorm as it rumbled along the old buffalo trace in the late 1780s. "God was present in the storm," he wrote about the family's perception of the tempest. "We might be destroyed; but another and purer emotion blended with our fears—a feeling of reverence converting terror to awe. We were in the midst of a great and sudden visitation of Divine power." The invisible world of the sacred was made visible in the natural world, inspiring reflection and emotional response. (A few lines later, Drake confessed that, as an adult educated into a more "enlightened" worldview, he had come to recognize the family's convictions as mere superstition!)[26]

Place assumed supernatural quality. Springs became healing sites; streams became baptismal fonts. Pioneers mythologized Blue Licks as a place of supernatural sterilization, a holy ground where the blood of heroes desolated the land. In the woods east of the old buffalo trace in Bourbon County, revivalism transformed the forests around a meetinghouse into sacred space where even natural law was seemingly suspended, as when settler Peter Houston witnessed the specter of "negroes flying thro' the air."[27]

As the most familiar way in which early Americans conceptualized the supernatural, Biblical allusion provided descriptions and expectations for the new West. Most people knew the Bible less as text than they did as truths conveyed through hymns, sermons, and worship. One Backcountry farmer compared himself to "Abraham of old" upon deciding to "go and see myself" the grandeur of trans-Appalachia. Equating western migration to the Israelites' exodus, pioneers proclaimed Kentucky "the godly inheritence," "the Land of Milk and Honey," and the "Promised Land." Colonization of British North America had occurred with similar Biblical reference: English colonists and their sponsors envisioned America as an Edenic land of fertility and botanical bounty long before American colonists breached the Appalachians. As late-eighteenth-century Americans severed their political relationship to the mother country, however, they could not so easily undo the shared enthusiasm about the lands of North America. Drawn by a combination of natural abundance and apparent providential destiny, people swarmed into the region. "Absolutely infatuated by something like the old crusading spirit to the holy land," wrote one anonymous commentator, Americans headed westward to find "what the Israelite discoverers said of the promised land" and to experience what their own ancestors had found in previous Edens on the North American continent.[28]

Place was more than where one lived. It provided the stage for human, natural, and supernatural activity. Selfhood was circumscribed not only by one's relationship to others but by the memories and values shaped by the natural setting. Identity, in that sense, meant belonging to a locality, a point drawn in the broadest sense by Daniel Drake about his family's migration to Kentucky in the late 1780s: "We ceased to be Jerseymen, and became Virginians"—classifications that had little to do with political boundaries.[29]

By transforming the wilderness into an agrarian environment, settlers immersed themselves in natural and supernatural spaces that proffered identity. Domestication of fauna and flora empowered them over the creation in ways that hunting and gathering could not. Adam, after all, exhibited purity, innocence, and guardianship over the simplicities of a *garden;* and John Bradford's success at promoting the "garden" ideal in his almanacs indicates widespread receptiveness. Eden-as-garden meant appropriating the wilderness rather than adapting or submitting to it.[30]

In contrast, pioneers interpreted human life as erratic, fragmented, and teetering toward disarray. Society required a moral economy that protected the individual's rights (to barter, to gain social rank, to own the fruits of one's labor, to pursue happiness) and required his or her participation in the community (to lead or follow, to fulfill obligations, to repel external dangers and expel internal threats, to guard against oppression, to place no unreasonable burden on the group, to abide by common behaviors). Failure to meet communal standards meant loss of goodwill. Individuals expected neighbors to be hospitable and cooperative.[31]

But they did not want neighbors to be intrusive. Pioneers did not expect community to be pervasive, which would have made it incongruent with the individualism so integral to personal success and the autonomy so central to the workings of the household. Occasional gatherings like house raisings, road openings, log rollings, corn huskings, militia musters, dances, weddings, and other frolics kept individuals attuned to their moral responsibilities to each other. At such frolics, as Daniel Drake remembered, "the neighbors were notified, rather than invited, for it was an affair of mutual assistance." The gatherings reassured residents of the moral obligations that bound neighbors together, but when they ended, so too did the communal presence, and each pioneer returned to the isolated, autonomous family farm at the center of his or her world.[32]

Without a strong sense of community, pioneers conceptualized order as good citizens maintaining peace and harmony through mutual respect and responsibility, and meting out justice to those who disrupted society. Through pioneers' cultural lens, Indian attacks on isolated homesteads symbolized less the struggle for regional control than crimes that demanded retribution. Without much attention to evidence or due process, settlers responded violently. From the perspective of outsiders, pioneers warranted a label of lawlessness, for they appeared to have little use for sheriffs or structured codes of law. Farmer Henry Duncan remembered how, in the "early times, about Paris, they used to fight with their fists and clubs." Violence and retribution contributed to the stereotype of the uncontrolled pioneer. As one visitor claimed, when a courthouse arose in a neighborhood, lawless men followed the bison and deer farther into the wilderness, wishing "to have nothing to do with justice." Quarrels and brawls resulted from accusations of cheating, theft, or some other crime. The expectation of justice in the face of such deceptions supported a small industry of thugs. "Fellows would come in from the outskirts of the country, determined on having rows," Duncan explained. "There were bullies, who for pay, would espouse your quarrel, and do your fighting for you."[33]

Those fellows may have been seeking more than just a good fight. Risk-taking through personal bravery and physical skill defined masculine identity. In an egalitarian sense, any and all men were engaged in survival, and each could aspire to become a Big Man, the larger-than-life local leader who attained the admiration and loyalty of the group. Through tall tales of courage and deed, citizens embellished his hunting prowess, cunning in land acquisition, courageous action against Indians, and physical dominance in contests. Of course, the biggest man was Daniel Boone, but other men like Robert Patterson of Lexington and Simon Kenton of Limestone also represented the ideal for their neighbors. John Campbell, an early settler of Lexington and an Irishman of "herculean form and strong intellect but rough in his manners though much esteemed," was such a character, enjoying a local reputation along the southern end of the old buffalo trace for defiantly surviving close calls with Native Americans and British. Living out his life in a rustic log cabin, he also withstood the material refinement that many early pioneers thought would emasculate men.[34]

To outsiders, the availability of this rough-and-tumble masculinity to anyone brave enough to grasp it made settlers appear "democratic." Yet, all pioneers were not considered equal and did not share in decision making. While emphasizing individualism through self-interest and self-preservation, pioneer culture also required competition and risk. There

Simon Kenton. Courtesy of the Filson Historical Society, Louisville.

were winners who gained status and power, and there were losers who lost rank and authority in political life. It was a struggle of all against all in which the least impressive conceded, and not always gracefully, until the next opportunity to prove themselves.[35]

Although many Euramerican men arrived in Kentucky and worked to establish their manly reputations alone, they inevitably sought to transplant families and secure landed inheritances for their children. Once in trans-Appalachia, however, family ways became distorted. Having left kinship and neighborhood networks behind, and restricted financially from bringing elders and other relatives on the westward trip, families became more nuclear and autonomous, consequently revising conventional gender roles. In regional mythology, the wife of a pioneer was never ignorant of her capabilities, displaying a pluckiness and industriousness to his aggressiveness and tenacity. Folklore of the early 1800s celebrated the perseverance of the brave women of Bryan's Station who, in 1782, supposedly continued to milk the cows and carry water from nearby springs despite the Indians who surrounded the fort. Lore also held that women occasionally assumed male roles as gun-toting and axe-wielding fighters. During an Indian raid on McConnell's Station, for example, two Eur-american women poured boiling water on their attacker as he tried to crawl through the floorboards. As the isolation and perils of western life fostered more nuclear pioneer families and redefined female gender roles, masculinity became exaggerated as well. If other male kin did not travel westward with the family, fathers carried more of the physical and psychological burdens of providing for and protecting families. Because women and slaves could display courage and combativeness in the heat of Indian war, greater pressure was on men to define and demonstrate manliness in other ways.[36]

The Limestone Road provided multiple venues for white men to prove themselves. Along the old buffalo trace, men hunted bison to demonstrate sporting skills. Along the warpath, men pursued and fought the Indians who threatened their families and homesteads. And as the 1780s proceeded, the trail took on yet another role that reinforced pioneer conceptualizations of masculinity. One of the two required activities levied upon citizens by their government was roadwork (militia duty being the second). Clearing and maintaining the Limestone Road, typically a family or individual effort, symbolized a man's participation in the larger society, his contribution to diminishing the wilderness, and his manliness. He was a citizen, an identity deprived white women and blacks despite their industry. White men alone could claim responsibility for road clearing or militia service, obligations that could be negated only by assuming an equally responsible public function: for example, Michael Fishette received appointment as a courier, releasing him "from Militia duty and exempt from working on the highways."[37]

By the mid-1780s, opportunities to demonstrate and claim manliness through road clearing were, like the competition of Big Man status, available to all white men. In 1785, Virginia's legislature assigned three days of annual road maintenance to all males at least sixteen years of age. Road overseers had to conscript neighbors to clear and level the path with hoes and horse-drawn scrapers. In 1789, the Mason County Court at the northern end of the road ordered public election for road overseers, transferring the competition of frolics to public service. Residents voted by district, usually selecting neighbors of status to supervise the work. After statehood in 1792, Kentucky's legislature continued to demand similar civic attention to the conditions of roads. Laws of the early 1790s made all heads of household living within one-half mile of a road responsible for its clearing and upkeep. Despite the aberration of gender roles that often resulted in the wilderness, roadwork allowed men to serve manly purposes by fulfilling civic and moral obligations. A man was one who provided for his family, cleared the wilderness, prepared for militia service, and opened the road to the future.[38]

MANY SORTS OF PEOPLE

Pioneers who lived along and cleared the Limestone Road between the 1770s and early 1790s had few roots in the anglicized, cavalier world of Virginia. They were peoples of Scots-Irish, Scottish, and Irish ancestry who migrated from the backcountries of Pennsylvania, Maryland, Virginia, and North Carolina. Their Backcountry culture was one of forts and stations, big men and brave women, hunting and farming, competitive frolics and communal roadwork. By the mid-1780s, however, peoples of different cultural backgrounds began to migrate along the old buffalo trace as well: settlers of Welsh heritage from northern New Jersey, Yankees from New England, poorer families of English heritage from the Mid-Atlantic and lower South, African Americans from the Southeast, and immigrants from France, Germany, Ireland, Scotland, and England. Their folkways and expectations challenged Backcountry culture, but they too were pioneers.[39]

As Virginia's western county, Kentucky should have become a replica of the Old Dominion. Certainly, that is the assumption from which most historians have approached the region. But the villages and isolated homesteads scattered across the countryside housed a variety of peoples from many origins who were not interested in re-creating Virginia in the West. Along the Limestone Road, approximately 49 percent of pioneers emigrated from Virginia. A little less than 25 percent came from Pennsylvania. The final 26 percent was a hodgepodge of peoples from other origins. By 1790, populations were multiethnic and multicultural, and the menagerie of folkways contrasted with the conformity imposed by a demanding wilderness less than a decade earlier. Villages became heterogeneous—a majority of residents traced ethnic roots to England, but neighborhoods filled with peoples of African, French, German, Irish, Scottish, and Scots-Irish heritage as well. The countryside was also diverse, sufficiently so that settler Edward Harris remarked in 1797, "we are a mixture of many sorts of people," a point reiterated by Presbyterian minister Adam Rankin, who declared in 1801 that Kentuckians were "as complete a mixture of all nations, as ever met in equal length and breadth, since the first planting of man upon the earth." Many pioneers moved en masse, anxious about Indian attack and determined to maintain familial bonds. Along the old buffalo trace, eighteen families from Carlisle, Pennsylvania, established Millersburg; a party of Backcountry North Carolinians inhabited Bryan's Station; Mayslick was a colony of settlers from New Jersey; and a neighborhood of upcountry Georgians formed in Lexington. As diverse as the region became, people often segregated themselves and their cultures from those who spoke different languages or dialects, who cooked and ate different types of food, and who conceptualized the world differently. Consequently, pioneer groups tried to guard their cultural identities as the regional population increased, trusting no one different from themselves in market, court, or politics.[40]

Before moving westward, peoples in eastern states had already begun to form sectional identities. In the mid-1780s, Thomas Jefferson elaborated:

In the North they are	In the South they are
cool	fiery
sober	voluptuary
laborious	indolent
persevering	unsteady
independent	independent
jealous of their own liberties, and just to those of others	zealous for their own liberties, but trampling on those of others
interested	generous
chicaning	candid
superstitious and hypocritical in their religion	without attachment or pretentions to any religion but that of the heart[41]

Jefferson's taxonomy of North and South is particularly insightful when exploring the manner in which Americans began to spread along the old buffalo trace. Mason and Nicholas Counties, the two northernmost along the road, filled with pioneers from Pennsylvania, Maryland, and New Jersey, and were appealing to New Englanders and New Yorkers as well. The two southern counties—Fayette and Bourbon—were settled by Virginians, North Carolinians, Georgians, and southern Backcountry peoples.[42]

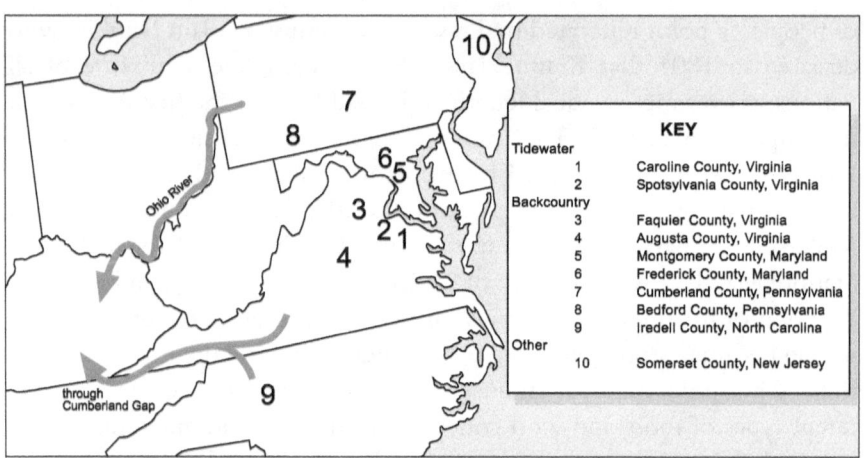

Group Migrations into Kentucky in the 1770s and 1780s. Map by Mapsmith etc.

Distinctions were clear enough that settler John Hodge "could tell where a man was from, on first seeing him." Judging appearances and mannerisms was the easiest way to label neighbors and strangers. "As to religious and moral refinement & a knowledge and use of the domesticants of civilized life," explained Daniel Drake, a New Jersey settler in Mason County, "the Jersey emigrants, as a body, were superior. Next came the Virginians, and last and lowest, the Marylanders." The backwardness of Marylanders along the northern end of the road was particularly disconcerting: "Whenever I am writing of our ignorance, the Maryland element of our population comes to my mind.... They were not only extremely ignorant compared with the Jersey, and most of the Virginia immigrants, in all school learning, but likewise in the domestic arts." John Hill found his Mason County neighbors appalling; most were "Northwardly bred" who "care for little more than a little whiskey, vinison & bread from hand to mouth." The image reinforced Drake's characterization of the Backcountry folk (particularly the Marylanders) among whom he and his family settled as "all country people by birth or residence—all were illiterate, but in various degrees—& all were poor, or in moderate circumstances."[43]

In cultural contrast to Marylanders were New Englanders. One of Drake's schoolteachers was "Kenyon, a Yankee at that time a *rara avis* in Kentucky.... a man of some personal appearance, and, in point of manners, not less than attainments, much superior to [the previous schoolmaster, Jacob] Beaden" who, by the way, was a Marylander and "an ample exponent of the state of society in that benighted region" as he could teach "cyphering as far as the rule of three; beyond which he could not go." During the War of 1812, the image of New England Yankees as sophisticates faded in trans-Appalachia: "The Corse that the yankeys took in opposition to the war," explained Ebenezer Stedman, "Condemned the whole yankey Race." Still, well into the late 1810s, New Englanders continued to hold some standing as commercially and culturally advanced Americans. Transplanted Rhode Islander George Corlis explained to his father in 1816, "these tuckahoes are afraid of the sight of a yankey in money concerns."[44]

As Virginians from the Piedmont and Tidewater, Tuckahoes seemingly embodied the fieriness, independence, and zeal for liberties of which Jefferson wrote. François Michaux commented on how their "passion for gaming and spirituous liquors is carried to excess, which frequently terminates in quarrels degrading to human nature." Yet, among the Tuckahoes there was a distinction to be made between more cultivated gentry and their less cultured neighbors. The Virginian gentry carried an air of erudition unmatched by others. As Drake remembered, "there was in the country, except among wealthy emigrants from Old Virginia (of whom, however, there were none about Mayslick), a great deficiency of books." By 1790, the gentry and their libraries were just beginning to arrive along the Limestone Road.[45]

As pioneers of all backgrounds swarmed into Kentucky, the region along the trail became a field upon which assumptions, suspicions, prejudices, and even animosities between eastern regional cultures extended westward. Just recently concluded in Virginia was a cultural collision between Tuckahoe Old Lights and Backcountry New Lights, and the conspicuous gentility of wealthier gentry families continued to be repugnant to Backcountry settlers who resented and resisted them. The Whiskey Rebellion in Pennsylvania, the Carolina Regulations movements, sides chosen in the Revolutionary War: contests of eastern life manifested in the West through suspicion and animosity.[46]

For the Bryan family, lingering questions about Backcountry loyalties during the Revolutionary War destroyed their trans-Appalachian dreams. In 1779, four Bryan brothers led a contingent of Backcountry North Carolinians to Kentucky, settling along the southern end of the old buffalo trace and building Bryan's Station for protection. But even before the

Indians' 1782 siege, the original family had fled. Details are few, but as neighbor Joseph Ficklin recalled, not soon after their arrival, "the Bryans rested under the imputation of being tories—and all went back to North Carolina."[47]

More typically, cultural friction evolved gradually as pioneers of differing backgrounds settled near each other, as in Mayslick. The Shotwells, Morrises, and three branches of the Drake family arrived in the late 1780s from Plainfield in northern New Jersey where most were of Welsh ancestry and had been members of the Scotch Plains Baptist Church. Upon relocating to Mayslick, they attempted to segregate their community from the Backcountry culture that predominated in Mason County. "Had our visitors . . . been so numerous as to prompt and permit selection," conceded Daniel Drake about the neighborhood in which he grew up, "we should have had no great range of choice." The early ethnic homogeneity of the hamlet meant that strangers stood out. Typically, these were schoolteachers like the Yankee Kenyon or Master McQuitty, a Scot, "but whether he was from the 'land o' cakes,' I can not say."[48]

The New Jerseyites who created Mayslick settled in close proximity to the lick, reflecting a communal ideal different from Backcountry patterns of dispersed cabins and secluded farms. Residences clustered in relative closeness so few families felt detached, making community a more pervasive ingredient in their folk culture. As they settled the fourteen-hundred-acre tract, homesteaders agreed to "live so near each other that no house . . . would be unsupported by some other," building their cabins on the slopes of the hills, not in the creek valley where lands were more fertile and level. Visitors frequently commented on the neighborhood where a brook crossed the old buffalo trace near the lick, and a series of houses strung across the hillsides for mutual protection and support.[49]

As the New Jerseyites privately complained about the customs and habits of those Marylanders, Pennsylvanians, and Virginians whom Drake found so backward, illiterate, and poor, they publicly adapted to and even assimilated many Backcountry folkways. Original residents of Mayslick used frolics—for example, log rollings, corn shuckings, barn raisings—to reinforce the ideal of the good neighborhood. At quilting bees, the men did not segregate themselves to pursue masculine competitions. Rather they "whetted their appetites by drinking whiskey and looking on," and the gathering often ended with "plays of various kinds, interlarded with jokes and bursts of laughter." Backcountry frolickers, in contrast, used community gatherings as moments of rare sociability and contests of risk and individualism. In time, Backcountry ways overran the community-reinforcing frolics found among the New Jerseyites. Horse races, cockfights, wrestling matches, eye-gouging brawls, and other contests of strength and superiority supplemented and eventually replaced Mayslick's less competitive frolics.[50]

Religious culture underwent similar revision. Backcountry religious services, particularly those of the Methodists, were typically field meetings with the same sense of frivolity, revelry, and opportunity for masculine aggressiveness. "Their camp-meetings in the woods," recalled Drake, "presented scenes of fanatical raving among the worshipers, and of levity and vice among the young men who hung about the camp." Those Backcountry ways similarly seeped into the New Jerseyites' religious culture. The little Baptist cabin-church where the Drakes worshiped adopted an open-air meeting style: the congregation gathered outside the building so that the singing of "Old Hundred" mingled with "the notes of birds in the surrounding trees" before Deacon Morris presented a sermon on election, reprobation, or predestination.[51]

The intrusion of these new folkways into Mayslick's frolics and religious services "excited disgust" from the New Jerseyites, including Daniel's mother who appeared "by nature and religious education, a non-combatant" and father who issued his "strongest condemnation." Even Daniel rejected the competition of Backcountry frolics as "the mother of cheating, falsehood and broils." Original settlers of Mayslick had understood that their transplanting of New Jersey culture would not go unadulterated; recall Drake's realization that "they had ceased to be Jerseymen, and became Virginians." Additionally, the wilderness begat "an exaggerated idea of the distance from place to place" that weakened the cultural integrity of any group. But the ways of the Backcountry settlers remained strange and unappreciated by people who had certain expectations about the good life and the proper ways to live it.[52]

Despite condemnations of Backcountry culture, the Drakes succumbed to the cultural blending under way. "My first business in the morning was to pull, and husk and silk enough for breakfast," remembered Daniel, "and, eaten with new milk, what breakfast could be more delicious?" Corn mush, previously a minor fare in the family diet, became a standard dish even as his mother made a conscious effort to retain New Jersey foodways, especially in the form of desserts. Pumpkin pies and mince pies: "A Jersey housekeeper could never neglect or forget"; and "dough-nuts" and "wonders" resembled the "pop-robbins" of the East, except they were boiled in fat rather than in milk.[53]

Even the family's notion of masculinity became compromised. Drake recalled that "it was held by the whole neighborhood to be quite too 'gaal-ish' for a boy to milk; and mother, quite as much as myself, would have been mortified, if any neighboring boy or man had caught me at it." But since he *did* milk the cows, the New Jerseyites' restrictions on masculinity did not prove strong enough to deter him. The demands of wilderness life and the influence of Backcountry culture had begun to reform the New Jerseyites' folkways.[54]

Other groups also interacted, condemned, and in many cases, assimilated to the culture of the Backcountry pioneers. John Johnston was among the first Yankees to settle along the old buffalo trace, relocating from Salisbury, Connecticut, to Washington in 1785. Before the turn of the century, most Yankee immigrants settled in Mason County at the northern end of the road. In the 1810s, Ebenezer Stedman and the Corlis family led a new wave of Yankee migration into the southern counties. The degree to which their backgrounds dictated identity in early Kentucky remains difficult to discern, although there was a general consensus that business acumen defined Yankees as different. Certainly, Yankee culture and Backcountry culture were incompatible. Backcountry peoples celebrated Christmas with bonfires and gunplay. Among New England Congregationalists, the day demanded solemnity and worship. In Washington, at five o'clock on Christmas morning in 1817, tourist James Flint awoke to gunshots and complained of the revelry that he came upon in the streets. Arriving in Maysville later that afternoon, he found similar festivities under way; "Every sort of labour without doors was suspended." Similarly, Lexingtonian Susan Yandell was shocked that "on Christmas eve, while we were at the Episcopal church, & Mr. Chapman in the middle of his discourse, some unprincipled wretches broke a pane of glass, & fired a gun immediately under the windows.... And the next day (sunday,) there was still a repeated firing of guns." Still, Yankees interacted and assimilated. As Drake's memoirs indicated, they were welcomed as teachers. Backcountry folk and others appreciated the education that such men brought into the communities. Yankees like John Johnston adapted to their new neighbors, establishing business ties and personal relationships. Many found that by maintaining identities as Yankees they garnered some respect.[55]

Migrants of German heritage were also conscious of their folkways and the emerging cultural patchwork that threatened to absorb them. John Chambers arrived in Washington speaking English corrupted by the Low Dutch of New Jersey Germans. Lawrence Protzman, also from New Jersey, founded Paris where the local burying ground was known as the Dutch Graveyard. John Seitz early became a prominent Lexington merchant. By the turn of the nineteenth century, a community of German and German American residents formed in the southwestern quadrant of Lexington. Among them were John Peter Schatzell (merchant), John and Jacob Springle (bricklayers), Abram Howe (wagonmaker), William Clarke (hatter), Melchoir Myer (butcher), John Shedil (baker), Francis Krinkle (tanner), and Joseph Hostetter (butcher). German American settlers far more successfully preserved the integrity of their folkways than did the Welsh of Mayslick. In 1789, for example, a group of Dunkards—German American pietists from Backcountry North Carolina and Germantown, Pennsylvania—formed a community and congregation just east of the old buffalo trace in Mason County where they continued their ritual of triple immersion baptisms. In 1795, Bourbon County resident Peter Smelzer appealed to his wife in his will to raise their children "in a Christian manner and give them such learning as may be done amongst us Dutch people." And in 1800, a German Lutheran church arose in Lexington. Part of the reason for their success in preserving cultural patterns was that Backcountry peoples avoided social interaction with German-speaking individuals. Joseph Ficklin, a Backcountry migrant, recalled how it "was the amusement of the wits to make fun of the Dutch."[56]

As evidenced by Lexington's German neighborhood, villages along the Limestone Road appealed to European immigrants. John Brand of Scotland became a successful rope and bagging manufacturer in Lexington. John Maxwell (an original settler of Lexington), Alexander McGregor (merchant), and Hugh Crawford (bluedyer) organized the St. Andrew's Society to assist fellow Scots immigrating to the Bluegrass. Irishman Cornelius Coyle and nephew John Coyle established a tailor shop. Other Irish immigrants included James Campbell (merchant), Joseph Charless (bookseller), James Weir (merchant and hemp manufacturer), Robert and Alexander Fraser (watchmakers), and William West (merchant). English settlers included Jeremiah Neave (a Quaker merchant and "zealous Democrat"), John Wiggleworth (merchant), Robert Wilson (cabinetmaker), John Smith (rope and bagging manufacturer), and Harry Toulmin (farmer). Waldemarde Mentelle fled his native France for Lexington in 1792 to escape the Terror and found work first as a house painter and later as a teacher of French. In 1805, his wife Charlotte capitalized on the stereotype of French refinement and opened Mentelle's for Young Ladies, providing instruction in French, dancing, literature, and social etiquette. Other French immigrants included Henri Terasse (coffeehouse owner), Mathurin Giron (confectioner), and John Delisle and Loftus Noel (metalworkers). Louis Vimont made a reputation in Paris as a leading merchant. John Savary, a Millersburg merchant, became a successful exporter. The arrival of French immigrants in the early 1790s coincided with Jeffersonians' enthusiasm for the French Revolution, inspiring a new wave of fashion among younger clerks and students who wore "coats waistcoats and pantaloons fitting to the skin."[57]

One of the most prominent European immigrants was Englishman Harry Toulmin, whose writings offer perspective into the purposes behind migration. Republican denunciations of the debasement of European society, including a rejection of monarchy, hereditary aristocracy, and established religion, echoed in Toulmin's writings. The American West liberated "Ye poor Europeans, ye, who sweat, and work for the great: ye, who are obliged to give so many sheaves to the church, so many to your lords, so many to your government, and have hardly any left for yourselves,—ye, who are held in less estimation than favourite hunters or useless lapdogs,—ye, who only breathe the air of nature, because it cannot be withholden from you; it is here that ye can conceive the possibility of those feelings I have been describing; it is here that the laws of naturalization invite every one to partake of our great labours and felicity, to till unrented, untaxed, lands!" Trans-Appalachian emigration was figuratively and literally a breath of fresh air. "An European when he first arrives, seems limited in his intentions as well as in his views; but here suddenly alters his scale: two hundred miles formerly appeared a very great distance, it is now but a trifle: he no sooner breathes our air than he forms schemes, and embarks in designs he never would have thought of in his own country."[58]

European idealism contrasted greatly with the expectations of African American migrants. Most had little choice in moving to the West, little ideological interest in it, and little hope in breathing that fresh air. Spurred on by masters eager to secure a prime tract of real estate, the earliest black pioneers lived in the forts and stations, laboring for land surveyors and clearing a wilderness where a slave's life was considered far more expendable than that of a Euramerican. Through the reward system that characterized Backcountry definitions of masculinity, male slaves who demonstrated character and courage in the face of Indian threat sometimes received emancipation, and, to that extent, the pioneer culture of risk and competition applied as well to African American manliness. Given that blacks had far more experience at clearing and planting than many whites, those who remained enslaved often assumed leadership roles in the fields.[59]

Nevertheless, their race prevented full membership in pioneer society, and their isolation from other African Americans hindered attempts to create community. If an entire plantation did not uproot and travel westward, slaves faced a loss of kinship networks and the consequential emotional and psychological support that was important to people who had limited individual choice in their lives. "Tomorrow the negroes are to get off," explained a correspondent of the migration of Robert and Ann Harrison's household from Virginia in 1804, "and I expect there will be great crying and mo[u]rning, children Leaving their mothers, mothers there children, and women there husbands, although but for a year; for the ensuing Faul I suppose whoever lives to see it both black & white will Leave this State." For many blacks like the Harrison slaves, the move to Kentucky did not instill idealism and optimism. Instead, migration further agitated already disrupted lives.[60]

Still, if we can believe his master, at least one enslaved African American sensed possibilities in Kentucky. In 1790, Samuel Ayres's slave, Godfree, apparently sent instructions to the Virginia plantation from which the two men had migrated to "inform his Fellow servants that he is much pleas'd with the Countrey and thinks this is a much better part than Wheir he came from for the black people, provided thay have Good Masters, and if not they are no worse off than they were there." Ayres probably wrote the letter to persuade his other slaves not to be frightened by the move, but possibly, Godfree did find slavery more relaxed in Kentucky.[61]

If so, it was because most white pioneers gravitated towards family farms, dabbling little in crops that required large labor pools. Their needs for slaves were seasonal at best. In the nascent businesses of the villages, in contrast, a demand for labor was constant. "The want of hands excites the industry of the inhabitants of this country," remarked François Michaux. Black pioneers took on positions in shops and smithies because whites interpreted such jobs as subordinate and weakening individual independence. Some slaveowners hired out their chattel, occasionally sharing profit with the slave. For hired-outs and free blacks, positions as laborers, skilled artisans, or domestic servants were always available. In Paris, Frank Bird, who billed himself as "formerly owned by the great and good Washington," served as hostler at an inn. If Bird had been George Washington's slave, but even if he had not been, his presence along the beaten path testified to some blacks' anticipation of a West from which opportunity beckoned. Godfree, if a hired-out slave, may have found some hope in that.[62]

Certainly, Peter Durrett did. In 1784, he and his wife settled near the headwaters of Boone's Creek east of Lexington. Like other pioneers who relocated along the old buffalo trace, they harbored deep anxieties about the move, but as slaves they had little say in the matter. Baptist minister Joseph Craig had insisted on keeping husband and wife together as he moved his household to Kentucky and bartered successfully for Peter's indenture. With Craig's consent, Durrett and his wife hired out in Lexington where they established a home on the lands of a friend and one of the town's founding fathers, John Maxwell. Durrett soon made a name for himself as "Old Captain," the leader of African American spiritual life in Lexington. At a meetinghouse on a local farm, he preached regularly to both black and white Baptists and Methodists. In the mid-1790s, he applied to the South Kentucky Baptist Association for ordination, a request they refused to honor because of his slave status. To ordain Durrett meant to challenge the racial structure and the recent legitimization of slavery in the state's 1792 constitution. To deny him the right to preach, however, could spark anger and possibly revolt among the fifty enthusiastic converts whom Durrett had brought into his fold. In the end, the association "gave him the right hand of Christian affection, and directed him to go on in the name of their common Master," without ordination. Old Captain continued to preach actively to free and enslaved blacks, entreating them to embrace salvation and opening his home to weekly services.[63]

By 1790, over forty-eight hundred slaves and thirty-two free African Americans lived along the Limestone Road. Nearly all were from the tobacco-producing regions of Virginia and Maryland. Although on the margins, they were nonetheless pioneers. Few could aspire to live as members of an autonomous black family household, but many sought ways to create and maintain communal support, as Old Captain did in Lexington. All other migrants enjoyed a luxury, however, that African Americans could not: the decision to segregate themselves. Whites were ubiquitous in black life. More often than not, blacks were trapped in work and living situations that forced them to interact with others different from themselves. As subordinate members of American society, African Americans proved the most adaptable of early westerners, constantly negotiating the shifting cultural circumstances in which they found themselves.[64]

As Old Captain was no doubt aware as he preached to racially mixed congregations, the greatest common bond between African American and Euramerican pioneers was the traditional worldview that shaped their perspectives and their faiths. But while most white pioneers were deeply grounded in the Judeo-Christian tradition and most black pioneers converted to it, the majority of Christians was not actively religious. Churches along the Limestone Road in the 1780s and early 1790s were weak institutions, constrained by the lingering Indian threat, the absence of permanent clergy, and Backcountry inclinations toward dispersed settlement. In fact, reiterating the traditional notion that human society was fragile and unstable, Daniel Drake recalled how, before 1790, spiritual life along the old buffalo trace appeared one of competing polarities, "a state of society which presented the opposing elements of virtue & vice, piety and profanity, in many of their most lovely and hideous forms." Pioneer culture always seemed on the verge of pandemonium: the masses engaged in secular frolics that allowed cursing, drunkenness, gambling, brawling, and sexual looseness. But a pious minority determined to stop them. The Drakes and other churchgoing pioneers, most of whom lived in close proximity to each other and to a meetinghouse, conceptualized their world dichotomously: "The village church and the village tavern did in fact represent two great opposing principles: good & evil,—the spirit and the flesh." With its drinking, cavorting, and brawling, the tavern symbolized the supposed depravities characteristic of secular life, exemplifying to many how a human-controlled order, devoid of adherence to divine law, could easily degenerate into chaos.[65]

This lesson was particularly relevant to the settling of the early West. Whereas republican theorists argued that the wilderness would erode civilization, pious Christians believed that human sinfulness was the greater threat, the enormity of which was evident in meager church memberships. By the turn of the nineteenth century, only 10 percent of Kentuckians had joined an organized congregation. Preachers complained that without established religion, true Christians would become lax in their faith, and the faithless would be lost. "I found scarcely one man and but few women who supported a credible profession of religion," criticized Reverend

David Rice. The solution was not more meetinghouses, however. Many settlers, in particular Backcountry Virginians who were suspicious of established religion, distrusted churches and clergy. In 1798, an anecdote from the *Kentucky Gazette* mused about an "honest bluff country farmer" who met the local parson but refused to deferentially step aside. The preacher rebuffed the farmer that "he was better fed than taught." "Very true, indeed sir," replied the layman, *"for you teach me and I feed myself."* Within the process of settling a wilderness, religion was not a necessity. Faith could be sustained individually or in familial settings beyond church walls.[66]

Yet, while proportionately small in number, church members wielded significant influence over cultural development along the Limestone Road. Biblical studies and theological debates gave them opportunities to evaluate their lifestyles and those of people around them. The tenets through which Christian pioneers attempted to structure society originated in their understanding of what God demands of his flock, and consequently, they hoped to order their lives and their culture by integrating a pious ethos (how "to serve God in a proper and acceptable manner," as one congregation put it) with how they perceived life to be. Their panacea to the secular pleasures that Backcountry pioneer culture seemed to promote was a society ordered in accord with the Spirit: one that followed scriptural dictates and was moralistic, exclusive, literalistic, and afterlife-oriented. In order to implement this ideal, many advocated a "church of law" that aggressively and unapologetically demanded strict adherence to the Bible.[67]

Christianity has always entailed a concern for moral purity, although those of Calvinist bent have been more attentive to the stringent and visible application of its principles. For Christian purists along the old buffalo trace, moral law was absolute, more so than the natural laws trumpeted by the Enlightenment and Scientific Revolution. Pure values and beliefs did not depend on human interpretation. They existed as an expression of divine will, and those who violated them were destroyed either in the earthly realm or the afterlife. Salvation was experiential within those values and beliefs, forming a connection between good thoughts and good deeds. Righteous behavior characterized a pious community; distraction from those accomplishments branded one impure.[68]

Ray's Fork Baptist Church—one of the "Old Calvinist Baptist, or, as they were then called, the Hard Shell or Iron Side Baptist," as settler James Ireland recollected—was a typical church of law, nurturing a Christian faith concerned with moral purity and spiritual law. Situated west of the old buffalo trace near Paris, Ray's Fork Baptist Church became the site of the witch trial of Nancy King, "an old decrepit woman of not very preposessing appearance." Although the origins of the complaint against her remain hidden from the historic record, during her inquest at the church, King's accusers claimed that her influence was so strong "that they could not do a good deed or have a good thought." Billy Davidson, "a big pursy, lazy old fellow," provided the testimony that proved the old woman's undoing. "She often rode him a great many miles to dances and other sinful places," he claimed, "and would sometimes carry home on his back sacks of potatoes and turnips . . . and next morning he would be so stiff and sore that he could scarcely walk, and that it kept his mind in such confusion that it was impossible for him to serve the Lord in a proper and acceptable manner." When King refused to renounce the demonical vows of which the congregation accused her, they dismissed her, and the "old lady left the church crying." While the purists of Ray's Fork had suspicions about King's association with "dances and other sinful places" (the frolics so central to Backcountry definitions of the good life), even the old woman's physical appearance may have been considered a superficial stamp of an impure soul.[69]

Within scripturally pure congregations, accusations of occultism were unexceptional. As explanations for the inexplicable, witchcraft and other superstitions continued widely accepted convictions. "Beliefs in ghosts, wizards, and witches prevailed to a considerable extent," as Daniel Drake admitted:

> My superstition, and that of the people of Mayslick, in the days of which I am writing, extended to other things than heaven and hell. It embraced omens, ghosts, and even the self-motion of dead man's bones. Some cabins were startled by strange sounds; a night or two before the death of my cousin Dr. John Drake, some member of the family heard the sound of a plane, as in preparing boards for a coffin; the barking of dogs during the severe illness of a person was ominous of death; the inmates of a cabin, about a mile from father's, saw a piece of white drapery moving on the snow in the moonlight near their dwelling; and the arm or thigh-bone of a man who had been buried on a spot which was afterwards cultivated, was exhumed. I do not remember how or why, but it was reburied, and afterward appeared on the surface of the ground. For myself, if not a firm believer in these specimens of the supernatural, they were so

established in my imagination, that I was always, when alone in the dark, in a kind of expectation or fear that something would show itself from the world of mystery.[70]

In response to the mysteries that surrounded them, Euramerican pioneers employed potions and folk medicines. Their African American neighbors turned to various forms of "conjuring" that not only targeted unearthly foes but oppressive whites as well. Unlike whites' superstitions that cultivated fear among believers, blacks presumed that some ghosts and spirits actually aided the conjurer: remnants of an African heritage that venerated ancestors and assumed continued guidance even after death. Despite the suspicions raised by witchcraft and conjuring, the practices were merely exaggerated forms of the traditional belief that it was possible to tap into and use the known *and* unknown creations.[71]

Supernaturalism was central to the traditional worldview because it attended to the most common and intimate of human circumstances: birth, sleep, sustenance, and death. Symptoms of King's supposed witchery manifested as attacks on production and reproduction. As Ireland recalled, "If any of the stock died or was afflicted in a peculiar manner, they were bewitched; or if a child came into the world deformed, hair-lipped or any peculiarities about it, old Nancy was blamed." Her connection to natural calamities made King as great a threat to survival as smallpox or drought, and she became the scapegoat for the difficult challenges that characterized pioneer life. The congregation viewed her dismissal as requisite to restoring "peace, harmony and prosperity."[72]

That the trial occurred at Ray's Fork Baptist Church, then, is significant. Its congregants embraced a Calvinistic ethos in which God controlled the actions of the visible and invisible realms. The Almighty set the rules, established the patterns of life, and predetermined fates: order was imposed divinely. In the minds of such Christians, a person practicing witchcraft proposed to interfere directly with that Calvinist cosmos. King's participation in the invisible world empowered her to manipulate the visible one, circumvent God, execute her own judgments, and exploit supernatural powers to upset the created order. If unalterable natural laws could be suspended through witchcraft, then how dangerous were such conjurers to the supposedly immutable laws of Scripture?[73]

In the years before the Revolutionary War, beliefs in the occult had become largely confined to the lower classes where more educated

Americans could ignore them. Revolutionary ideologies in trans-Appalachia had unleashed egalitarian forces, however, that not only gave a democratic flavor to pioneer life but elevated the supernaturalism of the folk. As Daniel Drake confessed, embedded in the psyches of individuals and communities were superstitions rooted in Christian faith and inherent to traditional views of the natural and supernatural worlds. Because Scripture referred to God, angels, witches, spirits, conjurers, folk healers, miracle workers, demons, and the devil, such were the beliefs among those of a scripturally pure faith.[74]

Concerns over purity were not limited to rural Baptist churches or defined through supernaturalism. When, in 1784, Adam Rankin took the pulpit at the Mount Zion Presbyterian meetinghouse in Lexington, he determined to eradicate the use of *Watt's Psalms,* a revision of the Psalms of David. Strife over psalmody had convulsed Presbyterianism for years. At Mount Zion, the debate flared into outright secession. At issue was the purity of Scripture versus a liturgical device that represented a human-filtered understanding of the Word. Rankin censured the "Psalms and Hymns of human composition, and of human authority" for hiding the true message of God. By 1789, he refused communion to congregants who accepted *Watt's Psalms.* In turn, they brought charges against him before the presbytery. Within three years, the church had split. Rankin and his followers remained in the meetinghouse. His less orthodox opponents constructed the First Presbyterian Church adjacent to Lexington's public square.[75]

The two Presbyterian congregations epitomized extreme visions of spiritual order and, ultimately, of societal relationships that were common to traditional worldviews along the Limestone Road. Those who had led the exodus from Mount Zion included Robert Patterson, John Maxwell, James Trotter, Robert Megowan, and Robert Steele—all town trustees, church elders, and, with the exception of Patterson, merchants. Their First Presbyterian Church, a frame structure with a cupola, bell, and gallery, flourished under James Welsh, a clergyman who moonlighted as a language professor at nearby Transylvania University. *Watt's Psalms,* as "human authority," echoed an increasingly rationalized version of the mechanics of life: God may be an all-powerful deity, but man abided by reason and directed the world through his own endeavors. A few blocks away, Mount Zion Presbyterian Church remained under Rankin's leadership until 1825 when it died quietly, as did the preacher on a trip to Philadelphia two years later. His influence spread widely, however. Years later, Daniel Drake remembered his mother as a "Rankinite," her admonition still ringing in his ears: "God has said it! The Bible forbids this. How simple & yet how sublime." Like the Hard Shell Baptists at Ray's Fork, Rankinites called for a scripturally pure ethos uncorrupted by human intercession as the only sound basis for faith and everyday life.[76]

For many traditional Americans, insistence on scriptural purity buttressed an assumption of rewards and punishments: "whatsoever a man soweth, that shall he also reap." If one did not succeed or prosper, apparently one had lost God's blessing by straying from the path of righteousness. Faced with the secular pull of taverns and frolics, purists reacted with admonitions that only through cautious attention to scriptural dictates would impure neighbors avoid hell. In purist strongholds, pleasures like dancing and drinking contradicted the dictates of Scripture, as did rituals like infant baptism and baptism by sprinkling. A witch trial and the banishment of non–literal believers from Communion were reasonable responses to impure activities, making each congregation an exclusive community of believers that dismissed and labeled apostate any person who did not interpret eternal salvation as a reward for the righteous. "Election, reprobation, and predestination were the favourite themes," explained Drake. "They were all held strongly affirmative, and the slightest doubt was branded as tending to heresy." Similar to their perception of a finger-wagging God who admonished and punished the wicked, purists criticized and attempted to penalize Nancy King, the Wattsites, and others who doubted the verity and supremacy of Scripture. For many purists, Kentucky likened to John Winthrop's "City on a Hill." But unregulated Backcountry folkways and frolics impeded the development of a truly sacred space in the American West. As one purist, attempting to chastise the wicked and hypocritical, lamented in 1793, "we Christians are not contented, unless we are committing evil every day of our lives. Oh sad degeneracy! To enjoy the light of the sun, and yet to act infinitely worse than those that were enveloped in the most impenetrable darkness! Oh my country! Oh pity! Oh Kentucky! Christians in knowledge, and worse than heathens in practice. The favorite people of God, and yet devoured by Satan! Eminently favored in order to be superlatively disgraced."[77]

While many pioneers hoped to create a scripturally pure society, pioneer culture was not monolithic, and the diversity of peoples and cultures made the region a true American melting pot. Cultural segregation and integrity continued to characterize community life, particularly in congregational enclaves where people of like mind could actively regulate group membership. Ironically, the road by which the pious went to church was also the route that frolickers traveled. In the 1790s, when summer droughts limited baptism by immersion in Mason County because "there was *not* 'much water,'" Preacher William Wood of Washington led Baptist men, women, and children along the old buffalo trace. They moved northward to Lee's Creek or traveled southward to Johnston's Fork, always seeking baptismal pools so as not to violate scriptural dictates of baptism by immersion. As they traveled, they passed inns and taverns boisterous with frivolity. And occasionally, they were run off the road by young men from Washington, racing on horseback to eat, drink, brawl, and pursue sexual pleasures at Mayslick's frolics.[78]

THE VILLAGE WEST

While Backcountry peoples settled in dispersed patterns, their reliance on forts and stations for early protection inspired compact settlements. Later pioneers, wanting to live collectively to maintain cultural integrity or religious identity, formed neighborhoods around those old stations and forts. By the end of the 1780s, a series of small, densely populated villages sat along the winding dirt road. Consequently, as a new wave of migrants—families from Tidewater and Piedmont Virginia with expectations to subdue the effects of the Kentucky wilds—began to settle along the beaten path in the 1790s, they found neither a foreboding wilderness nor an Edenic garden. They discovered a "village West."[79]

That it happened along the old buffalo trace is a bit surprising. After all, better known and more heavily traveled was the Wilderness Road. It was a narrow path through the Cumberland Gap upon which hundreds of families from Georgia, the Carolinas, and Virginia pushed into southeastern Kentucky and then moved northward into the Bluegrass or continued westward into the Green River region. Just the sheer volume of migrants intimated that this would become the predictable path to a densely settled western landscape. Yet, while the Wilderness Road's identification with Daniel Boone conferred celebrity upon it, it was a dreadful course to negotiate. Even after state-sanctioned improvements in the 1790s enabled "Waggons loaded with a ton weight, . . . [to] pass with ease, with four good horses," the trail remained substandard. "What a road have

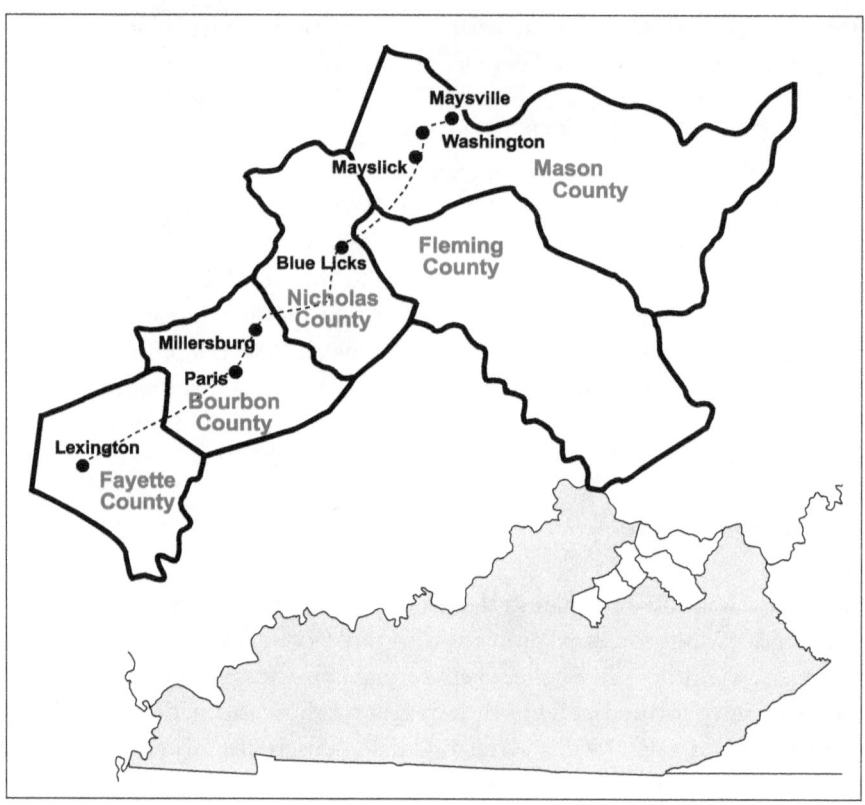

The Village West. Map by Mapsmith etc.

we passed!" complained Methodist Bishop Francis Asbury as late as 1803, "certainly the worst on the whole continent, even in the best weather; yet, bad as it was, there were four or five hundred crossing the rude hills whilst we were." The ruggedness and tribulations of the Wilderness Road not only represented the arduous westward journey but inhibited concentrated settlement.[80]

While the Wilderness Road was an obvious route for southern populations moving westward, it was out of the way for northerners and overly demanding for peoples accustomed to less strenuous transportation. European tourists, residents of mid-Atlantic and New England states, and wealthy Virginians looked to two northern routes as the best westward passages. Both began at the headwaters of the Ohio River, where Pittsburgh stood as the gateway to the West. In "well-peopled" regions around the town, "all the comforts of life are in the greatest abundance," most of which were offered at inflated prices. People crammed into Pittsburgh and waited for supplies, for boatmen, for guides, for traveling companions, or for the river to thaw. Land speculator John May recalled temporarily sharing a small log home outside Pittsburgh with "not less than 50 souls . . . bo[u]nd to Kentucky."[81]

A great but apprehensive optimism animated potential settlers. "We are about to enter a grand field," wrote Virginian David Meade in 1796 as he waited for his boat down the Ohio, "which neither writers nor ordinary travelers have in my opinion given a faithful account of." Pilgrims rejoiced at the rumors of the region's beauty and fertility, and they wanted to experience Kentucky for themselves. More cynical observers puzzled over the westward movement. Ironworks entrepreneur Moses Austin wondered about the "hundreds Travelling hundreds of Miles, they Know not for what Nor Whither, except its to Kentucky, passing land almost as good and easy to obtain'd." Disgustedly he reacted, "Can any thing be more Absurd than the Conduct of man?" To some, it did not appear so at the time.[82]

The thousands who temporarily bunkered in Pittsburgh had two choices: they could meander in flatboats down the Ohio River or they could travel overland along a "prodigious thorofare," the winding post road through Wheeling in western Virginia and Chillicothe in Ohio Territory that in time would be known as Zane's Trace. Before the waning of the Indian threat, westbound travelers commonly moved along the road "forty or fifty in a party." By 1796, tourist Isaac Weld was relieved to find farms flanking the trail so that few Kentucky-bound pilgrims had "to sleep more than two or three nights in the woods going there"; and he was pleased that, with the Indian menace alleviated, "five or six travel together they are perfectly secure." Migration along the post road appealed to many who knew that shelter and food could be found at intermittent farms and taverns. Yet, even as the Indians retreated, white thieves posed a new danger, ambushing unsuspecting journeyers and stealing the few belongings that migrants carried with them.[83]

So, many elected to travel the Ohio River. When the Drake family boated the waterway in 1788 en route to their new Kentucky home, anticipations of Indian attack kept them from landing before their destination at Limestone. Unlike the post road, as late as the mid-1790s, there were few farmsteads along the Ohio's banks to provide refuge, and most travelers quickly left Gallipolis and other jerkwaters where morals seemed loose and the laws of civility unacknowledged. Contributing to anxieties were Kentucky flatboats: great oblong vessels between thirty and fifty feet in length, twelve to twenty feet in width, and about four feet deep that, because of their bulk, were difficult to maneuver and exposed passengers and crew alike to the whims of the river current. As late as 1816, settler George W. R. Corlis discouraged his father from bringing the family "in one of those Hog pens" aboard which "insolent boat men ... demand high prices."[84]

Imagined perils shaped voyages along the Ohio River as well. Migrants always suspected sinister natives peering from the underbrush. In 1796, settler John Breckinridge had "as pleasant a journey as it was possible where any danger is apprehended." Paranoia over Indian attacks tempered settlers' enthusiasm. Still, even though trepidation proved a constant companion, one that turned many migrants back, it also made a successful journey all the more meaningful.[85]

For those who rode the current of the Ohio, their destinies began to unfold at "The Point" where the small port of Limestone had arisen. Trapped by a steep hill that stood only eight hundred yards from the river, marking the southern edge of the floodplain, Limestone was a compact village. In 1784, at the mouth of Limestone Creek where the old buffalo trace met the mighty Ohio, early settler Simon Kenton built a blockhouse for the security of new arrivals; populations thereafter remained huddled around the juncture of the river and the trace. Three years later, the Virginia legislature officially designated the hamlet as the seat of Mason County and renamed it Maysville for John May, original claimant to the lands on which the village developed. State recognition did little to boost prospects, however, and by 1795 when Frenchman Victor Collot began his tour along the Limestone Road, there was general agreement that the settlement, "from the narrow space between the hill and the banks of the river, can never be very populous." Despite its redesignation, the name Limestone lingered, a nostalgic reminder of the first step new settlers had taken on Kentucky soil. In fact, the post office did not officially change to Maysville until 1799, and many westerners continued to call the town Limestone well into the 1820s.[86]

Although individuals and families arrived in Maysville determined to find their futures in Kentucky's interior, the village enticed some who were willing to forego further adventure. The appeal was merchandising, overland shipping, providing accommodations for travelers, and a nascent river-

Arriving at the Point. From a mural on the Maysville Flood Wall, Maysville. Photograph by Craig Thompson Friend.

---

based economy. By 1790, ninety-six residents occupied approximately thirty houses. "The depôt of whatever goods pass from Baltimore and Philadelphia to Kentucky," the settlement became a crowded and busy little village, communicating "an appearance of liveliness and bustle which might induce a stranger to think it a place of more consequence in itself than it really is."[87]

Like other river towns, however, Maysville just did not enjoy good repute. Virginian gentleman David Meade, heading to his lands outside Lexington, left behind most of his belongings at the river port because only one wagon could be acquired. Lodging, while usually available, was substandard: "The inns are wretched public houses without provisions," complained Victor Collot, "and the little that can be obtained is procured with difficulty and at an exorbitant price." By 1789, the "Amazing immigration into Kentucky had stripped all the country around Limestone of every kind of provision in such manner that nothing can be bought at that neighborhood under three times the Lexington price for the same article." Villagers' greed made them seem a despicable lot. Pioneer Needham Parry regretted that he had arrived in a place void of "people of principle." "The inhabitants of this town live in idleness and poverty," complained settler John Heckwelder in 1792, "and for support depend upon what they can lay hold of from the travellers to whom they occasionally lend assistance when in difficulty." Even residents recognized the deficiencies of their village which, as one observer related, they derisively labeled "the fag end of Kentucky; you are glad to get away from it as fast as possible."[88]

Escaping Maysville proved nearly as arduous a task as arriving there. The trace wound up steep bluffs, making the hike difficult and often injurious. New settler Asa Farrar and his family took two days to climb the hill with their belongings. Victor Collot remarked how "large stones, many of which were loose, filled the uphill roadbed." Coaches packed with people and wagons laden with goods overturned frequently. Riders walked their horses up the incline "for fear of blowing him in the early part of his journey."[89]

Once arrived at the peak, however, most travelers deemed the uphill challenge worthwhile. Rolling hills covered with woodlands and intermittent grasslands stretched toward the horizon. The panorama was "more waveing than any other part of Kentucky." The soil seemed extraordinarily rich: "When you take it between your fingers," wrote settler Edward Harris in 1797, "you cannot perceive any more grit than in butter," and he proclaimed it "the best Country for Corn, wheat, Rye, Oats, Barley, flax, hemp & grass." "Black as the bottom of your dung heaps," the soil revived expectations that had been momentarily forgotten in Maysville. But it was also a region "badly watered" by inconsiderable streams and springs, a problem that few settlers realized.[90]

Four miles south was Washington. In 1785, Baptist minister William Wood and Virginia surveyor Arthur Fox purchased a tract from Simon

Kenton and designed a village. Officially established in 1786, Washington quickly enjoyed a better reputation than Maysville, largely because the bluffs that frustrated travelers also hampered Indians. Shawnee attacks usually caught Maysville's residents off-guard, but warnings reached Washington before the Indians did. Many settlers left The Point and relocated to the inland village because "the danger of Indians from the other side of the river was great."[91]

By the early 1790s, Washington emerged as the dominant village of the region and the third largest settlement in Kentucky after Lexington and Frankfort. Its neighborhood contained fifty limestone houses and log cabins sheltering approximately three hundred inhabitants. Washingtonians' sense of the potential of their village manifested in an early improvements movement, beginning in 1786 when residents laid flagstone walks along Main Street and opened several public wells. On his 1795 map, Joseph Scott identified Washington as the northern terminus of the road, symbolic of the rising prominence of the village, and by 1801, Washington replaced Maysville as the seat for Mason County.[92]

The distinction between Washington and Maysville had much to do with topography. "One peculiarity in the quality of lands, and which perhaps does not exist in any other part of the United States," wrote a French visitor in 1795, "is, that those situated on the summits are much better, and have greater depth of loam than these in the vallies." Washington's situation atop the bluffs advantaged it with richer soils. Yet, the bounteousness of nature was not to be relied upon: "these fine lands have, however, one very bad quality; they produce naturally no herbage, or very little, fit for pasturage." In this northern part of the region, the twitch grass and clover on which livestock fed in the East were rare, forcing the earliest farmers to lay "artificial meadows, which is attended with great expense, and a loss of time which is peculiarly precious to new settlers."[93]

As travelers breached Washington's southern boundary, they faced a landscape heavily forested with ridges, rocky ledges, and unpredictable streams. Occasional crossroads, inns, and farmsteads appeared, but for the next seven miles, migrants and visitors struggled to make their way to the hamlet of Mayslick. French tourist François Michaux complained that the distance did not "afford the least vestige of a plantation." The landscape was "thick Woods, where nothing was to be seen but Trees, except now & then two or three Wild Dear & frequent flocks of Wild Turkies." The condition of the road was equally troublesome. According to early settler Ned Darnaby, the trace from Washington was "nothing but a tolerable cart-road."[94]

Making the journey even more difficult were creeks, streams, and rivers—some fordable, others not. Three miles south of Washington flowed North Licking Creek or, as locals called it, the Northfork. Like so many of Kentucky's waterways, the Northfork fluctuated between low and high water levels. "A very good bridge" spanned the stream, but when the bridge washed out, the river could not be forded because, as Collot noted, "the banks are steep, the bottom muddy, and the land on each side marshy." A mile beyond was Lee's Creek. Summer and winter fluctuations dramatically affected streams like Lee's Creek, and often the rocky bed of the rivulet served as the road. Here was one baptismal pool for William Wood's Baptist congregation. Here also was Lee's Mill, forced to close operations in summers when the waters evaporated. Five miles farther coursed Johnston's Fork of the Licking River where, on other occasions, Wood led his congregants and where Clark's Mill operated when the waters flowed. Even though owners constructed dams to direct the flow over the water wheel, their efforts could not sustain power with unpredictable water levels. Because both perched alongside such minor streams, Lee's Mill and Clark's Mill became "feeble operations that only run in wet weather."[95]

Between the two mills, some seven miles south of Washington, the road wound through Mayslick. In 1788, five New Jersey families purchased the tract from William May, and, within two years, approximately fifty people lived in the neighborhood. Without the commercial activity of Maysville or Washington, Mayslick was merely a small cluster of farmsteads. The community was the first and most concentrated of the truly *rural* neighborhoods that arose along the Limestone Road, a pattern illustrated on Collot's map by a number of farmhouses in the four miles between Mayslick and Johnston's Fork.[96]

But after travelers crossed Johnston's Fork, they entered a topography more angular and rugged. The road itself became a track of immense flat and chalky stones that ran atop hillcrests and along creek bottoms, making travel even more difficult and uncomfortable. On his 1784 "Map of Kentucke," land promoter John Filson labeled this region "Fine Cane Land," and for those who arrived before the 1790s, Filson's proclamation proved reliable. *Arundinaria gigantea,* growing three to four inches in diameter and seven to sixteen feet in height, formed thick canebrakes. "Continuing in verdu[r]e all the winter," cane was the draw for the great bison that initially blazed the road, and it proved attractive to many settlers who sought natural pasturage for their livestock. "Perhaps, the most nourishing food for cattle on earth," heralded Gilbert Imlay, "no other milk or butter has such flavour and richness as that which is produced from cows who fed on cane." By the mid-1780s, however, the canebrakes along the road were already disappearing. Exaltations of Kentucky's flora misled thousands who came expecting forests of natural pasturage and found fields of weeds.[97]

As they approached the main branch of the Licking River, travelers came upon Blue Licks. Named in 1773 for the blue-gray limestone that surrounded its saline spring, the place was a "desert" country, "dry and open, strewed with rocks, and consequently barren." Through the crevices of a rocky landscape overspread with mosses and lichens, a few small pines and cedars sprouted, but the area had an appearance of "sterility, desolation, and sadness." Here was the site of the 1782 Battle of Blue Licks. The myth of the battlefield—that soil saturated with the blood of heroes would never again be fertile—added to the impression of a sterile and eerie land.[98]

Both for its historic significance and because it was a major crossroads where several traces joined, Blue Licks became central to settlers' mental maps of the region. In a legal deposition in 1810, Robert Patterson, an early settler of Lexington, recalled how the road south of Blue Licks was "a plain marked path to Limestone, with apparently very old and plain marks" that stigmatized it as a warpath. Just north of Blue Licks, the road became wider and had the "appearance of a Buffaloe Road." Whether buffalo road or warpath, the old trace both literally and figuratively metamorphosed at Blue Licks.[99]

At Blue Licks, the trace intersected the Licking River, the largest watercourse in the region and the northern boundary of Bourbon County. The Licking wound tortuously through steep, rocky ravines. Its banks were difficult to maneuver, no bridge forded the river, and not until the late 1780s could travelers rely on an operating ferry. The river had political implications in 1786 when Maysville residents seceded from Bourbon County because the "Intervention of a Mountainous tract of Barren Land running down on each side of the main branch of Licking Creek that cannot be inhabited" made court attendance extremely difficult. The result was the birth of Mason County to the north. Still, the Licking promised trade potential, although when the waters subsided, navigation dried up for two or three months at a time. Five hundred yards downstream from the old buffalo trace, a bank of stone jutted into the Licking, making travel impossible during droughts. In good seasons, however, the river was navigable "one hundred and fifty miles for the largest boats."[100]

Before 1795, sojourners recalled only two farmsteads in the stony and barren eight miles south of the Licking River. Still traveling across shale hills, migrants began questioning whether, in moving beyond the northern villages, they had left behind Kentucky's best lands. The road hugged hillcrests that produced little herbage for grazing and required major improvement for farming. Victor Collot pondered settlers' judgment: "We came to a desert country composed only of masses of rock; we journeyed eight miles along a road which was almost impracticable, from the immense quantity of ravines and enormous stones with which it is encumbered, and found on our way a wretched hut but inhabited by woodmen and hunters."[101]

Eventually travelers left the hills and came upon Millersburg. In 1778, a party of Backcountry Pennsylvania farmers led by John Miller preempted the lands and constructed Miller's Station. Although not officially recognized until 1798, the village was well established with some 240 citizens and 50 houses. It sat along the banks of Hinkston Creek where a sawmill, fulling mill, and two gristmills operated, "one dam answering the whole."[102]

A change in soil quality delighted travelers. In contrast to the topography around Blue Licks, the southern lands along the old buffalo trace comprised a vast plain of gently rolling, fertile hills that became known as the Bluegrass region. The land, "if possible, increases in richness," proclaimed Imlay. Several yards of topsoil covered a limestone bedrock. "This portion of Kentucky was once the paradise of the paw-paws," recalled regional botanist C. W. Short. Well into the 1790s, woods flanked the seven-mile stretch of the road from Millersburg to Paris. One summertime journeyer slept every night save three "under two blankets as the nights are very cool—the days about as hot as in Pennsylvania." As in the hillier lands to the north, cane had once flourished in the Bluegrass. In 1776, Levi Todd proclaimed "nearly one half of it covered by cane" and "between the brakes, spaces of open ground as if intended by nature for fields." And, as in the hillier lands to the north, indigenous flora disappeared quickly in the Bluegrass.[103]

Seven miles beyond Millersburg, travelers came upon yet another branch of the Licking River. Stoner's Fork was "a considerable creek [that] ... supplied falls for two mills, and water of good quality for domestic and other purposes." Collot described how its "banks in general are low and firm and its bed excellent." Here arose Paris, the seat of Bourbon County. In 1785, New Jersey farmer Lawrence Protzman purchased lands and laid off lots for what was then Hopewell. Situated in the lush Bluegrass and alongside a fordable river, Hopewell quickly bloomed: "There is constantly that stir and bustle which denotes a place of business." In 1790, the Virginia legislature renamed the village Paris. Between 1793 and 1795, the number of residences grew from eighteen to sixty, many of which were brick, stone, and frame houses, "not one of which are finished." The post office began using "Bourbontown" as the official address, although Paris remained the more common name. Still, unlike Washington whose demographic development exceeded that of Paris, the latter appeared to be more urbane with its brick courthouse and less rudimentary architecture, which may account for settler Valentine Peers's surprise upon his arrival in Bourbon County that "however flattering the descriptions, my Ideas of it were far short."[104]

The nineteen miles south of Paris comprised "a great and extensive plain, sometimes grouped with woods and at times interspersed with farms, equal for the construction of the buildings and the cultivation of the land to any in Europe." Travelers crossed branches of Elkhorn Creek on several occasions. At one point, they passed what remained of Bryan's Station. The road became "very wide and fine, with grazing parks, meadows, and every spot in sight cultivated," and the broad, treeless plain "insensibly assumed the appearance of an approach to a city."[105]

Refinement of the landscape announced Kentucky's largest town. Named by a surveying party in 1775, Lexington did not have a true physical presence until 1779 when Robert Patterson led construction of Fort Lexington alongside a small, shallow branch of the Elkhorn. The following year, pioneers joined together to plan a town grid. Agreeing to share expenses, the forty-seven heads of households organized their town into half-acre lots in three rows with access to a commons situated along Town Creek. On the hill south of the creek ran High Street; on the slight incline to the north was Main Street. Both were parallel to Water Street, which ran alongside the creek. Intersecting these streets were Main-Cross and Mulberry Streets (the latter evolved into the old buffalo trace one mile north of the courthouse). Convenient to inhabitants' in-lots were five-acre out-lots set aside for farming. It was a common, middle-American urban grid surrounded by an open-country neighborhood of farm lands.[106]

In 1782, when the Virginia legislature recognized the community as the seat of Fayette County, residents erected a log courthouse. By 1791, 834 people resided within town limits; five years later, 1,500 citizens lived in nearly four hundred houses. Whereas Thomas Chapman scoffed at Paris's unfinished condition, he thrilled at the genteel brick and stone edifices of Lexington.[107]

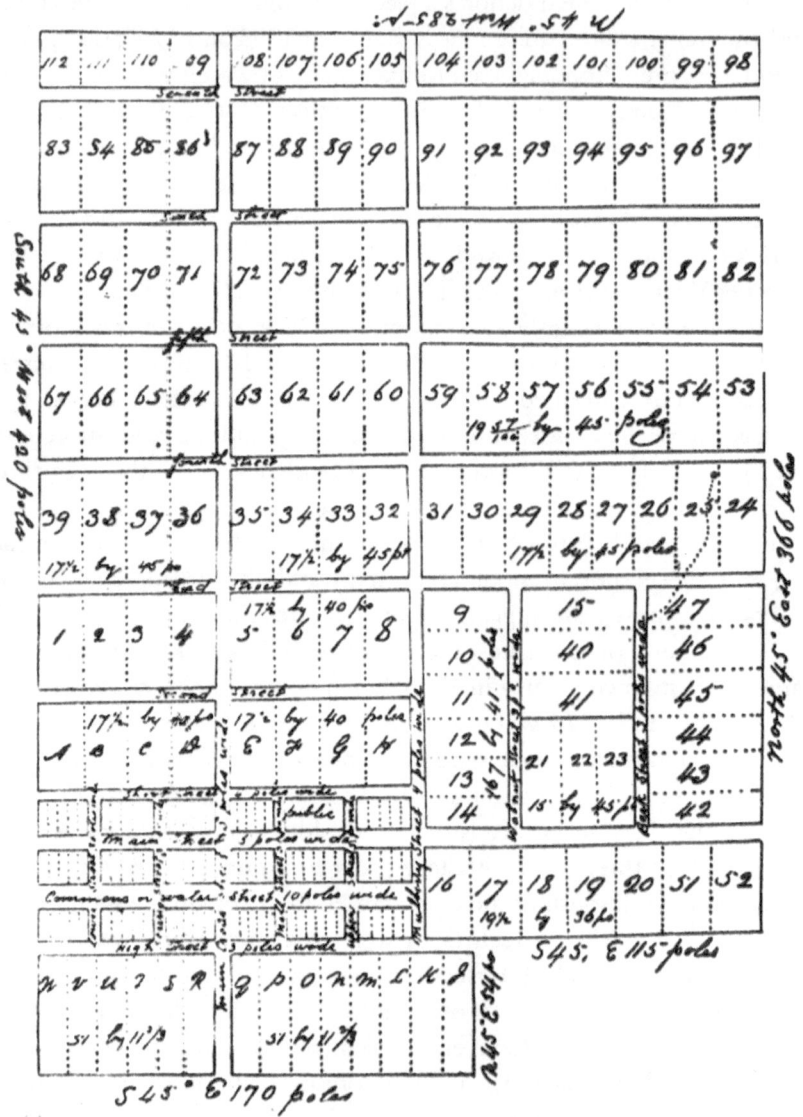

Lexington Town Grid, 1780. Reproduced from Doct. Luke Munsell's Large Map of 1818 *(Cincinnati: Carey G. Fairbank, 1834).*

When Collot arrived in Lexington in 1795, he assumed that without navigation the village would "not be great, and that Frankfort would be the real place of commerce." Time would not bear out his prediction. Still, Collot echoed the opinions of many. In 1792, the state assembly located the state capital in Frankfort, some twenty miles to the west along the Kentucky River. Lexingtonians complained that the decision "deprived the citizens of the corner stone of republics": access to the seat of government "in the center of the population of the State." Having anticipated their town as the cultural, economic, and political center of the new West, Lexington's trustees had begun construction on a statehouse. By the summer of 1794, with the state assembly meeting in Frankfort, the "statehouse" was folly. Trustees transformed its first floor into a market, an unsuccessful enterprise that pushed them finally to lease the building to local printer John Bradford.[108]

From Limestone to Lexington, as the central thoroughfare of the pioneers' world, the Limestone Road assumed a role that had not existed prior to Euramerican colonization of the region, integrating a town, villages, hamlets, and neighborhoods into a nascent *urban* corridor of the early American West. As determinative elements in trans-Appalachia, such places have elicited less historical attention than the rural qualities of the region: the "frontier" and the Indian threat, the hunting world of Daniel Boone, the struggles to validate landownership, the commercial agrarianism of Henry Clay. Still, new settlers could not travel the old buffalo trace without moving through these places and having their experiences and perceptions influenced by village economic, social, and political activities.[109]

Circa 1793, Elihu Barker drew a map of Kentucky that details this village West. Roads emanate from Lexington and connect dozens of little places into a constellation of village life: one road extends northward through Georgetown toward Fort Washington (soon to be Cincinnati), another heads to Frankfort and on to Louisville, another traverses westward to Versailles, one stretches southwestward to Harrodsburg and Danville, still another—the Wilderness Road—leads towards the Cumberland Gap, one spans southeastward to Boonesborough, and finally the old buffalo trace winds to the northeast through Bourbon Courthouse and on to Washington. Places and place-names, cartographically distinct from rural spaces, dotted the Kentucky countryside. But to separate them from that countryside is impossible; they were intimate features of it.[110]

Like historian R. S. Cotterill in 1916, we could accept that each dot, each place, was "but a tavern or a roadside inn without prospects or hope." We might even dismiss them altogether. "To the unindustrialized, nonurbanized regions of the new United States, to a region such as Kentucky," historian Raymond Betts wrote on his way to slighting the towns and villages, "European eyes were turned, now alert to the urban-rural dichotomy and given to a consequent appreciation of the rural environment." Nonurbanized? Certainly, these villages and towns were not of the same category as New York City, Boston, Philadelphia, or the European cities that Betts's subjects called home. Still, characterization of early Kentucky in this manner does little justice to the aspirations and endeavors of many of the region's pioneers who understood urban and rural not as inimical but as symbiotic, and who strove to build villages, towns, and even a city to complement the farmlands around them.[111]

The people who early populated these villages were pioneers themselves, trying to re-create businesses like they had in small towns east of the Appalachians. And many rural pioneers from the eastern backcountries expected locales like Maysville, Washington, Millersburg, Paris, and Lexington to furnish processing, storage, artisanal, mercantile, and transportation services. Thus, quite early, pioneers played a crucial role in the "urbanization" of the old buffalo trace. Even initial settlement patterns had established a framework for the village West. In their efforts to defend themselves, pioneers had erected stations and forts that served as communalcenters for social, economic, political, and military organization. As thousands poured from behind the walls in search of land, the places remained as county seats and commercial entrepôts. Ironically, a foundation forurban development had been laid by pioneers whose culture emphasized autonomous households and dispersed settlement.

Decades later, when enough time had passed to reveal the consequences of pioneer settlement patterns, the 1810 census certified Kentucky's urban frontier. Enumerators in Kentucky, Georgia, and Indiana identified citizens by place, and, although the percentage of Kentuckians living in towns and villages was much smaller than those in Georgia and Indiana, fifty-three hamlets, towns, and villages dotted Kentucky's countryside, compared to Georgia's twenty-seven and Indiana's four. Settlement in the Deep South and Old Northwest was truly rural, generating large expanses of lightly populated countryside and occasional centralized populations. In comparison, Kentucky was composed of moderately pop-

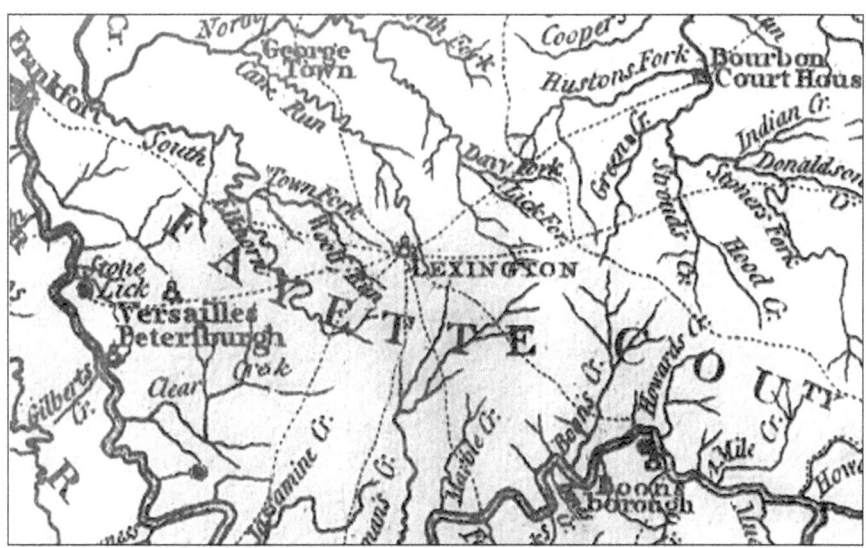

Elihu Barker, "Map of the state of Kentucky, from actual survey," 1795. Courtesy of King Map Collection, University of Kentucky Libraries, Lexington.

ulated towns and villages offering services to a moderately populated countryside.[112]

The villages also benefited from their situation along the Limestone Road. New migrants traveled the old buffalo trace where they could restock provisions before moving on to their final destinations. Thousands of people squeezed into Lexington and Maysville, "the halting-place of all travellers who visit these countries." Many took up permanent residence; others poured into the other villages and hamlets where concentrated populations demanded food, pressing local farmers to clear more of the wilderness and produce more for village markets.[113]

The rise of this village West took place within a nationwide urban boom that transformed the Early American Republic. Between 1790 and 1810, the population of the United States increased 84 percent, and urban populations in most places grew much faster. Lexington expanded at nearly five times the national rate and, by 1810, was second only to Cincinnati in the West. The rapidity and orderliness of Lexington's development earned it the distinction "the Philadelphia of Kentucky," at least according to visitor Lewis Condict. And the attention that Lexington garnered spilled over into Paris, Millersburg, Washington, and Maysville. It was truly an *urban* corridor that arose along the Limestone Road.[114]

Simultaneously, this village West became more intricately connected to agricultural life, blurring distinctions between urban and rural. The roadside along the beaten path was a landscape shaped by the natural environment. Rows of townhouses jammed along a main street looked little like the singular farmhouses that freckled the countryside, but strangely, observers noted repeatedly how the rural landscape blended into the villages and vice versa. Unimproved roads, barns and stables, spacious town lots: all elicited comment about the rural character of urban life. In a very real sense, villages were intimately related to the countryside around them. Where the environment was difficult, where the rocky ridges around Blue Licks and the flats of the salt licks impeded rural development, so too was village growth stymied. Where farms flourished atop hills along the northern end of the road or in the lush Bluegrass, there stood the larger villages and towns. An environment favorable to extensive agricultural development guaranteed similarly favorable growth for the village West. Court and market drew rural residents into these hubs of trade. Sociability, revivalism, a need for foodstuffs, and the basic agricultural nature of early Kentucky drew village attentions to the countryside. Both patterns dulled distinctions between town life and country life, and both contributed to the shaping of the pioneers' world.[115]

If previous historians were accurate, *if* the constellation of place-names on Barker's map meant little more than clusters of cabins and possibly a tavern or store that visitors viewed no differently than the features of a rural landscape, then why did people live in these places? Why were they not out pursuing the more "virtuous" agrarian lifestyle of the Early Republic? What role did such places play in the development of an agricultural region? These questions cannot be sufficiently answered if we accept the urban-rural dichotomy, for it assumes the two were divorced in pattern and purpose. Instead, we must appreciate how, along the old buffalo trace, town and country were integrated into one agri*culture*—a continuum with Lexington at one end, a river port at the other, and many farms and a variety of places betwixt. The presence of a village West not only evidenced and reinforced the diverse cultural patterns forming in the pioneers' world, it inspired theorists and future settlers to believe that the framework existed to transform pioneer culture and civilize society along the Limestone Road, or as it was more often called in the 1790s, the Maysville Road.

 2 • GREAT SETTLERS

The Gentry and the Rage for Republican Order

In 1795, as he wandered southward along the old buffalo trace, Victor Collot scribbled notes that he would one day convert into a map and a narrative of his travels. A veteran of French forces in the American Revolution, Collot became governor of Guadeloupe in 1783. Ten years later, he found himself exiled by a British naval invasion and impeded from returning to France by its own violent revolution. He turned his sights to Philadelphia, where he arrived the following year determined to find commercial opportunities for his troubled nation. In the spring of 1795, the quest took him into the early American West where, after several weeks aboard a flatboat, he debarked at Maysville.[1]

What he found was the village West and a rural landscape that held pockets of pioneers creating small farms here and there, culling from the earth a subsistence that they supplemented with herding, hunting, buying, and selling. This was the world that he re-created on his map, "Road from LIMESTONE to FRANKFORT in the state of KENTUCKY." The natural contours and wilderness setting of the trace took center stage, and Collot spent much of his journal musing about how the road could be transformed into a dynamic economic thoroughfare. The most unique trait of the map is that it reads upside down, exactly as settlers and visitors envisioned the road when they landed at The Point: laid out before them, coursing across the countryside towards the center of the promised land. At the bottom of the map sits Limestone (Maysville). Geographically south, but cartographically up, the eye travels through Washington, Mazeleak (Mayslick), Blue Licks, Millersburgh, Paris, and Lexington. Here and there appear signs of a rudimentary economic order: mills, salt works, small farm after small farm, lesser roads that fade into the blankness of the page, and towns, villages, and neighborhoods that comprise a nascent urban corridor.[2]

Collot's map was outdated by its 1826 printing; and even in 1795, it was incomplete, omitting locations like Mefford's, Kenton's, Bryan's, McConnell's, Houston's, Owings's, and Grant's Stations, and other places that would have been easily overlooked by a European in search of markets. Still, they were places that reminded many Kentuckians of a pioneer heritage under assault by the 1790s. Missing as well is Ellis's Station on the banks of Stoney Creek which, in 1805, became Ellisville, the seat of Nicholas County. As the 1800s wore on, other new features that failed to appear on Collot's map—taverns, inns, cattle farms, hemp plantations, tobacco warehouses, and other institutions of an increasingly commercial economy—became more prominent along the dirt trail. And the road itself would be recast: straightened in some places, rerouted in others. The old buffalo trace was a world constantly in revision. As Collot wandered the beaten path in 1795, of course, all of that was yet to come through the designs of Americans who had been watching cultural developments along the old buffalo trace, disturbed by the lack of agricultural production, the weaknesses of community and polity, and the lawlessness of pioneers.

IMAGINING CIVILIZATION

Victor Collot's journal and map belie the expectations that he and others like him had for economic and political growth in the American West. But possibly the objects that he omitted are most intriguing: the log stations that had once provided sanctuary for the earliest settlers and had persisted as historical markers in pioneers' mental maps of their world and as physical monuments along the dirt road. In his journal, he noted their architectural distinction: "a kind of small block-house, larger at the top than the bottom, with crannies above and below, and surrounded with a great palisado twelve feet in height: these block-houses are built with trunks of trees, the intervals between which are filled up with clay mixed with chopped straw; the roof is covered with bark or boards: the chimney consists of a pile of stones placed at the extremity of the appartment, in the

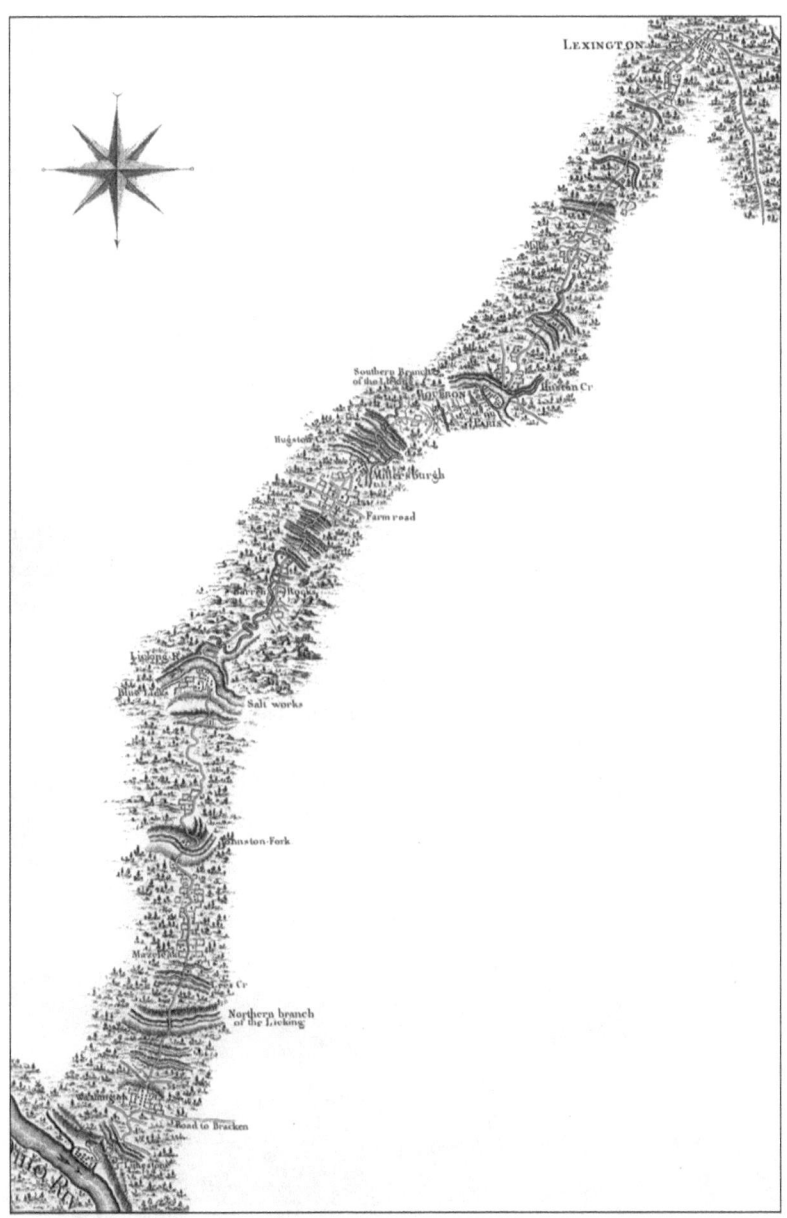

Victor Collot's "Road from LIMESTONE to FRANKFORT in the state of KENTUCKY," 1795. Courtesy of King Map Collection, University of Kentucky Libraries, Lexington.

roof of which is a hole for the smoke; and another hole is made in one of the sides of the house, which serves for the admission of light, and is of course the window."[3] Sketches of these stations were not to be found on his map, for a simple reason. Collot drew his map in 1826, having never expected the stations or the pioneer culture they symbolized to survive those thirty intermittent years.

Like many educated Europeans and Americans of the late eighteenth century, Collot believed in societal evolution. Simply put, theorists proposed that societies matured from rude hunting cultures to pasturage, then to agriculture, and finally into commerce. By the late eighteenth century, the idea had become widely accepted among European and American intellectuals. Ranking settlers by "occupation, fortune, and particular character," Collot described a three-tiered transformation that began with Forest Men, nomads who neither cultivated the land nor engaged in "other employment than hunting." In the shale hills around Blue Licks, he had come upon such men in a small hut covered by tree bark. Theoretically, Forest Men lived crudely beyond organized society and inevitably moved farther into the wilderness as game diminished and population increased to the degree that a courthouse arose in a neighborhood. In their places arrived First Settlers, pioneers who anchored themselves to a place, cleared land as they needed it, and herded domestic animals. The very term *pioneer* exposed prejudices towards such early settlers; derived from the Latin *peon,* a pioneer was one who pre-

The Station at Lexington. From Z. F. Smith, History of Kentucky *(Louisville, Ky.: Courier-Journal Printing Co., 1886).*

pared or opened the way for others. Collot viewed them in this manner, explaining how, after four or five years, this second wave of people gave way to Great Settlers, "good farmers" who created homesteads by clearing large spaces of ground, laying out meadows, planting orchards, and living "in security, plenty, and happiness."[4]

The Great Settlers were the only crowd who could tame the wilderness because they were the only ones with money and time, without which a farmer could not secure the tract of land that would bring success. One had to invest heavily and without certainty as to outcome: "The dearness of land, and especially the uncertainty of tenures, which keeps purchasers in endless lawsuits, and frequently exposes them to be put out of possession after the expenses they may have incurred in clearing and cultivating, have prevented emigrants from settling in this part of the country, and led them to prefer the north-west territory ... where the titles are indisputable." The success of securing land and title (an objective, incidentally, beyond the reach of most pioneers), would facilitate an agrarian promised land: "It must, however, be observed, that when once these artificial meadows are in crop, they produce a third more than others, and those especially which are sown with trefoil are extremely fertile."[5]

Collot's assumptions about social evolution were not unique. Over twelve years earlier, J. Hector St. John Crèvecoeur promoted a similar vision of the "hideous parts of our society" and how they were to be replaced through the evolutionary process. Along any frontier, within a dozen years, "prosperity will polish some, vice and the law will drive off the rest, who uniting again with others like themselves will recede still farther; making room for more industrious people, who will finish their improvements, convert the loghouse into a convenient habitation, and rejoicing that the first heavy labours are finished, will change in a few years that hitherto barbarous country into a fine fertile, well regulated district. Such is our progress, such is the march of the Europeans toward the interior parts of this continent. In all societies there are off-casts; this impure part serves as our precursors or pioneers. . . ."[6]

Collot and Crèvecoeur declared what had become conventional wisdom among formally educated men: that there were predictable patterns of social change, that the continuum ran from rude simplicity to civilized complexity, and that even as late as the mid-1790s, the trans-Appalachian wilds were populated with the lowest states of society. They could imagine an evolution from the primitivism of Forest Men to First Settlers' "forlorn hope" (as Crèvecoeur described it) to the permanence of Great Settlers, making stations mere symbols of the transitory character of pioneer life.

As the 1790s opened, many men envisioned themselves as Great Settlers who, upon their arrival, would advance western settlement through more refined manners, habits, and morals. In 1793, Harry Toulmin and his family arrived in Kentucky. Months earlier they had departed Lancashire, England, for Virginia, where Toulmin sought out James Madison and Thomas Jefferson who in turn equipped the Englishman with letters of introduction that guaranteed immediate access to Jeffersonian Republicans in trans-Appalachia. With the assistance of John Breckinridge and other transplanted Virginian gentry, Toulmin began a whirlwind political career in the early West: he became president of Transylvania University in 1794, resigned in 1796 to became Kentucky's secretary of state, followed that appointment with a seat on the superior court of Tombigbee District of Mississippi Territory, and in 1819, helped fashion Alabama's first constitution.[7]

Toulmin too had a theory about the peopling of the old buffalo trace. In *A Description of Kentucky* (1793), written to inspire other Englishmen to migrate to the American West, he depicted the "character" of western settlement which read, in part, "Emigrations to this country were mostly from the back parts of Virginia, Maryland, Pennsylvania, and North Carolina, until 1784; in which year many officers who had served in the American army during the late war came out with their families; several families came also from England, Philadelphia, New Jersey, [New] York, and the New England States. The country soon began to be chequered after that area with genteel men, which operated both on the minds and actions of the back woods people, who constituted the first emigrants. A taste for the decorum and elegance of the table was soon cultivated; the pleasures of gardening were considered not only as useful but amusing." Toulmin's model elaborated on the final stage of social evolution: the genteel men of unstated origin who would direct the region's cultural development. As he understood it, the supremacy of backwoods pioneer culture had already begun to crumble as a more diverse population of New Jerseyites, Yankees, and Europeans settled along the Maysville Road. All that remained was a final blow by a refined citizenry, migrating from more heavily populated and more commercially developed areas, that would refashion both the village and rural Wests into a civilized society. In 1793, he already found cultured people embracing fashionable refinement and genteel manners along the beaten path.[8]

At the heart of these prophesies were anxieties over the Backcountry brand of pioneer culture that predominated in Kentucky. It troubled many Americans and Europeans who imagined an American West where, once conquered, republicanism could be realized. In creating his narrative and his map, Collot was not content with pondering the mere topography of western settlement. He assumed that the old buffalo trace offered a stage upon which a republican society would arise and that the descendants of pioneers, people whose restlessness had been reformed by civility, could lead the way: "It is easy to conceive that the children of such men, accustomed early to hunting, to distant courses, to felling trees, opening roads, and braving the inclemencies of the seasons, become themselves soon in a state to form establishments, and to acquire that love of liberty, that honorable pride, which belongs to every man who owes his happiness, and that of his family, only to his own industry and labor. . . . It is, therefore, only among such that we find traces of the austere and simple manners of their ancestors, that hospitality which heretofore formed the ornament of the Old States, and where we might dare pronounce the name of liberty."[9]

More than description of the natural landscape emanates from Collot's map, then; his descriptions of social development along the old buffalo trace revealed his own ideological biases. The most important of these preconceptions was a concern over the fine line between "savagery" and "civility." The terms were laden with specific meaning to those who shared Collot's values. As it pertained to humans, savagery was an uncivilized condition triggered by an uncultivated, wild environment; and its derivation from the Latin *silva,* meaning "a wood," suggested too great a role for the natural environment in human development. Civility, in contrast, was the human condition freed from the natural state of savagery, demanding conformity to accepted principles of social order and civic organization.[10]

Collot delineated the conceptual distinctions between savagery and civility on his map. A cursory review of "ROAD from LIMESTONE to FRANKFORT in the state of KENTUCKY" confirms the power of nature over human settlement and development. From impediments of watercourses to his several portraits of urban disorganization to the immediacy of farms and villages to the road, it is not difficult to discern that Collot considered the natural environment a powerfully negative force upon humans, seemingly threatening order and civility.

Certainly, for republicans like Collot, conquest of the environment signaled a step toward civility. The West was wild precisely because it was a place distanced from eastern and European civilization. Humans needed not only to separate themselves from the degenerative powers of the wilderness but from their own animalistic passions and fantasies. Hence, there was plenty of reason to label some fellow humans—those who were seemingly comfortable in an untamed wilderness, overly involved in supernaturalism, or deficient in technology, intelligible language, or moral principle—as beastlike, savage, and uncivilized.[11]

If places were as remote and the corridor along the road as heavily forested as Collot portrayed, the urgency of extensive civilized development would have been apparent to all who feared that westerners, separated from the advantages of the East, faced deprivations that would first debase them and then possibly extend eastward to erode American civilization. By the 1790s, reversion of westerners to a state of savagery seemed a universal concern. In 1795, Toulmin attempted to ease the apprehensions of English friends about his new neighbors: despite their isolation, "I do not know that they are rendered more savage."[12]

By the 1790s, Thomas Jefferson took the lead in promoting an ordered, agrarian empire to civilize the West. "I think our governments will remain virtuous for many centuries," he wrote James Madison in 1787, "as long as they are chiefly agricultural; and this will be as long as there shall be vacant lands in any part of America." The Jeffersonian view advocated that savage, chaotic, natural, available space had to be converted to civilized, ordered, agrarian, occupied space in order to sustain republicanism. Benjamin Rush, republican physician of Philadelphia, perceived a definite correlation between citizenship and the control that man exerted over the environment. He encouraged "*cultivating* a country," by which he did not mean simply gardening but transforming a region: "draining swamps, destroying weeds, burning brush, and exhaling the unwholesome or superfluous moisture of the earth by means of frequent crops of grain, grasses, and vegetables of all kinds." Clearly, much of his concern reflected contemporary ideas about miasmatic origins of disease, but Rush also promoted an environment adaptable to the agrarian emphasis of republicanism. New settlements (like those along the old buffalo trace) needed to exude a "rural elegance," retaining the most ancient of trees, "rows of them standing, to adorn lanes and walks." Independence and affluence became embodied in an improved, orderly environment, one whose appearance reflected the purpose of humans rather than nature.[13]

Collot anticipated that the villages and rural elegance that manifested along the beaten path offered a corrective to the uncivilized wilderness of the early West. Cartographically, he clearly differentiated between larger places like Maysville, Paris, Washington, and Lexington—all of which he drew with neatly ordered rows of buildings—and the smaller places of Millersburg, Blue Licks, and Mayslick, which appear more irregularly patterned. His cumulative portrait of the road's places suggests that Collot rationalized the distinction between cultivated/civilized and wilderness/savage neighborhoods as an issue of order and security versus dispersion and fragility.[14]

Collot's map also reveals the village West. As early as 1771, just two years after Daniel Boone's first excursion across the Appalachians, republican poets Philip Freneau and Hugh Henry Brackenridge's "The Rising Glory of America" foretold the demise of the natural forests and the ensuing urbanization of the western wilderness: ". . . I see, I see / A Thousand Kingdoms rais'd, cities and men / Num'rous as sand upon the ocean shore; / Th' Ohio then shall glide by many a town / Of note: and where the Mississippi stream / By forest shaded not runs weeping on, / Nations shall grow and state not less in fame / Than Greece and Rome of old. . . ." Accepting the infallibility of natural law, many observers anticipated an inevitable conversion of the wilderness to urban civilization. The Age of Reason acknowledged the uncompromising rules of the natural and social worlds, including urbanization. In 1796, when David Meade arrived in Kentucky, he resigned himself to what "in the nature of things must inevitably come to pass . . . an immense population on the banks of the Ohio—Cities, Towns & Villages never out of the travelers view." For Meade, the rise of western towns and cities was imminent "unless nature should violate he[r] own laws which is impossible." Likewise, Thomas Ashe contemplated in 1806 that "the day is not far distant when its [the Ohio River's] whole margin will form a continued series of villages and towns." Even John Filson, champion of the republican Stoic, predicted that "where wretched wigwams stood, the miserable abodes of savages, we behold the foundations of cities laid, that, in all probability, will rival the glory of the greatest upon earth."[15]

Inevitable as they believed it to be, urbanization was also a condition that most republicans feared would degenerate the young nation. "When they get piled upon one another in large cities, as in Europe," Thomas Jefferson fretted to James Madison about the American people, "they will become corrupt as in Europe." Jeffersonians thought that the Revolution and Constitution had temporarily suspended America in a more virtuous stage of national development, and they wished to postpone advancement into European degeneracy. As the citadel of Europeanization (and therefore, antirepublicanism), urban society flew in the face of republican thought. Benjamin Franklin memorialized the farmer as he condemned the "Great Man" of the city. Hector St. John de Crèvecoeur contrasted the more honorable, classless farming societies of America's countryside with the caste structures of its towns.[16]

Thus, western urbanization, while certainly not predetermined, was predicted. But it raised another concern about Kentucky's development: how to civilize both rural people living in a "state of nature" and urban people teetering on the edge of European-like corruption. According to Lockean thought, humans gave meaning to life through their own actions, and when guided by reason and science, they actively reconstructed social environments and social institutions. Even though many Americans were seemingly unaware of the dangers of wilderness and civilization, a social compact in the form of a written constitution promised to incorporate isolated individuals under the republican vision. If the majority of those individuals were farmers, it would also bind human society to the stability represented by the conquest and cultivation of the natural environment. As Jefferson, Madison, and others directed the flurry of constitution writing during the 1770s and 1780s, these conceptions of society and nature, civility and savagery, and the social compact became infused into the American political mind.[17]

Alongside Rush's rural elegance and the Lockean notion of a social compact was Collot's own idea of how to transform western life. When commerce took place along routes like the old buffalo trace, stringing centralized places together as potential pearls of commercial profit, urban and rural residents alike engaged the markets of an expanding commercial economy as they retreated from barbarism. Even those who refused to leave their farmsteads for occasional ventures to villages and markets would, if they lived along the road, reap benefits. As Daniel Drake experienced, "I saw aspects of things and people, which I should not have seen had we lived off the road, and the sight of which was no doubt intellectually beneficial." Whether the actual structures sat along the beaten path or not, Collot drew his farmsteads always contiguous to the road, where their occupants seemingly had opportunity to encounter the benefits of commerce.[18]

Commerce was crucial to civility, and the free operation of markets became an important component of western plans, since in its genesis, capitalism set forth a new sense of the market that complemented republican civilization. In 1776, Adam Smith's *An Inquiry into the Nature and Causes of the Wealth of Nations* elaborated on the theory of social progressions, concluding that manners, habits, and morals were shaped by the particular forms of economic and social development that characterized each stage of civilization. In an agricultural society, "the inhabitants of the town and those of the country are mutually the servants of one another," Smith explained. "The town is a continual fair or market, to which the inhabitants of the country resort, in order to exchange their rude for manufactured produce." The economic symbiosis between farmers and townsfolk—smiths, carpenters, wheelwrights, masons and bricklayers, tanners, seamstresses, shoemakers, tailors, butchers, brewers, bakers, and merchants—was natural: "Had human institutions, therefore, never disturbed the natural course of things, the progressive wealth and increase of the towns would, in every political society, be consequential, and in proportion to the improvement and cultivation of the territory or country." Within Smith's scheme, every marketplace offered some profit to an entrepreneur willing to invest: "His capital is continually going from him in one shape, and returning to him in another."[19]

Thus republicanism, refinement, and market orientation were potential plans for American development. But the theories of Smith, Rush, Jefferson, and other grand thinkers were formulated with a mind towards nation building. It would be left to individuals who wandered into trans-Appalachia to imagine how those theories could be applied in microcosm. In the early 1790s, men who had the wealth, authority, and inclination to revise the pioneer West began arriving along the Maysville Road determined to conquer the natural landscape, construct an economic order, and civilize society and polity. The American Revolution may have inspired an egalitarian impulse and freed the religious beliefs of the unorthodox, but it also swept away lingering constraints on capitalist development and made the social compact achievable. Resolving that the age of the pioneers along the beaten path was complete and a new era of Great Settlers was about to begin, gentry families packed up their households and headed westward, determined to address a pioneer culture that had allowed too much democracy, too much wilderness in which savagery could harbor, and too few markets to support a fully agrarian republic.

REFINING THE SOCIAL LANDSCAPE

In the 1790s, Great Settlers began to arrive along the road. Most were from the Tidewater and Piedmont regions of Virginia where society had become fairly bifurcated. "We have two sorts of people in this country," wrote one resident of the Old Dominion, "one called tuckyhoes, being Generall. of the Lowland old Virginians. The other class is Called cohees, Generally made up of Backwoods Virginians and Northward men, Scotch, Irish, &c., which seems, In some measure, to make Distinctions and Partitions amongst us." Part of the cultural baggage transported westward with Backcountry folk was an understanding that they were different from Virginians of the tobacco-growing regions. That may have been the primary reason that many of them migrated: to escape the political and economic dominance of the Tidewater and Piedmont gentlemen. While land was most often mentioned as the reason for migration, that thousands passed good land in western Virginia and refrained from moving into the federally controlled Northwest Territory suggests that it was land without regulation that they sought. They had it, briefly, before the "Tuckyhoe"-dominated Virginia legislature attempted to structure Kentucky's land distribution in 1779.[20]

With the promise of a more regulated land distribution, families of wealth or status or both from Tidewater and Piedmont Virginia moved along the Maysville Road in search of land. Frustrated by exhausted soils in the East, hundreds of Jeffersonian men and women wanted, as settler John Breckinridge explained, to provide "*good* lands here for my children, & insure them from *want,* which I was not certain of in the Old Country." Unlike earlier pioneers, the gentry had the wealth and connections to find and secure land quickly. As Virginia's economy struggled, the richness and fecundity of Kentucky's fields beckoned. Esteemed Virginians like John Randolph contemplated as late as 1813 how "In a few years more, those of us who are alive will have to move off to *Kaintuck,* or the *Massissippi,* where corn can be had for sixpence a bushel, and pork for a penny a pound. I do not wonder at the rage for emigration. What do the bulk of the people get here, that they cannot have for one-fifth of the labor in the western country?"[21]

The pull to the West was often counterbalanced, however, by gentry desires for luxuries, their emphasis on strong communal ties, and their reliance on family for social support and advancement. It was particularly difficult for women who, like Elizabeth Preston Meredith, opined that "I have just parted with my poor Dear Mother in a great deal of distress about my going to Kentucky indeed I have never met with anything in my life that give me so much unea[si]ness as parting with my Mother." David Meade determined not to sacrifice any of his lifestyle and relocated his entire plantation. The cavalcade of "the coachee, my chair, our baggage wagon, two hired waggons, . . . altogether twenty-one horses and about fifty souls" was a magnificent sight as it moved northward to the Monongahela River, where all were loaded onto flatboats for the trip to Maysville.[22]

Most migrants, despite their wealth, could not transport all of their possessions. Not surprisingly, upon settlement they worked to assuage their uneasiness by creating order within the chaos along the old buffalo trace. The gentry had to open and maintain a reliable trade route to the East. The beaten path no longer just connected neighbors and facilitated migration; it would bind citizens to the social, economic, and political phenomena of the Early Republic. By improving the natural track, the gentry wanted to make possible conveyance of commercial goods, distribution of news, extension of ideas about gender and status, processions of militiamen, transport of slaves, sale of domestic and agricultural productions, traffic to markethouses and courthouses, and most importantly, a distinction between the savage state of the 1780s and the civilized culture which they offered. Consequently, as gentry gained prominence on county and state levels, the purpose and egalitarianism of road maintenance changed. A 1797 "Act Concerning Public Roads" excused from road clearing those with two or more slaves who could do the work for them. The state assembly also transferred the selection of overseers from democratic election to county courts, the least democratic branches of Kentucky's nascent government, in which a small and elite corps of justices selected their own successors and perpetuated an oligarchical structure that determined everything from tax collection to ferry rates to road crews. The materializing highbrow intentions of courts and legislature became obvious by 1801 when an "Act to Amend the Act Concerning Public Roads" instructed overseers to grade roads "so smooth that carriages may pass with convenience."[23]

Those carriages traveled to and from genteel homes that arose convenient to the road. The gentry's rural and village retreats reiterated the republican vision so prevalent in their homeland: agricultural, patriarchal, and deferential structures that perpetuated the social values of Tidewater and Piedmont society. Architecture became an early delineation of genteel status. In 1796, when David Meade arrived in Lexington, catching a glimpse of a town still evidencing the vulgar architecture of the pioneers, he sheepishly informed his sister of the "many (I will not say elegant) handsome brick houses—and some few of stone—framed houses likewise & many of logs." The log residences of Lexington led many late-century arrivals to imagine that "the people live in cells, and are barbarous and uncultivated in their manners." Meade understood the bias and took pains to explain to his sister how the architectural distinction between a log house and a log cabin was a subtle but important one: "the former has a shingled roof—whereas the other has a roof of slabs put on without nails." The "uncultivated" more often resided in log cabins; in log houses lived "many opulent and some Genteel people . . . & I am told that they are by the latter made more than barely comfortable."[24]

Log cabins and log houses: the differences were minor but increasingly important to the gentry as they settled among the pioneers. Six miles north of Lexington, in a two-story residence named Cabell's Dale, one of the most prominent of the newly arrived Virginia families settled in a log house. Having migrated from Piedmont Virginia in 1793 as members of a large kin-group driven by wealth, large property ownership, political influence, education, and family heritage, John and Mary Hopkins Breckinridge cultivated an ideal of elite status by entertaining guests in a reception hall with carpets, polished pine floors, and yellow wallpaper. Compared to Meade, John Breckinridge was a man of moderate wealth when he arrived, but his ambitions quickly elevated him. Previous political experiences in Virginia had taught him that he, as a man of means, could command deference and respect even without direct, personal appeals to more common folk. He established himself as a political juggernaut: leader of the Lexington Democratic Society, associate to Thomas Jefferson and primary proponent of the Kentucky Resolutions, father of the state's 1799 constitution, and, as evidenced in that document, guardian of the gentry's interests in slaveholding, more oligarchical systems of local government, and the institution of voice voting by which the gentry heard exactly who voted for and against them.[25]

Like Breckinridge, when the gentry wanted to demonstrate growing regional and national affluence to their less refined neighbors, they took care to situate their houses in republican environments. Rush's accolade of "rural elegance" reverberated across the landscape. Long avenues of ancient

John Breckinridge. Courtesy of the Filson Historical Society, Louisville.

trees lined roads and paths leading up to Cabell's Dale and other country estates. Vistas were cut through woodlands, making genteel homes visible from a distance. Elms and chestnuts flourished as markers of status. Indoors, yellow poplar was recommended for floor planks since it polished into a golden hue. The dark colors of cherry and walnut made them common materials for fireplace mantles and other interior ornaments. Genteel country estates bolstered the ideal of elite status and brought republican civility to the countryside, particularly in the neighborhoods surrounding Lexington and Paris.[26]

Like Breckinridge, David Meade aspired to rural elegance. His depiction of log houses prefaced the construction of his own home south of Lexington. Situated on one hundred acres of rolling hills, La Chaumière des Prairies ("little thatched cottage on the plains") epitomized the transference of architectural gentrification to Kentucky. It became the most formal of Bluegrass estates. The house proper—an entrance hall, four chambers, and a dining room with black walnut wainscoting and a table adorned with decorated Chinese porcelain and printed ceramics—was of log construction. A year or so later, a frame building that accommodated two more chambers arose behind the log house. In 1800, a final addition was a brick octagonal pavilion attached to the side of the original building. Meade situated his home in a romantic landscape with a Chinese bridge and pavilion that brought a bit of the exotic to the meandering footpaths. A low stone fence covered with honeysuckle and climbing roses surrounded the estate. With a kitchen, dairy, smokehouse, distillery, servants' cabins, and assorted outbuildings all located behind the frame addition, Meade boasted "the first lordly home in Kentucky." In his mind, however, the stigma of log construction remained. Despite the size and opulence of his estate (the dining room alone, capable of seating up to one hundred guests, was larger than most log cabins), Meade still felt it necessary to defend log houses to his sister back in Tidewater Virginia.[27]

Meade's home also exemplified a gentrification of building materials, moving from log to frame to brick as the 1790s progressed. As the architectural symbol of Virginian gentility, the brick great house drew the greatest attention from visitors and neighbors alike. It typically was a two-story residence with elaborate gardens and outbuildings for stables and servants' quarters. At Levi Todd's Ellerslie, built east of Lexington in 1787, the first floor had a central hall that drew air through the building and ushered guests into the flanking parlor and dining room. Five years later, Todd doubled its size. Patterned after the great houses of the Tidewater, Ellerslie's symmetry and proportion announced the status and gentility of its owner.[28]

Brick exteriors, ornamental gardens, imported glass windows, brass knobs and locks: luxuries defined genteel architecture. But the most symbolic feature was beyond the view of less refined neighbors. A parlor, usually adjacent to the front hallway, was where gentry displayed upholstered furniture, libraries, and pianos. It was the entertaining venue for the cohort of self-styled elite who regularly gathered for social affairs. John Breckinridge, who politicked often, entertained guests in a carpeted parlor. Levi Todd, Fayette County clerk and guardian of the county's land records in a region where land titles proved indispensable to status, regularly used his parlor at Ellerslie for conducting business. The parlor was the most public space in the gentry's private homes.[29]

Stamping a genteel architectural imprint on the landscape was necessary as the gentry observed around them the persisting markers of pioneer culture. Log stations and portions of forts still stood: Bailey's Station just north of Washington, Irish Station south of Blue Licks, McConnell's Station in southern Bourbon County, McGee's Station in eastern Fayette County, and even remnants of Fort Lexington remained visible near the public square. There were neighborhoods of vulgar architecture: the rustic log houses in Mayslick, the crude frame shanties of the Irishtown neighborhood in Lexington, and hundreds of log cabins in various stages of deterioration along the length of the road. In numbers alone, pioneers in central Kentucky could not be ignored, and that was worrisome to the gentry because the inhabitants of such crude residences had seemingly succumbed to the wilderness, living in log cabins constructed from cheaply acquired common materials. An occasional pioneer might aspire to a more permanent structure of frame or stone construction, but generally even the finest of pioneer homes was less permanent, less genteel, and less commanding on the landscape. In contrast, brick great houses were luxuries, symbolically standing diametric to the log cabin: multiroom versus one-room, manufactured brick versus unhewn logs, permanence versus transience.[30]

Robert Patterson's Cabin (c. 1780), Lexington. Photograph by Rod G. Turner.

As they built their great houses, then, gentlemen and gentlewomen accepted the role of a permanent, self-appointed meritocracy that would order society and promote stability and cooperation between themselves and the masses. Social hierarchy and collective interest were key to the type of society that the gentry hoped to construct. Only in knowing one's place did the true republican citizen, regardless of economic status, fulfill moral and social obligations and benefit the political welfare of all. The contrast was simple: In the egalitarian culture of the pioneers, each individual pursued private interests, in the republican culture of the gentry, "each individual gives up all private interests that is not consistent with the general good, the interest of the whole body," as Pennsylvania republican John Dickinson once put it. The vision of social hierarchy also brought a new dynamic to cultural development along the old buffalo trace: socioeconomic class.[31]

Class plays a minor role in historians' ideas about early Kentucky. Conventional wisdom holds that, in the 1770s and 1780s, the struggles of settling a wilderness under constant threat of Indian attack leveled class distinctions. "Frontier democracy" resulted from a wilderness that gutted social and political aspirations. While Thomas Jefferson and friends foretold a republican society west of the mountains, a democratic one supposedly emerged for two reasons: the mood of the new nation teetered dramatically toward egalitarianism, and an abundance of land eroded the power of gentlemen over their "inferiors," who always had the option to set out on their own.[32]

This view of early Kentucky evolved in order to explain why, in 1792, Virginian George Nicholas would incorporate "universal" manhood suffrage into the state's first constitution. It rests on two inaccurate assumptions: that the pioneers were democratic, and that the suffrage was universal. We have already seen how little democracy actually characterized pioneer life. The degree of egalitarianism that characterized Backcountry culture did not manifest in "a widespread participation in public affairs, a diffusion of leadership, a widespread sense of personal competence to make a difference," as historian Eric McKitrick and Stanley Elkins defined democracy. Pioneers, both Backcountry types and the many sorts that followed, demonstrated little collective interest in political action. Engagement in public affairs was limited to small-scale petitioning and usually dealt with the locations of mills and warehouses, court schedules and locations, and personal complaints over lost land titles. Frolics more clearly evidenced widespread political participation, but in a most localistic sense. The selection of Big Men concentrated leadership in the hands of a few white men. It also demonstrated that a ready-made, nondemocratic structure of leadership migrated westward with pioneers.[33]

Pioneer culture along the Maysville Road, in other words, was too individualistic to facilitate a collective, manipulative attitude toward government. Culturally, groups tended to be competitive both internally and with other groups. And the few pioneers who had a strong sense of personal competence to make a difference operated through the religious sphere rather than the political one. From west of the Appalachians, government had become a fuzzy, distant institution that no longer warranted vigilance. Thus, democracy as a political or even social reality never germinated.[34]

When, in the late 1780s and early 1790s, Kentucky statehood was inevitable and the gentry began debating the form for a new state government, there was significant discussion and concern not about the problems with pioneer democracy but about the political ignorance of the masses. Lacking a tradition of political participation, pioneers were a dangerous crowd in which to invest power. "There is a wide difference between power being derived from the people, and being seated in the people," fretted "Sidney" in his *Kentucky Gazette* editorial of 1788; "Disorder and tyranny must ensue from all power being seated in the bulk of the people. . . . The opinions of the people at large, are often erroneous." On the eve of the first state constitutional convention, Fayette County lawyer Harry Innes expressed concerns privately to Thomas Jefferson that the "Peasantry are perfectly mad—extraordinary prejudices and without foundation have arisen against the present Officers of Government—the Lawyers and Men of Fortune. They say *plain honest Farmers* are the only men who ought to be elected to form our Constitution. . . . They have given a very serious alarm to every thinking man." To stave off a polity of pioneers whose cultural backwardness and ignorance threatened to unravel society, the gentry worked toward a republican society in which "thinking men" ruled.[35]

Even George Nicholas, who among the thinking seemed most attuned to democratic ideals unleashed by the Revolution, had reservations about the pioneers. The inclusive language found in Kentucky's first constitution reflected Nicholas's own identification with Jefferson's intellectual circles. But privately he expressed anxiety: "The peculiar character which belongs to our citizens in general will contribute for a time at least to our unhappiness. They were formerly citizens of other countries, and a great proportion of them have been induced to come here by a spirit of discontent or adventure. Citizens generally consisting of such men must make a very different mass from one which is composed of men born and raised on the same spot. Our people are all wise and *ought to be* great men; they see none about them to whom or to whose families they have been accustomed to think themselves inferior." The isolation characteristic of Backcountry settlement patterns had inhibited deference based on wealth or lineage, and Nicholas recognized that he and his "unhappy" peers might have to wait for the day when certain citizens understood "themselves inferior." The situation was truly a predicament for the gentry, a very small number of families with landed wealth who could not call upon tradition to legitimate their power, for tradition in Kentucky did not situate them at the pinnacle of a social pyramid. Much to their chagrin, there was a possibility that the throngs who lived and traveled along the beaten path would always remain problematic.[36]

George Nicholas. Courtesy of the Filson Historical Society, Louisville.

With this in mind, Nicholas decided to contract all settlers to the common weal by including in his writing of the 1792 constitution that "all men when they form a social compact, are equal, and that no man or set of men are entitled to exclusive separate public emoluments or privileges from the community, but in consideration of public services." Under the new constitution, settling in Kentucky would signify voluntary participation in society, a contractual association that acknowledged the autonomous individual but also bound him to the moral obligations of living in community. If the gentry could define and control those obligations, the social compact could lift unthinking men out of their states of nature. With the possibility of such a society, Nicholas felt secure in opening suffrage to all free white males of at least twenty-one years of age.[37]

It was a fleeting moment of democratic political opportunity. In the villages, and particularly in Lexington where an aspiring gentry was most influential, trustees successfully petitioned the state legislature to restrict local suffrage. In 1796, only Lexingtonians owning £25 worth of in-town real estate could vote for or serve as trustees. Paris, Washington, and Maysville followed suit, although they did not require a given level of proppertied wealth. Still, the prerequisite that a resident own a town lot eliminated many potential voters who rented. In 1800, the state assembly modified its "Acts concerning Towns in this Commonwealth," granting suffrage to all free male town lot owners and dropping the age requirement to eighteen years of age. Yet, trustees in Lexington, Paris, Washington, and Maysville again petitioned for and received exemption, continuing their exclusionary practices. By 1804, for example, this restriction eliminated 165 of Lexington's 298 heads of household. Not until 1812 did legislation harmonize local voting rights with state suffrage. At least for the first two decades of statehood, the village West was neither democratic nor representative.[38]

Nor was the rural West. Republican theorists surmised that only where men owned land could they exert enough independence to vote and serve disinterestedly. Fear of the poor and unpropertied had a long history in British and American political thought. By the 1780s, republicans had even begun to accept the usefulness of manufactures in providing work for the poverty-stricken, an economic concession that, in their opinions, threatened to push America into a more advanced and degenerative stage of civilization. As the foremost political theorist along the old buffalo trace, George Nicholas maintained an epistolary friendship with eastern republicans who shared his belief that poverty was detrimental to the creation of a republican polity. Idealistically, Nicholas hoped that, in a region with high landless rates, he could dismiss landownership as a prerequisite to good citizenship and expect that liberty would provide enough incentive for responsible voting and selfless representation.[39]

Such idealism proved unpractical. Nicholas's constitution actually left poorer residents—people who, without land, sought out work on tenant farm after tenant farm and from town to town—outside the political process. Their mobility became a means by which their political participation could be checked. Article 3 of the constitution required residency "in the state 2 years, or the county they offer to vote one year next before the election." While the results prove impossible to quantify, it stands to reason that large numbers of poorer and younger Kentuckians in search of land and labor were denied political participation regularly. Whether in local or state politics, premeditated or not, the political restrictions faced by large numbers of less established settlers exhibited a trend on the part of empowered gentry to eliminate the remnants of pioneer society and distinguish between civilized and savage, permanent and temporary, "us" and "them." The gentry could vote; less wealthy or sedentary neighbors often could not.[40]

In dictating the terms of citizenship, the gentry also introduced their own ideals about masculinity and femininity. Dismissing pioneer qualifications of bravery and physical skill as prerequisites for leadership, the genteel worked to establish social rank, evidenced by wealth, public virtue, and moral principles (such as temperance and devotion to the common weal), as the defining characteristic of leadership and masculinity. For example, in 1783, Isaac Shelby and wife Susannah Hart Shelby migrated across the Appalachians where he quickly established a political reputation based upon his military experience in Lord Dunmore's War and the War for Independence. In 1792, Kentucky's electors unanimously chose him as the state's first governor. When another war with Britain loomed in 1812, citizens returned Shelby to the governorship, and in the following year, he organized and delivered over three thousand volunteer troops to William Henry Harrison's army. Shelby, at sixty-two years of age, joined in the Battle of the Thames, one of the most decisive American victories of the war. With status, entitlement, political loyalties to the Jeffersonians, determination to eradicate the Indian menace north of the Ohio River, and a role in the organization of a republican government, Shelby exemplified civic virtue and republican masculinity. He was a pioneer when he arrived in Kentucky, but he rose above that station in the opinions of the gentry who regularly praised him in Independence Day toasts and early histories. If others followed Shelby's example, both public and private life could become energized with an assertive model of republican manhood.[41]

Travails of the western wilderness served the republican ideal of masculinity well. As in Shelby's case, the pioneer was often symbolically transformed into a republican, reinforcing arguments that those crass and uncivilized settlers of the 1770s and 1780s could become "thinking men." As with the pioneer ideal of manhood, Indians provided the perfect foils for republican masculinity. Jefferson himself drew a clear distinction between civilized manliness and barbarian cowardice in the Declaration of Independence, accusing George III of bringing "on the inhabitants of our frontiers, the merciless Indian Savages, whose known rule of warfare, is an undistinguished destruction of all ages, sexes and conditions," a military tactic that civilized society did not condone. Consequently, as patricians wrote the first histories of Kentucky in the 1810s, there was little shame in the defeat at the Battle of Blue Licks because the opponents acted more as savages than as men, making the defeated civil by default. Civility—the very attribute that republican theorists considered requisite to the survival and success of western development—became a quintessential quality of manliness.[42]

The gentry used public occasions to reinforce republican masculinity, civility, and citizenship. When news of Washington's death reached Lexington in late 1799, the town trustees organized a memorial procession. Despite a severe winter storm and a relentless cold spell, residents turned out in late January 1800 to commemorate their fallen hero in a melancholy march across several blocks to the First Presbyterian Church, into which participants crowded to hear an eulogy.

> Military, arms reversed
> Music, playing a solemn dirge
> Chairman and Trustees of the University
> President and Professors of the University
> Students
> Masonic Lodges, dressed with the insignia of their order
> Clerk of town
> Trustees of town

Clergy
Justices of the Peace
Citizens

Like frolics, processions were male-oriented events, but these were different men. Paraders were not competing to determine the most masculine and deserving of many. From military to education to government, those who filed through Lexington's streets were community patriarchs (or, as in the case of students, patriarchs-in-training), demonstrating appropriate masculine and deserving places in the republican social compact. Even in the final group of the procession, successful artisans and merchants joined in, expressing their collective identity as independent, republican men. The title of "citizen" reinforced paternalistic features of republicanism, eliminating from its own ranks those who could not display independence: white women, free and enslaved blacks, and white men with less than £25 in landed property.[43]

Despite their marginalization, women were important to reinforcing this republican ideal. Unlike pioneer frolics where wives often became insignificant to the ultimate purpose of the event, in the special moments that bolstered gentry notions of masculinity, such dependents were crucial. As audience, they furnished the reason for public ritual. These were white women of middling and upper rank who watched as the parade passed by. That they were wives, mothers, and daughters made difficult men's decisions to exclude them from full political participation. By conceding women's indirect involvement in patriotic moments such as Washington's memorial procession, however, men recognized female patriotism while restricting women from the slippery slope of partisanship. As white women observed parades, dined with guests of honor, and danced at patriotic balls, their participation remained circumscribed by the men who surrounded them.[44]

While the Revolutionary War had a substantial impact on white women, empowering them with duties and responsibilities previously considered manly, the patriotism that arose in the wake of revolution equally affected women, redirecting their duties and responsibilities towards a new public role that blended domestic virtues with an emphasis on contribution to community and polity. As mothers, republican women were to instill in children (especially sons) the duty, fortitude, and commitment central to patriotic participation in a republic. As wives, they were to sacrifice to the family, allowing husbands to concentrate on duty and freedom more acutely. As members of society, they were to display charity and virtue, fulfilling the family's communal obligations. Even as potential wives, they provided the same civic-minded purpose, as a toast at an 1808 dinner in Lexington elaborated: "Fair Daughters of America, in their choice of protectors, give a preference to men of virtuous and republican principles."[45]

The ideal of the republican woman enabled white females to express patriotism even as they remained outside politics. Otherwise, a woman's engagement in partisan bickering threatened to demean her and her husband, as when Rhode Island transplant Mary Ann Corlis visited the Lexington home of woolen manufacturer James Prentiss in 1816. She initiated conversation with another guest, William Henry Harrison, whom she later deemed *"no great things."* But the moment of great embarrassment for the hosts came when Harrison commented to Corlis, "I hope that you are a *good* Democrat," at which she quipped back whether "it was *possible* for *such* a thing to *exist*." Mrs. Prentiss dismissed Corlis's indelicacy as the haughty remnants of New England Federalism, but the damage was done. Overstepping the bounds of conventionally accepted ideals of genteel womanhood, Corlis returned to her Paris home with a severe admonishment. Only when properly executed did republican ideas of femininity promise to bridge the public and private spheres in ways that genteel women had seldom experienced in the past.[46]

By the mid-1790s, efforts to replace pioneer ideas of gender with appreciation for genteel ideas were pervasive. The pages of John Bradford's *Kentucky Gazette* and the *Kentucky Almanac* made sacrosanct the themes of republican government, values, and gender. So, too, did speakers at Independence Day and Washington's Birthday celebrations who advocated a more republican version of masculinity and, through their toasts, femininity. Breckinridge, Nicholas, and other seasoned politicians articulated the greatness of the day and, before recessing the genteel crowds to heavily attended balls or picnics, exhorted men to engage in agrarian republicanism and national development.

Yet, like the pioneers, the gentry were far from monolithic. More democratically minded gentlemen disapproved of this republican agenda. "A would-be Lexington Franklin," William T. Barry (future member of Andrew Jackson's "Kitchen Cabinet") derisively nicknamed Bradford. It was less a commentary on Franklin than on Bradford who, like most printers, claimed an openness to all views but was staunchly partisan and filled the *Gazette*'s columns with editorials that reinforced a republican vision, including the need to strengthen the social hierarchy. For Barry, the *Gazette* bordered on becoming an elitist rag with its subtle reinforcement of a deferential society and polity.[47]

Also standing along the parade route, adding symbolic weight to the march of independent white republican men, were enslaved black Kentuckians. During and after the Revolution, slavery stood in ideological juxtaposition to freedom: one was necessary to define the other. As John Adams bluntly stated, "There are but two *sorts* of men in the world, freemen and slaves." Many Euramericans accepted the brutal reality of slavery in order to benefit ideologically from its stark example as the absence of independence and materially from its labor. Reinforcing republican paternalism, the marchers impressed upon blacks their own political and social insignificance, thereby confirming further the independent status of white men.[48]

Symbolism was not sufficient, however. As part of their larger effort to order western culture, the gentry set rules for racial interaction. One unavoidable consequence of the unchecked character of pioneer society was that it gave both free and enslaved blacks liberties that were unacceptable in a slaveholding society. The respect for individualism that characterized pioneer culture opened opportunities for those who normally would have been marginalized and even oppressed in eastern politics and economy. Curiously, in acknowledging that all men were indeed created equal, Nicholas neglected to insert a racial qualifier in the 1792 constitution. Gentry responded with an erosion of African Kentuckians' economic and political opportunities. As early as 1787, Maysville's trustees prohibited slaves from "hiring himself or herself in the City." If a slaveholder acquiesced to a slave's endeavors, he faced a fine, as did any person who employed the slave. Additionally, slaves could not "sell any commodity," unless accompanied by an "authorized free person." In 1792, after the new constitution made slavery official, the state assembly expanded Maysville's law, passing "An Act to prevent dealing with Slaves" which restricted all white Kentuckians from trading coin or commodity with enslaved blacks. Two years later, sixteen citizens of Lexington implored fellow whites to "decline trading with our ... domestics, except it be in a lawful manner" (which was no trade at all!). Then came a 1798 slave code that applied weapons restrictions and courtroom policies to slaves, free blacks, mulattos, and Indians. At the 1799 constitutional convention, an overwhelming majority of proslavery advocates, desperate to regulate race relations and secure slave labor, muted debate on the topic and strengthened the institution. When the convention altered Nicholas's statement of community to read "all *free* men, when they form a social compact, are equal," the delegates literally removed enslaved blacks from a voluntary role in communal life and drew a clear distinction between citizens and noncitizens. The 1799 constitution also politically disfranchised all nonwhites, a change specifically targeted at propertied free blacks. By 1807, Lexington's trustees further reduced the economic privilege of enslaved African Americans by forbidding the practice of hiring out.[49]

While race and economics were central to shaping these laws, the ideological paradox between slavery and independence was also at play. The gentry *had* to dominate black slaves in order to demonstrate fully the distinction between those who had won freedom and the right to resist despotism and those who had not. How capably could slaves resist authority if they were "raged, foolish, and, in appearance, miserable," as one visitor described in 1819? Another tourist of the early 1830s noted Kentucky's slaves so broken that "if it were not for importation, the race would hardly be continued in many places." Slavery in pioneer Kentucky had not portended such devastating results. With the arrival of the gentry and their desires to transform the region from a society with slaves to a slaveholding society, the institution moved in a different direction. Racial lines were legislated. By the turn of the nineteenth century, slave patrols wandered the beaten path nightly, enthusiastically carrying out their missions to detain and flog slaves found without written permission to wander beyond their plantations. In 1810, Lexington's board of trustees directed its slave patrol to arrest "all verry suspicious blacks who may be found on the streets in doubtful situations." In private homes, racial lines were drawn equally rigidly: "negroes are not seated at the same table with the master of the family." Even in seemingly meaningless ways, the gentry worked to force blacks into deference. In naming slave children, owners who "ransack the Christian volume, that they may find fit names to their children" thought they humbled blacks with "heathenish appellations, such as Pompey, Nero, &c. usually given to dogs." Just as the pioneers proposed to appropriate lands with designations of ownership, so too did the gentry lay claim to black laborers with classical names that had come to represent servility.[50]

As they did in defining gender and delineating race, the gentry also worked to establish a new religious order along the old buffalo trace. Many embraced a spiritual consciousness based on an organic, hierarchical concept of society as reinforced by Enlightenment rationality. Among mainstream denominations, Episcopalians and some Presbyterians most ostensibly espoused this vision. Like Christian purists, members of these churches were moralistic, exclusive, and afterlife oriented. But they also employed formal rituals and abided by a series of creeds and liturgies that were heresy to the Rankinites. Their "Old Light" pietism accepted ideas of eternal damnation, confession of sins, and doctrinal truth as much out of tradition and theology (the mind) as out of faith (the heart). They also appreciated the patriarchal structure of their denominations and their doctrines: a deferential order that they expected to be replicated throughout society. The Wattsites' First Presbyterian Church and the American Episcopal congregation served Lexington's wealthiest and most renowned citizens, whose prosperity and charity suggested God's favor towards them.[51]

The Episcopal Church, partly because of the individuals who peopled its pews and partly because of its association with Anglicanism, drew the most vocal criticism from scriptural purists. "It was regarded as the persecuting ecclesiastical arm of the British government," Daniel Drake explained about the Episcopalians, "an organized body of Arminians, enlisted in the service of despotism." Purists interpreted Anglican persecution of John Bunyan and the church's liturgical appeal to the *Book of Common Prayer* as disdain for the literature of traditionalists who "read the *Pilgrim's progress* more than any other book, except the Bible." Possibly most disturbing was that in the pews of Lexington's Episcopal congregation sat gentlemen and gentlewomen (and after the turn of the century, a new middling class of lawyers, merchants, and bankers) whose actions seemed to undermine the pioneer culture that had nurtured purist Christianity. The result was public condemnations of gentry self-interest, avarice, and vanity.[52]

In 1803, a new brick Episcopal church arose just north of Lexington's town square, and over the next five years, the congregation worked to cultivate a parish. The list of twenty-four communicants and noncommunicants who subscribed to and rented pews reads like a who's who of the local elite. By 1813, the congregation outgrew the building, and a stucco-covered brick church capable of seating eight hundred people arose in its place. Its size conveyed the socioeconomic weight of its congregants. On Christmas Day, when Episcopalians "handsomely decorated with evergreens, cedar and box vine," the elegance of their denomination appeared most striking to purists who, if they celebrated the feast day at all, did so with guns or in unadorned manner. Traditionalists like William Price rejected such pretensions, as when he remembered his last Christmas dinner in Backcountry Virginia where he and his friends gathered for fellowship and revelry: "No Episcopalians has been invited. Such people are too aristocratic and overbearing. The people who are communicants of that church try to imitate their aristocratic brethren of England. . . . What we want in the church as well as in State is plain, practical, naive, devoted men who know and mingle with the people as one of their Selves." The attitude migrated with many Backcountry settlers who suspected that Episcopalians enjoyed too much materialism and status, and that purists should remain leery of the ostentatiousness.[53]

Old Lights did not take the critiques of their denominations well. "I should often be made to wonder at the extraordinary havock of superstition," wrote lawyer William T. Barry in 1801, "did I not reflect that it most commonly affects the more illiterate part of mankind, whose feelings are easily worked upon, and whose passions instead of being subjected to reason, are suffered to pre-dominate." In their attempts to refashion religion along the old buffalo trace, gentry gave little consideration to the superstitious and dismissed notions "that such ignorant proselytes, no matter how numerous, can add much dignity to this Christian religion." Disapproval of scriptural purists and the impropriety of continued superstition arose occasionally in the *Kentucky Gazette* and the *Kentucky Almanac*. In one story relating the near execution of a witch in Virginia, the editor blamed "the ignorance and prejudices of an illiterate people." Still, superstition remained commonplace, a clear indication to most of the gentry that a philistine culture remained all too influential.[54]

The contest between worldviews might have remained purely rhetorical if it had not been for the rise of yet another group. Harry Toulmin and other former Old Lights advocated a faith even further removed from literalist traditionalism. Theirs was a theology mediated by Enlightenment values: extending universal benevolence to all who abided by Jesus' teachings, and rejecting the Trinity, the reality of Satan, and the divinity of Christ. Although purists interpreted Unitarianism as a betrayal of Christianity, as long as the practitioners were few and unorganized, there had been little need for public condemnation.[55]

In 1794, Toulmin went before the Fayette County Court for a license to perform marriages. He could not produce evidence of his Unitarian "ordination," and a "bigoted" Presbyterian justice led the bench's charge to dismiss the request. Suddenly eager to defend Toulmin, a handful of attorneys (all members of the local Episcopal and First Presbyterian congregations and "fired with indignation") argued for five hours that the court's inclination threatened "a step towards religious domination." In the end, the justices and lawyers "ordained" Toulmin as "Christian Minister to the Independent Society."[56]

Toulmin's ordination was just the first salvo. Lexington, once christened the "Philadelphia of the West," became "this Sodom" among disheartened purists, and in the mid-1790s, the sale of Jeffersonian-Jacobin political texts became the next battlefield. Irish book-dealer Joseph Charless promoted sales of works by Thomas Jefferson, Mary Wollstonecraft, David Hume, Bolingbroke, Montesquieu, and Thomas Paine that inspired republican gentry and disconcerted the purity-oriented. The last author, in particular, raised eyebrows. In *The Age of Reason,* Paine denounced organized religion and with it the core of American spiritual life. Among the town's republican leaders, the book became a sensation, selling out almost upon arrival. In 1796, Thomas Barr reminded Samuel Trotter, his merchant friend who was on business in Philadelphia, "before you start from the city, don't fail in getting the second part of Age of Reason, and if possible an answer to it." After he read the book, Robert McAfee abandoned his own faith for a year; William Bledsoe, a Baptist minister, "quit preaching and has renounced his Bible."[57] Among the gentry, Christian doctrine seemed to be thinning out: Old Light Presbyterians and Episcopalians were among the most vociferous readers of Paine.

Purists compensated by renewing their own religious fervor and condemning the gentry's rage for Paine as synonymous with infidelity. In 1795, preacher David Barrow lamented how Kentuckians had "no true vital religion" because the "Deists as the disciples of Bolingbroke have been much strengthened as they have thought by a late publication of Thomas Paine." Deism's emphasis on individual reason and conscience sharply contrasted with the idea of divine purpose so popular among purists. It made faith irrelevant to the present and, because so many political and social leaders expressed interest in its tenets, threatened to undermine Christian influence, especially that of literal believers. A visitor to Lexington in 1798 complained that "the writings of infidels, and particularly Thomas Paine's *Age of Reason* was extensively circulated, and his principles imbibed by the youth particularly, with avidity; so that Infidelity, with all its concomitant evils, like a mighty tide, was desolating the land, with respect to religion and morals."[58]

If he had explored further, the critic might have uncovered the genesis of a reactionary movement. In 1802, during his Transylvania University valedictory address, Joshua Wilson warned fellow graduates to beware the "infidels who are excellent in public business. . . . They are men of natural and acquired abilities—tolerably skilled in Politicks and jurisprudence—they reject Christianity more from inattention, than from stupidity or wickedness—free from dissipation—lovers of liberty—haters of licentiousness—promoters of learning—attentive to business and possessors of common interest with yourselves." Wilson assured his audience that, as political and economic leaders, the gentry had satisfactorily molded Kentucky into a republican state and upon "these men you may with safety bestow your suffrage." As religious exemplars, however, they had failed. The "Blackest Atheism seems to be prevailing here," he fretted. Many of these "ungodly" met as "a devilish society," the Free and Easy Club which took its name from the local tavern in which it assembled every night for drink and song, twice a week for cards, and once a week when "an infamous scoundrel preaches them a sermon."[59]

The infidels, along with Paine and Satan, became the enemies of the scripturally pure, at least in the minds of many Methodists who sang: "The world, the Devil, and Tom Paine / Have try'd their force, but all in vain, / They can't prevail, the reason is / The Lord Defends the Methodist." Preacher David Rice articulated an emerging theological continuum along the old buffalo trace as one "from Calvinism to Arminianism to Universalism to Pelagianism to Semipelagianism to Arianism to Socinianism to Deism to Atheism." Situated opposite from Calvinists, Deists posed a frightening challenge to purity-based doctrine. Arminians, like the Episcopalians, voiced less concern because they shared a social vision similar to Unitarians and many of the "infidels." Attacks on Old Light orthodoxy did not stream from the pens of Deists or Unitarians; it was the purists who actively derided the Episcopalians and Old Light Presbyterians.[60]

The perceived swelling of infidel ranks exacerbated the theological turmoil. Josiah Morrow noted upon his visit in 1797 that Christianity had been given "a deadly stab hereabouts" but hoped that "the glory of scriptural religion, tho obscure for the present, will shine forth hereafter with redoubled luster." One critic decried how the gentry commonly moved with the latest intellectual and religious fads "in order to secure popularity, and promote their interest." Just as the region's upper sorts sought to exclude the lower sorts from political voice and social opportunity, purists barred gentlemen, merchants, and lawyers from eternal paradise. One purist took exclusion to the extreme, proclaiming "that the Evangelical doctrine is neither preached nor believed by any denomination on earth, except the Regular Baptist" whom, he continued, were the "people among other denominations that he will save."[61]

Of course, we have already seen how events in Adam Rankin's Mount Zion Presbyterian Church led to schism and the creation of a gentry congregation at the First Presbyterian Church. But that was only the beginning of the war between purists and gentry to control Lexington's Presbyterian community. In 1798, James Blythe accepted appointment as professor of natural philosophy at Transylvania University. As a stalwart Calvinist and scriptural purist, he reproved the Wattsites at the First Presbyterian Church for Arminian notions that every individual could attain eternal salvation. There may have been some personal impetus for this, given that James Welsh, minister to the First Presbyterian congregation, was a colleague to Blythe at Transylvania. But Blythe couched his attack in religious rhetoric, arguing for orthodoxy and staging his own services for Presbyterians who appreciated his vision of divine order. When John Lyle, minister of the Salem Presbyterian congregation near Paris, journeyed to Lexington in October 1801, he "found the amnity between Blythe & Welsh was higher than ever & greatly increased by Blythes preaching at night without an invitation."[62]

Transylvania University became the arena in which Christians fought to control religious culture along the Maysville Road. Within a decade of its founding in 1783, the Presbyterian school became the scene of one controversy after another. In 1794, the state legislature appointed John Bradford, editor of the *Kentucky Gazette* and superstition-denouncing Episcopalian, as chairman of the university's board of trustees. When the college president resigned that year, Bradford led the move to hire that "disciple of [Joseph] Priestley"—Harry Toulmin. The Unitarian candidate and newly ordained minister of the "Independent Society" rattled the purists on the board. After two votes, he barely secured the post.[63]

Toulmin's term lasted only two years before his opponents persuaded the state legislature to rescind his election on technical grounds. In 1796, James Moore became president of Transylvania University, for the second time. He had served as the first president of the college before Toulmin but, overburdened by work and a poor salary, had resigned. Although a moderate Presbyterian in his first term, soon after assuming the presidency a second time Moore became rector of the local Episcopal congregation, affiliating himself with a less purist crowd. The university took on a decidedly less religious purpose, appreciating students as rational men who could direct their own intellectual and spiritual development: "Whilst the trustees announce to the public the solicitude with which they are determined to watch over the morals of the youth, they pledge themselves in the most unequivocal manner, that no influence shall be used to inculcate those principles of religion which are characteristic of the different sects. It is their unalterable determination that the students should be left at perfect liberty in the formation of their religious creeds." Later that year, Moore addressed students, emphasizing intellectual curiosity in all matters, including faith. Under his leadership and Bradford's watch, the school took a more "enlightened" direction: James Brown assumed the professorship of law; Frederick Ridgely and Samuel Brown became medical professors. The three men and Moore, Virginians all, so embraced rationalism and the moderate republican ideals which Toulmin promoted that Thomas Jefferson himself encouraged friends to send their sons to Transylvania University. "The citizens of Lexington will soon be invoking the muses," bragged Samuel Brown, "learning will soon be *all the rage.*" For the gentry—Old Lights and Unitarians alike—the freedom to think for oneself, to expand beyond the narrow confines of superstition and suspicion, was central to republican life. Transylvania University students provided a captive audience for this brand of theological and intellectual instruction.[64]

The ascendancy of gentry control was not complete until early 1801 when several students protested the teachings of James Welsh, of all people. The incident actually had little to do with theologies. Rather, Welsh's Federalist leanings angered students who were fervent about Jeffersonian republicanism. Purist Presbyterians did not like Welsh or his employment of *Watt's Psalms,* but they saw him as a potential ally in regaining control over the university. They tried to silence the students, but a majority of trustees agreed to hear the complaints. One student attested, "I recollect that on the day of the feast at the Factory Mr. W[elsh] observing covered dishes to be carried there, said that notwithstanding the boast of Republicans those dishes were sumptuous enough for George the 3rd, that true Republicans should eat out of wooden platters. At the same time I also heard Mr. W. observe as well as on other occasions, that most of the partisans of Mr. Jefferson in this place were so, because of his Deistical principles." Unwilling to take the testimony seriously, the board refused to dismiss Welsh, but when fifteen students withdrew the following morning, his professorship ended.[65]

Thus, as the century turned, the gentry was actively redefining the natures and roles of religion, gender, race, and political suffrage in this new American West. According to pioneer Joseph Ficklin, there had been a time and place, back in the Old Dominion, when "the common man when he went to the gentleman's house, didn't pretend to go in, but stood at the door and took off his hat." This was what the gentry desired along the old buffalo trace, a deferential and patriarchal order that would prop up their own status. When John Brown moved to Kentucky in the mid-1790s, his wife Margaretta arrived amid a pioneer culture that clashed with her expectations of social hierarchy. Despite his admonitions "for some of my Aristocratic notions," having to socialize with pioneers led her to complain that "This *equality*, my Love, is a mighty *pretty* thing upon *paper*," but "where none are *beaux*, 'tis vain to be a Belle."[66]

As the 1790s closed, it appeared as if the gentry's actions would transform Kentucky into a beau society. When southwestern lands opened in 1797, a refined gentry rejoiced that "the rage among the poorer class of people here appears to be for the Spanish settlements." When outmigration from Kentucky escalated in the early 1800s, they applauded the phenomenon because it would "improve the morals of the state, as it will purge it of many of the *pioneers*." Re-creating life along the old buffalo trace was not an easy task, but the gentry would demonstrate themselves quite capable of limiting social conflict and bringing most white Kentuckians around to their vision of community, including the creation of a refined world in which beaux and belles were in fashion.[67]

VIRTUES OF THE FATHERS

The Fourth of July was the most celebrated day of the year along the Maysville Road. In the late 1780s and early 1790s, militias paraded about to the music of an accompanying band before leading crowds to community picnics. Independence Day celebrations were large frolics, inviting citizens to forget ethnic and religious differences in order to celebrate their common American identity. Militia troops often carried before them a liberty cap symbolizing the unity of these western pioneers. After the arrival of the gentry, however, the day took on a different tone. Militias still mustered, bands still played, and residents still picnicked. But beginning in 1799, self-appointed men of stature began giving speeches expressing their versions of "the importance of the meeting, and the importance of the day." And after Thomas Jefferson's election to the presidency in 1800, a reading of the Declaration of Independence became part of the ritual as well.[68]

Independence Day speeches followed a common format. Reiterating political economists of the 1780s who had anticipated a western republic in Kentucky, orators lauded those who "sought for liberty, and found it in a wilderness," as Joseph C. Breckinridge (John's son) pronounced in 1812. Through the trials of a revolutionary war, an incompetent confederation, and the settling of the western wilds, Kentuckians had become the epitome of "the American," exhibiting a patriotism unmatched elsewhere. Then came the warning, always sternly put, that "the continuance of your freedom, and happiness, will be precisely commensurate with the continuance of your publick virtue." Of course, public virtue reflected one's reclamation of "native dignity," of achieving the ideal of republicanism.[69]

The historiography of early Kentucky seems to dwell on one question that has never been clearly enunciated by historians: Why did the settlement process *not* result in the violent social conflict typical of other contemporary backcountries? In quests to prove internal discord, scholars have made too much of a few Kentuckians' occasional fascination with secession, the radical character of the region's Democratic-Republican societies, and a "whiskey rebellion" that was little more than a general disregard of an unenforceable law. Even when discontent manifested, it was more rhetorical than actual: "These town people [referring to Lexington's Democratic Society] made so much noise last year, spoke so big, and opened so many correspondences," complained a reader to the *Kentucky Gazette* in 1795, "that many of us were ready to join our brethren insurgents, in the other parts of the United States particularly those in Monongahela." But they did not join in Pennsylvania's Whiskey Rebellion, and they did not secede with Aaron Burr or James Wilkinson. While it becomes easy to read socioeconomic conflict into these episodes, the evidence remains unconvincing.[70]

In contrast to eastern Backcountry settlements where discontent and social turmoil targeted an established hierarchy, the absence of a rigid social structure in the 1780s and early 1790s Kentucky made such antagonisms unnecessary. In Maysville, Washington, Paris, and Lexington, those who served as community leaders more often than not were pioneers themselves. Even as new men of wealth and political ambition arrived along the Maysville Road, a pioneer Big Man typically continued to share authority on local boards of trustees. In Lexington, Robert Patterson was a trustee from 1792 through 1803, serving as chairman his last seven years in office. He had been at the creation of Fort Lexington, and his was the first signature on the "citizens' compact" of 1780 that allocated in-lots and out-lots. But it was Patterson's reputation as a farmer and marksman that sustained his career; and he was revered for his extensive knowledge of the region's development. In Washington, it was Daniel Boone who fulfilled this role; in Paris, it was Henry Duncan.[71]

When the gentry began to arrive along the Maysville Road, a class revolt did arise, but it was one of a more conservative nature which, by 1810, turned pioneer Kentucky on its head. Determined to change the world in which they settled by tempering democracy and securing privilege, the gentry positioned themselves as the architects and guardians of a civilized society in a world that they considered most uncivil. They did this by appealing to pioneers' patriotism and, as July Fourth orators exemplified, their own positions as the perpetuators of Revolutionary ideals.[72]

Images of the American Revolution seldom include Kentucky or any western lands for that matter. Where the conflict did encroach upon the trans-Appalachian wilds, historians have described it more as a struggle between pioneers and Indians than an imperial contest over the destiny of a continent. Even after the Treaty of Paris in 1783, hostilities across the Ohio River Valley raged, suggesting a distinction between the War for Independence and the western Indian Wars. In truth, however, Kentucky was born in and shaped by the Revolutionary moment. As the war approached its final stages, leaders of the new confederation of states both officially and unofficially drafted plans for the western lands that they anticipated inheriting. Republican ideals that triumphed with American victory propped up schemes for westward expansion. In Kentucky, place-names like Washington and Georgetown (named for the commander of the Continental Army), and Louisville, Paris, and Versailles (named in honor of the French allies) became monuments in the wilderness, relating the designs of Virginia's gentry legislators to celebrate and ensure the Revolutionary heritage in the West.[73]

Pioneers, too, had linked the Revolution's goals to their own aspirations. William McConnell left the Pennsylvania Backcountry in spring 1775 to find opportunity for personal advancement in the American Eden.

In May, he and a party of fellow dreamers set out for their futures, departing from the western reaches of Pennsylvania, journeying down the Ohio to the Kentucky River and from there to the Elkhorn, up which they paddled until they reached the center of the Bluegrass. In June, the party cleared trails and began constructing a cabin that McConnell expected would secure his claims. They were not the first Euramericans in the region: Fort Harrod arose in the summer of 1774, and Boonesborough followed in the spring of 1775. Yet, on June 5, they made a vital contribution to the shaping of life along the old buffalo trace. As they rested around a spring, word arrived of a conflict erupting in New England that would engulf them and their families in a struggle for liberty. McConnell and his party—in fact, all of the long hunters, surveyors, and curious explorers who wandered trans-Appalachia—were violating Great Britain's Proclamation of 1763, and they identified with the risks of rebellious Massachusetts farmers, appropriating "Lexington" as the name of their settlement.[74]

The designation of Lexington represented another way in which pioneers laid claim to the new West. But unlike labels of ownership, patriotic memorializing reflected a sense of *national* connection and purpose, a desire to shape western identity through the filter of the American Revolution. Still, in comparison to natural labels and designations of ownership, patriotic names were few and far between in the pioneers' world. Not until the gentry appropriated the Revolutionary heritage in order to shape identity and republicanize Eden did patriotic place-naming become widespread.[75]

Through Virginia's assembly before 1792 and through Kentucky's assembly after 1792, gentry determined the legalities of land acquisition, the placement of post offices, and the names of polities. Legislators realized that, like pioneers who laid claim to streams and stations with personal names, harnessing the mind of the West began with place-naming. Virginia's assembly purposefully designated early polities along the Maysville Road: in 1780, the state carved Fayette and two other counties out of its western territory previously known as Kentucky County; in 1785, the assembly cast Bourbon County from Fayette and, in turn, divided Bourbon to create Mason County three years later. The names for these counties derived from the Revolutionary heritage: Fayette for the Marquis de LaFayette; Bourbon for the French royal family; and Mason for George Mason, one of Virginia's signers of the Declaration of Independence.[76]

When Kentucky became a state in 1792, its legislature similarly undertook efforts to place symbolic labels on new counties along the road. In 1798, when Fleming County emerged from Mason and, in 1800, when portions of Mason and Bourbon combined to create Nicholas County, the names were familiar to most settlers. George Nicholas fought as an officer in the war, served at Virginia's ratifying convention for the federal Constitution in 1788, settled in Kentucky in 1789, and drafted the state's 1792 constitution. John Fleming arrived in Kentucky in 1779, served as a deputy surveyor from 1786 to 1788, and constructed Fleming's Station in 1790. The two men had very different roles in trans-Appalachian development, but both served as models—one through promoting political continuity with the ideals of the Revolution, the other by facilitating the settlement of an agrarian population. Kentucky's assemblymen lauded a state born within those years of independence and national and regional formation.[77]

Legislators' actions might have been in vain, but the manner in which the McConnell party reacted to news of Lexington and Concord bespoke pioneers' receptiveness to Revolutionary imagery. To pioneers, Kentucky symbolized desires to escape governmental regulation, to own property without fear of confiscation, and to secure some degree of self-governance. The geographical distance alone would have made all three seem simple. Trans-Appalachian migration served political as well as economic purposes, particularly endowing settlers with the "separate and equal station to which the laws of nature and of nature's God entitle them," a claim Thomas Jefferson made for all colonists in his declaration of 1776. The naming of the five counties along the beaten path memorialized those political ambitions.[78]

Virginia and Kentucky legislators did not stop with county names. In 1786, the Virginia assembly officially designated the settlement of Washington in honor of the Continental Army's commander-in-chief. In 1790, it reversed an act of 1789 that granted Lawrence Protzman permission to name a settlement after Hopewell, his former New Jersey farmstead, choosing instead to rechristen the hamlet as Paris to reflect its semiotic relationship to Bourbon County. Before statehood, Kentucky's three most populous counties—Mason, Bourbon, and Fayette—and their largest towns—Washington, Paris, and Lexington, respectively—were semantically connected to the Revolutionary moment.[79]

Such place names suggest that while it did not replace loyalties to local life and traditions with a new national society and culture, the American Revolution did incorporate thousands of individuals and hundreds of local places into a relationship with the nascent structures of a national entity.

Naming patterns alone could not guarantee identification with the Revolution, however. If they were to be successful in countering the centrifugal forces of the wilderness and the pioneers, the gentry had to incorporate the Revolution into Kentucky's past, making it the apex not only of the new nation's short history but of Kentucky's as well.

The unofficial but very effective leader of this effort was John Bradford, editor of the *Kentucky Gazette* and the *Kentucky Almanac*. He migrated from Virginia's northern piedmont in the late 1770s, working as a land surveyor and fighting in campaigns against native villages north of the Ohio River. Without any printing experience, he began publishing the *Gazette* in 1789 and the *Almanac* in 1792. In the *Gazette*'s first issue, he apologized to his customers for a delay in publication caused by the conditions of the buffalo trace: "A great part of the types fell into pi [disordered printing type] in the carriage of them from Limestone to this office, and my partner, which is the only assistant I have, through an indisposition of the body has been incapable of rendering the smallest assistance for ten days past." Despite the tardiness, the newspaper immediately became the arena of political debate in Kentucky. Bradford printed pieces arguing both sides on the issues of separation from Virginia, the location of the state capital, and the "conspirators" in Wilkinson's secessionist movement.[80]

The editor's most important role came after statehood when the gentry began to articulate their republican vision. Both the newspaper and the almanac became vehicles for patriotic republican rhetoric. The timing of Kentucky's development held particular significance. Its history paralleled the formation of a nation, as evidenced in 1794 when Bradford printed in the *Kentucky Almanac* a chronology of "remarkable occurrences," a recapitulation of "national" events since 1774 (not coincidentally, the year of Kentucky's first permanent white settlement), culminating with Kentucky statehood in 1792 as if it were the rational conclusion to the American Revolution. Frontispieces in the almanac typically read "xth year of American Independence, xth of our Federal Government, xth of this Commonwealth," a practice in imitation of French Jacobins who dated history from their own revolution, thereby underscoring the notion that Kentuckians' ability to identify as Americans depended upon their participation in (or their inheritance of) the events that created the nation. He used the *Gazette* to announce forthcoming patriotic celebrations and afterward to report the importance of such events. After the 1788 Independence Day celebration at which participants joined in song, Bradford printed the ode to reinforce the association of region to American liberty: "When the Almighty Fiat gave / 'Creation's boundless range' a birth, / The choir of Angels hail'd our Land, / The Land most favour'd of the Earth. / Hail Kentucke! Kentucke, thou shalt be / For ever great, most blest and free." Fourth of July toasts and orations filled the paper's columns, reminding Kentuckians of their intimate relationship to the Revolutionary moment and the significance of their own land to the national future. Bradford's purposeful publication of orations, toasts, and pieces that encouraged a nationalistic and republican perspective certainly made him one of the earliest of the Early Republic's partisan printers.[81]

Bradford and other republican gentry emphasized patriotic events because, despite the inclinations of pioneers to appreciate nationalistic rhetoric, there was an absence of patriotic space along the Maysville Road. Unlike eastern states which sported battle sites, Kentuckians had no geographic wartime monument to confirm Revolutionary membership. Daniel Drake recalled how "father and mother emigrated from a densely settled part of New Jersey, and had passed through the stirring scenes of the Revolution, in which their native State so largely participated." Memory is anchored to places, but Kentuckians had no physical landmarks to stir their passions and excite their nationalism. The region's only Revolutionary conflict had been the Battle of Blue Licks and, although the deaths of nearly seventy pioneers and the routing of almost three times that number at the hands of Shawnees could be attributed to dishonorable tactics by the Indians and their British supporters, the catastrophe was hardly worthy of commemoration. Pioneers and gentry alike incorporated the battle as best they could into their identification with the Revolutionary moment. Local lore situated their continued engagement in the Indian Wars as a theater of the Revolutionary War, with the hated British manipulating the Shawnees from behind the scenes. Having the tales related to him time and again as he traveled from Maysville to Lexington, tourist Fortesque Cuming mused that "there was not a half mile of the road between the two places unstained by human blood!"[82]

While a meritorious battlefield was lacking, honorable battlefield soldiers were not. The war conferred on veterans, both military and political, immediate recognition. As one Fourth of July orator exclaimed, "by honoring the acts of our fathers, we call forth those feelings of patriotism and of valor which stimulated them, and which will perpetuate the inheritance they transmitted to us." Most were from Virginia; in fact, nearly half of the

Old Dominion's soldiers eventually relocated to Kentucky. At an Independence Day celebration in 1794, settler William Price sat with forty men "who had served in the late struggle for American Independence. It was a glorious Sight to behold." Politicians promoted nationalistic pride by championing their neighbors as men who sacrificed for the cause of liberty. "The spirit of '76—Woe to the sons that prove unworthy of the virtues of their fathers," warned Lexington's *Reporter* in 1811. In a fledgling society, the *Kentucky Gazette* proclaimed, the aging veterans served a significant cultural purpose as role models for "offspring of our old Revolutionists . . . [who should] bear in mind the toils of their Sires, and not suffer themselves to be imposed on by any Tyrannical power." Many lived rough-and-tumble pioneer lives, but the gentry were willing to ignore the rudeness of their circumstances in order to present a collective standard of morality. Hence, the chasm between gentry and pioneers often faded at Fourth of July festivities, promoting a collective sensibility that all white Kentuckians, regardless of status and no less than easterners, were Americans.[83]

Cultivating historical memory was the most direct route to binding Kentuckians politically, socially, and economically to the new nation. J. J. Polk never forgot his father, "still bearing implacable hatred to kings and kingly power," trying "to instill the same spirit into the minds of his sons." On long winter evenings and in leisurely hours, the elder Polk gathered his boys around him and related "the conflicts between the Whigs and Tories, and in giving an account of the hard-fought battles in which he and his brothers had participated." To "thinking men," it was expected that, even as western populations became "civilized," the threat of self-interest remained. Polk made his sons pledge "fidelity to the government, constitution, and laws of the United States, and especially to Mr. Jefferson's administration." In a general sense, republicans throughout the new nation worried about the effect of localism on an emerging American character. James Madison expressed how, as America expanded, a "spirit of *locality*" would distract representatives from "the aggregate interests of the [national] Community, and even sacrifice them to the interests or prejudices of their respective constituents." In trans-Appalachia, that possibility was magnified by distance. Even as late as 1803, an editorialist in a New York City newspaper applied Madison's concerns to Kentucky, explaining how the region's situation had been and, without reliable access to the port at New Orleans, would continue to be "highly distressing—Cut off from every communication with the Ocean; their timber will become useless except for fuel; their agriculture will decrease, their mills never used, but for the support of flour for domestic consumption; the nerve of the country will be totally unstrung, and they will conceive themselves deserted by the union, and left to work out their own temporal salvation, by and for themselves."[84]

If Kentuckians adopted the self-interested attitude that they alone carried the torch of liberty in their agrarian republicanism, a "spirit of locality" could actualize. The pioneers of the 1780s and early 1790s, for example, steadily became convinced that a federal entity would not address the Indian menace, land distribution problems, or the need for an open Mississippi River trade. For many, the Confederation Congress seemed more interested in lands north of the Ohio River, and after 1788, the Washington administration was clearly more interested in East Coast problems. Because greater attention was not paid to western needs, whispers of self-interest circulated that the region should sever ties to the Union. Thus, a social memory that bound citizens to the cause of liberty, the advance of an American nation, and the development of an agrarian republic became a delicate task aggressively pursued. The gentry knew that ties, real and imagined, needed to be maintained with the new nation in order to secure a western republic.

Besides forging a national identity for Kentuckians, the gentry also worked to create national status for themselves. Men like John Breckinridge imagined themselves members of a national "ruling class." The term seems awkward in a discussion of an era marked by egalitarian tendencies. Although historians have long recognized the existence of an economic and social elite, political leaders across the new nation held such diverse goals and interests that placing them into one ruling class seems absurd. But that is exactly what they interpreted as crucial to ruling: a diversity of men who would expressly support particular parties. Issues of nationhood and employment of revolutionary imagery to establish status and order bound them together. The U.S. Constitution reinforced the development of this ruling class, promoting an elite that would think nationally.[85]

Through their efforts to fashion society along the old buffalo trace, the gentry tried to create a majority identity based upon nationalism. The rhetoric of liberty, the Revolutionary War as a gauge along which Kentucky's development was measured, the contextualizing of the Indian struggle as an extension of the war: these components increasingly shaped the character of white society, especially among males. But we cannot assume that the interests of the gentry were therefore secure or unambiguous. They also had to facilitate an economy that would support their republican polity and revolutionary imagery. "Venerated be the Plough, and those that follow it," proclaimed a July Fourth orator. But a simple farming society would not be enough. In a toast to agriculture, manufacturing, and commerce, one participant warned, "May the two former be always found busy at home, and the latter be their handmaid, instead of their mistress." The challenge for the gentry was to maintain their own regional and national prominence even as they promoted the fortunes of farmers, artisans, and merchants.[86]

 # 3 · Upsetting the Balance

Crises of Economy and Ecology

In late 1815, Rhode Islander John Corlis brought sons Joseph and George to the family's new tract of farmland just east of Paris. At Maysville, as an afterthought on his return trip to Providence, John mailed them directions—two-thirds of the old ground to be planted with tobacco, one-third with corn, a strip of rye between the two, and "don't use the new ground"— and his sons soon went about trying to satisfy his demands. The two aspired to convert the acreage into a productive and profitable enterprise before their father's return in the late spring. John Corlis's acclaim of his western lands certainly gave them cause to believe they could. But their hopes shattered as their farm-making commenced. They fenced in the new ground but struggled to clear the old without sufficient labor or tools, and when they finally succeeded, they discovered that the soil was not as arable as the previous owner had claimed. Despite an abundance of woodlands that they rented to neighbors for pasturage, none of the trees proved good construction materials. Even the little things seemed overly difficult. In late May, George complained to his sister Susan in Providence, "We have had hard work to keep the squrils from pul'g up the corn, the Negro's from stealing the Eggs & the crows from eating the chickens." The men quickly discovered the agrarian promised land along the Maysville Road as infertile as the land they tried to sow, and their father eventually invested far more money in his venture than he had anticipated or could afford.[1]

Like the Corlises, generation after generation realized that being freeholders, having land, and settling as a family may have been virtuous aspirations, but they put very little food on the table. Faced with the difficulty of their endeavors, the Corlis sons turned to their father's pocketbook to solve their problems: "I find no difficulty in 'making brick without straw' but I can't get timber & plank without money," George continued in his letter to Susan. Hiring carpenters, stone and brick masons, and neighboring farmers' slaves, George and Joseph let others wrestle with an environment modified by years of settlement and agricultural use.[2]

DEFINING AGRICULTURE

The difficulties faced by the Corlises in 1815 as they created a family farm were not uncommon. As they arrived along the beaten path, second- and third-generation settlers were frustrated by the incongruence of expectation and reality in gleaning a living from the earth. They could no longer depend on the stereotype of agriculturally self-sufficient pioneers that had drawn so many Americans to the region in the 1780s. Neither could they trust the gentry's idea of an agrarian republic, an image illustrated time and again in John Bradford's *Kentucky Gazette*. In 1798, for example, he celebrated the agrarian Eden: "Look round your farms—how rich the prospect seems! / The orchard bends, the field luxurious teems! / Here Agriculture opens to our view, / A land of milk and honey, rich and new." It was a dream that would have fulfilled the most virtuous of motives. "To be a freeholder, to have plenty of rich land, and to be able to settle his children around him," noted visitor Timothy Flint. But it was only a dream because, like the Corlises, most settlers quickly realized that to meet the objectives toward which they strove and which structured the agrarian republic—to live full lives on independent farms, to cultivate the land and harvest the produce, and to bequeath property to their children—they needed more than virtue; they needed money.[3]

Money, rather than land, became the foundation of the republic along the old buffalo trace. As early as 1784, promoter John Filson hinted as much on his "Map of Kentucke," superimposed over a watermark image of a plow under which was scribbled "Work & Be Rich." Hardly the type to use words in any metaphorical sense, Filson certainly did not see the attainment of self-sufficiency or the embrace of republicanism as means to becoming "rich." Rather, Filson meant profit and prosperity, and, because he prefaced the words with a call to labor, he certainly did not target the gentry.[4]

In the 1790s and early 1800s, an agricultural society arose along the Maysville Road. Drawn by the rumors and rhetoric of profit, thousands of newly arrived poorer and middling sorts passed the rural retreats of the gentry and the unsophisticated cabins of pioneer families, seeking opportunities to "work & be rich." Most aspired to more than subsistence simplicity, hoping to participate in communal and regional markets. The house, barn, and fields provided security and sustenance; the market offered profit and comfort.[5]

Those who found land had to decide what type of farm they would create for themselves and their posterity. When he arrived in Fayette County in 1794, Harry Toulmin unwittingly represented thousands of his neighbors as he penned his vision of a yeoman enterprise: a thirty-acre homestead on which a farmer must "plough and do other things appertaining to a farm; and if his wife or his daughter can spin either flax or wool, or cotton, enough to clothe the family," then "unquestionably he may have an abundance." As might be expected from Toulmin's pen, it was the republican mantra of simplicity, domestic production, and satisfaction with one's natural position in society.[6]

When Charles Julian settled in Kentucky in the early 1800s, he advocated another vision. Writing to friends still in Virginia about how to migrate and succeed in the "new country," he encouraged them to "rase mony" by trading all lands in the Old Dominion for slaves, horses, hams, wagons, gin, and Kentucky lands, to invest in bonds with deeds of trust, and upon arrival, to "imploy your women & old negroes in spining and young in a nail manufactory" and "[s]ell your horses that move you." At least one thousand cultivatable acres were necessary, according to Julian, but corn was a poor choice to plant; yet "always have plenty & some to grind into meal for market." Rather, he encouraged animal husbandry: "sheep will do very well—wool sells at 3/ & a good mutton in Dec[embe]r sells at 3 dol[lar]s and lambs in July at 7." It was a new tune of profit-making in a market economy.[7]

While the contrast between Toulmin's and Julian's descriptions exemplified whites' varying definitions of a productive farm, it also reflected how quickly the republican dream for the Early Republic began to unravel. Toulmin's vision of a yeoman agrarian republic with thousands of small, independent farms is what many historians envision as the Jeffersonian ideal. Along the Maysville Road, however, where struggles over land titles and the trials of land clearing just were not worth the effort if a farmer

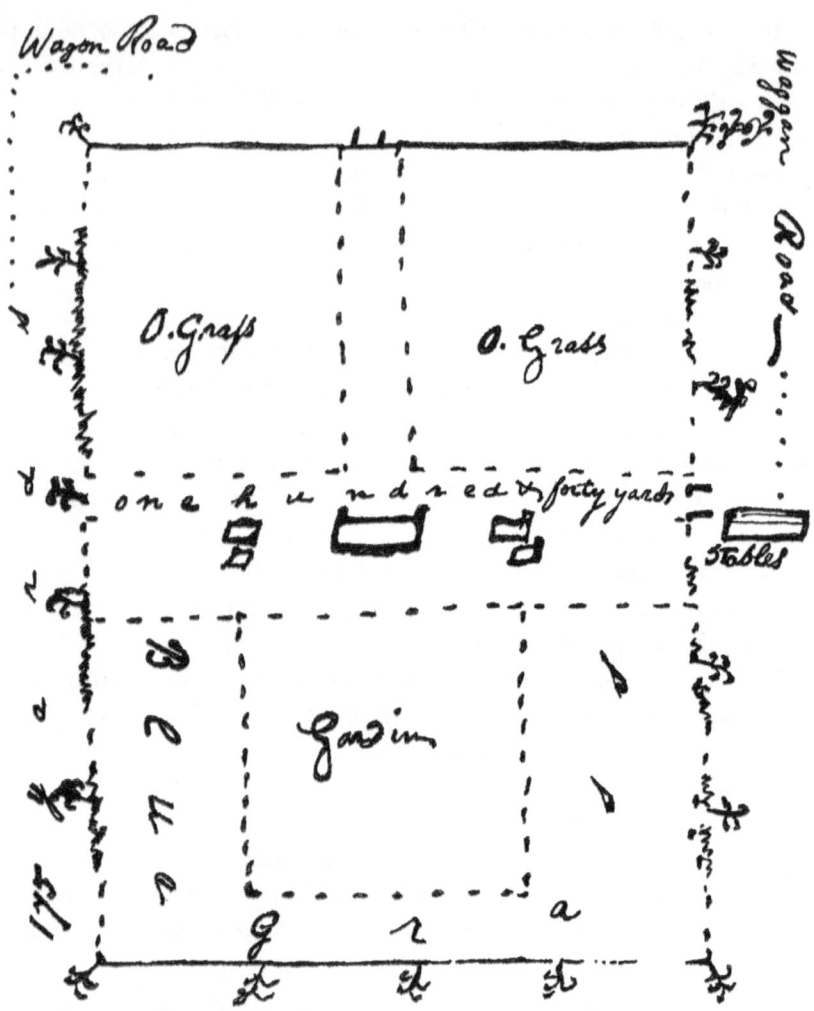

Charles Julian's Farm Plan. From Charles Julian Journal, Filson Historical Society, Louisville.

could not tap into the emerging regional market, the republican vision of agriculture was tenuous. There were cheaper lands to the north and less forested lands to the south for individuals determined to live meagerly and simply, to own and cultivate just enough land to confer a level of freedom and autonomy that differentiated the republican farmer from dependents: slaves, tenant farmers, propertyless laborers, and women. Julian's ideas represented another attitude toward economy and citizenship: a liberal impulse that promoted market participation for profit. Acquisition and cultivation of farms may have promised republican independence, but more importantly, they allowed farmers to pursue wealth, leisure, and refinement—to work *and* be rich.[8]

Before exploring the nature and consequences of the pursuit of agrarian wealth, it is important to recognize that Julian's and even Toulmin's ideals were never accessible to most settlers. Many arrivals of the 1790s and early 1800s faced the handicaps of youth, poverty, or both. Men in their twenties and early thirties were at a disadvantage. Virginia had originally granted lands in the counties along the old buffalo trace to veterans of the French and Indian War. (Only in southern Kentucky, south of the Green River, were tracts officially designated for Revolutionary War veterans.) Nearly 67 percent of Fayette County was awarded to older men who, through settlement or speculation, laid claim to status, wealth, and power at the expense of younger men, many of whom were Revolutionary veterans. By 1800, only 49.2 percent of the state's heads of household owned property. While its landless rates approximated the national average, Kentucky proved the least egalitarian of the western settlements in land distribution, and the state ranked with the more stratified societies of New York, South Carolina, and Georgia as having the most inequitable distribution of wealth and property in the nation. Along the beaten path, the statistics were more discouraging: ownership ranged from 46.1 percent in Bourbon County to 39.3 percent in Mason County. For the propertyless, Kentucky became more a quagmire than a garden. The high land-to-labor ratio that enticed settlers to the region quickly converted to a high labor-to-available-land ratio, creating large numbers of tenant farmers and urban workers.[9]

While speculation made widespread individual freeholding difficult, land nonetheless remained available to those with sufficient money. In every issue of the *Kentucky Gazette* from 1792 to 1812, at least two advertisements offered improved or unimproved in-state lands. "New comers can be at no loss (if they have Cash) to secure an Estate to their taste," related David Meade; but as a wealthy and genteel Virginian, Meade's idea of enough cash was hardly that of the average migrant. Compared to other developing regions at the turn of the century, Kentucky land prices escalated quickly. In the Old Northwest, the federal government sold an acre for $2; property in William Cooper's New York tract sold for $3 to $5 an acre. By contrast, only remote lands in Kentucky drew similar rates, between $2.75 and $3.50. By the mid-1790s, property near Millersburg already brought $6 to $8 an acre. Harry Toulmin warned, "the fact is, that in various parts of the state, land has risen 100 percent . . . and in some cases it has risen 3, 4, and even 500 percent." By 1811, John Melish observed along the beaten path that "very little good land is now to be had under 12 dollars per acre." Similarly sized improved tracts in the vicinity of Lexington soared to $200 each! Charles Julian's recommendations that settlers invest in horses, hams, wagons, gin, anything that could be promptly sold begins to make more sense in light of the immediate need for capital, especially as those items commanded higher prices in the West than in the East.[10]

Nevertheless, among men pursuing the dream of an agrarian republic, the clamor for land remained constant. Ownership of a farm held moral significance. Landed inheritance bound families together in webs of obligation that ensured that one's children retained the worldview and social standings of the parents. Without acreage, a settler could not fulfill this traditional moral obligation to family and society. Still, because land early became a commodity, farms also took on a monetary association that could not be ignored. Without ownership, a settler not only lacked the capital that land represented but the facility to produce and profit as well. In a society where less than half of the householders owned land, the need for hard money to purchase and create a farm compromised even aspirations to live in self-sufficiency. An unknown visitor warned the future settler that if he did not "bring a fortune with him, he will find he must at first live low and work hard."[11]

Such was the life of John Wallace who, in 1792, arrived penniless in Bourbon County with hopes of eventually buying lands, establishing a farm, and relocating his family. High prices impeded his aspirations, making him dependent on his brother, a Paris merchant. After nine years as an itinerant peddler and distiller of rye whiskey, Wallace saved only enough cash to purchase land at cheaper federal rates in Ohio. Lack of hard money not only limited settlers' fortunes in Kentucky, it often left them stranded without opportunity. If the wealthy did not "feel at home," they could return to the East, one observer noted, whereas "to the family of a poor man, woman, and children it is a dreadful, I may say, almost impossible thing to return."[12]

Because many of the unpropertied continued to hold tight the dream of a farm, but lacked the cash to pursue it, they opted to stay in the region and work as tenant farmers. Landlords considered them unreliable given that tenants little hesitated to sever commitments, pack up, and move on when opportunity for a farm beckoned. Robert Breckinridge warned fellow property owners that investment in tenant farmers would prove fruitless: those without land "will not remain tenants longer than they can procure 100 acres of tolerable farming land" in an undeveloped region. Yet, some evidence suggests that even after the opening of lands north of the Ohio River and farther west, many persons remained in tenancy, forgoing opportunities to migrate. Nearly 25 percent of Bourbon County's landless population in 1792 was still landless and still living in Bourbon County a decade later. Insecurities about succeeding on their own most certainly restrained some of the propertyless; landlords like John Breckinridge, Robert's half-brother, reminded tenants that if they "cannot make a living from the most valuable lands in the heart of Kentucky," they had little business moving to the wilds "where the land is poor, the Country unsettled & sickly."[13]

John Breckinridge was one of the handful of resident and nonresident speculators who laid claim to over ten million acres (a quarter of the state) at the turn of the century. In 1795, he advertised for twenty tenants to lease his Mason County lands for terms of seven to fourteen years. As a large landowner who depended upon others to clear his lands and risk the trials of the wilderness, Breckinridge, like so many other large-scale property owners, exploited the hordes of needy who sought shelter, food, and work. Having invested in the purchase of lands, a speculator increased his profits not by selling uncleared lots, but by renting at low rates to poorer settlers who cleared the land and cultivated the soil. On occasion, owners even suspended rent on the condition that tenants cleared enough acreage each year. Once the land acquired new and higher value, the speculator then rented at higher rates or sold, often to the very people who invested their labor in fattening his purse.[14]

Tenants, therefore, remained because the institution offered hope. "[M]any people come from Virginia & other States very poor & are strangers," remarked Edward Harris in 1797; after a four- or five-year tenancy "ordinarily if prudent they go off on land of their own full of stock & provisions." Tenancy operated as a way-station to ownership of one's own farm and therein addressed a fundamental cultural problem: how to incorporate the unpropertied poor into an agrarian republic predicated upon monied wealth. Rental agreements like those between two Lexington merchants, Thomas Hart and Samuel Price, and the fifty tenants living on their lands in western Jefferson County near Louisville hinted of the cultural rationale behind tenancy. The laborers could avoid paying rent for five years and then pay $2 per acre or four bushels of wheat per acre for the following two years; *or* they could live rent-free for three years, pay $.50 per acre the fourth year, $.75 per acre the fifth year, $1 per acre the sixth year, and $1.25 per acre the seventh year. Because there had to be a net profit to reward the landowner for his initial investment in the lands, the terms propelled lessees into the market economy. In order to pay their rents, tenant farmers had to acquire cash in exchange for their own produce or labor. The annual increase in rent found in the second option reflected expectations of fiscal improvement. Even those reluctant to abandon rudimentary farming had to grow a staple crop to meet rent demands. Tenancy, then, prepared future farmers for production of sustenance foods and commercial crops—and the pursuit of profit. The experiences of tenant farmers contrasted greatly with those settlers who owned their own farms. Rather than work and be rich, these poorer Kentuckians worked to survive. Still, both propertied and unpropertied learned that agricultural labor and commercial participation were inseparable.[15]

An indifferent group of poor whites seemed immune to this pull; they neither owned land nor sought out tenancy. While determining their numbers remains impossible, recognizing their presence is inescapable. In 1810, Philadelphian Alexander Wilson observed Mason County residents who lived in "miserable huts" but excitedly told him "with pride" of the rich soil, the abundance of production, and the healthfulness of the country. These boosters existed, nevertheless, in squalid poverty: "their own houses worse than pig-sties; their clothes an assemblage of rags; their faces yellow, and lark with disease; and their persons covered with filth"—conditions, along with their unproductiveness, that he attributed to their laziness. Harry Toulmin offered another explanation: the country "has hiterto been settled principally by the poorest class of people," he wrote, "whose finances were too low to enable them to proceed farther into the country." John Hill, a new settler in Mason, did not speculate on the reason but made clear his contempt for many of his less inspired neighbors, who cared "for little more than a little whiskey, vinison & bread from hand to mouth," none of which required much labor or currency. Daniel Drake remembered a Maryland man named Hickman and his family who squatted on his family's farm. They first lived in a tree house built with "a heavy garniture of green leaves," then moved into a small stable before finally settling in a cabin. Drake's father permitted the intrusion, eventually taking Hickman on as a tenant. Yet, as poor as the Marylander may have been, he eluded tenant labor by acquiring two slaves—"a negro man in middle life, and a woman rather old"—and forced them to do his work in the field "under the whip to the extremist degree," much to the consternation of the antislavery Drake family.[16]

Without legal claim to land and often without permission to live where they did, the very poor seemed the greatest threat to the ideal of "work & be rich." Their inclination to avoid work fundamentally clashed with the industry of persons determined to carve a profitable existence out of the West. John T. Lyle's father arrived in Kentucky with thirty thousand dollars, squandered the money in Lexington, and married and unemployed, "never increased his property." He failed to establish a farm because, like so many, he "didn't care about it." Critics interpreted such lack of industry, and the apathy and pessimism that seemingly accompanied it, as ultimately debilitating to the agrarian ideal itself. In February 1803, the *Kentucky Gazette* contained an "Ode to Poverty," a verse of which reinforced the point: "But not alone of this am I complaining: / Nature herself's so altered by thy power, / That fields and meadows, each gay tint disdaining, / No more to me display the gaudy flowers." Within republican thought, languor evinced the more primitive condition that theorists had feared would evolve in the early West. Through inactivity, this poor white class threatened to return the republic to a less civilized condition.[17]

Hence, this indigent crowd became a major concern to the gentry. In 1795, "A Friend to the Distressed" appealed to the compassion of central Kentuckians in their dealings with the unpropertied poor. The author encouraged establishment of county committees to assist settlers who arrived "without the means of purchasing even bread for their subsistence" and encouraged farmers to construct and open one cabin on their farms for a homeless family. Two years earlier, the state's General Assembly had initiated a similar program, requiring county courts to use funds "for the relief of such poor persons . . . incapable of procuring a living." Like the tenancy system, however, poverty relief incorporated hidden objectives. The law ordered justices to take poor children from their families and put them through indenture to "some art, trade, or business" where they would receive appropriate education, pay of "three pounds and ten shillings, and a decent new suit of clothes." By separating children from their work-shy parents and placing them in the homes of craftsmen and merchants, the legislature sought to assimilate them to "responsible" economic participation.[18]

Indolent squatters and tenant farmers provided a constant reminder to landed settlers that failure was an eager, albeit unwanted, visitor. Anxiety about losing one's property and livelihood simmered below the surface, and critics of the gentry's vision for a republican West took full advantage of the mounting tension. Samuel Taylor complained as early as 1789, "is not the unequal division of our landed property naturally calculated to promote an aristocracy?" In 1803, the small stone building on Levi Todd's estate that housed Fayette County's land titles burned to the ground. Although the arsonists escaped, landed farmers "believed the office was set on fire to destroy land claims" and quickly pointed accusatory fingers at unpropertied settlers, demanding a speedy reconstruction of property boundaries and claims.[19]

The benefit of landownership became evident by the turn of the nineteenth century. In 1790, 80 percent of propertied farmers along the Maysville Road were small landowners, owning too few acres to compose even Toulmin's version of a thirty-acre family farm. A decade later, however, nearly two-thirds of all landowners owned farms suitable to Toulmin's ideal, and the other third owned estates that met or exceeded Julian's ideal. Those who owned land were making money and investing profits in acquiring more land (and slaves) by which they could make more money. Still, landownership was not enough. Starting a farm and surviving for the first year or two presented a major obstacle. As George and Joseph Corlis learned, farm making demanded time, energy, and money often beyond farmers' resources.[20]

An "industrious settler" anticipated a "neat farm and snug cleanly habitation." For most pioneers of the 1780s, a simple house had provided identity and orientation. Living in rude cabins in an as-yet-uncleared wilderness "drew us more together, and compelled us to rely more intimately on each other," Daniel Drake commented. In comparison to the solitude of the wilderness, the intimacy of the farm house "enabled us to extract from the visits & company we *did* have, a high degree of enjoyment." Mayslick residents opened homes to travelers and each other on a regular basis. As we have seen, members of the gentry also invented identity through homes, using more sophisticated architecture and landscaping to differentiate themselves, but also opening parlors to business or social associates. Whether plain or ornate, private residences along the length of the road served as public spaces: "Every farmer's house was a home for all, and a temple of jollity," reminisced Mann Butler, a resident first of Lexington then Maysville. The house became the first of many ties between family and community and, on the occasions when travelers arrived with national and international news, the world.[21]

Some farmers, like the Corlises, bought properties with a house and other buildings already constructed. Still, in many cases, the preexisting residence was a log cabin or house well beyond its usefulness. For a year, Mary Ann Corlis and her brothers lived in the two-room cabin on their Bourbon County farm, but in 1816, their father sent plans for a new house with more windows, private rooms, and a piazza, features more appropriate to a family of their middling status.

Farm making did not stop with construction of a residence. In 1809, Nathaniel Hart projected the costs: six thousand dollars for a house, one thousand dollars for a distillery, five hundred dollars to construct a mill, another five hundred dollars for a barn. "When this is all done *if I do not break in doing it,*" he concluded, "I hope to be in a situation to make money." As a wealthier resident of Fayette County, Hart's plans were rather extravagant, at least in cost. But in intent, they were not unusual. Beyond the house and barn, farmers invested in such profit-seeking businesses as distilleries and mills as part of their economic and moral obligations to the community. As social and economic outlets for any given neighborhood, gristmills, tanneries, fulling mills, bakeries, distilleries, stores, and even

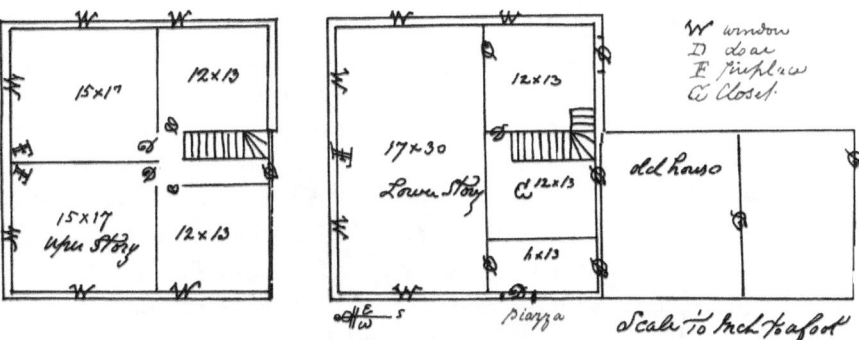

John Corlis's House Plan. From John Corlis to Joseph Corlis, 19 March 1816, Corlis-Respess Family Papers 1698–1984, Filson Historical Society, Louisville.

ferries connected individual farms to the increasingly commercial agrarian republic around them.[22]

Because these endeavors addressed social and economic communal needs, they tenuously straddled the line between private and public enterprise. The resulting complications became well illustrated in the late eighteenth century as the grinding of corn and other grains shifted from home to mill. Demands of mill owners, whose operations played increasingly crucial roles in local agriculture, began to clash with other communal interests. In 1789, for example, Laban Shipp's construction of a milldam on Stoner's Fork in Bourbon County elicited complaints because it blocked river access to the county's other commercial facility: the tobacco inspection warehouse. The ensuing controversy formed along lines of agricultural and economic interests. Larger and wealthier farmers hoping to create a tobacco culture generally opposed the milldam because access to the warehouse was central to such a goal. Allied with them were many poorer residents who fished the stream for food and shared the dream of a staple crop economy. Smaller farmers, more dependent upon home productions and needing market access, supported Shipp's enterprise. Without the mill, these residents faced an overland journey of nearly thirty miles to grind the quantities of corn that many of them anticipated selling. The Bourbon County Court compromised by requiring Shipp to build locks into his milldam so that navigation and fishing along the watercourse could continue.[23]

Caution should be taken, however, not to read into the story a conflict between premarket and market worldviews similar to those that seemingly erupted in eastern societies of the Early American Republic. In the late 1780s and early 1790s, neither the plantation agriculture that represented the market mentality in the South nor the proto-industrial enterprises that embodied that orientation in New England had planted solidly in Kentucky's soil. Mills and tobacco warehouses arose to serve communal, agricultural, *and* commercial enterprises. Grain production, like tobacco cultivation, offered market opportunity. To ensure exportation of superior quality produce *and* to protect the commercial reputation of neighborhoods, the state assembly regulated millers' fees just as it established standards for tobacco, hemp, and flour inspection. As Nathaniel Hart's intentions demonstrate, most millers were themselves farmers. Tension between farmers and profit-oriented entrepreneurs, therefore, seldom occurred since they were one and the same. The struggle between mill owner Laban Shipp and customers of the tobacco warehouse evidenced a clash between two components of one agriculturally commercial economy.[24]

In an agrarian republic where cash dominated, in an economy where commerce and agriculture intertwined, the question was not whether one would take part in the market, but how one would participate. Some residents chose the gristmill subculture of less established or newly arrived farmers who depended on corn to survive, finding profit in the sale of meal or whiskey. Others strove to expand into production of a staple crop like tobacco, one that would bring greater profits or could be used as "commodity money" in local markets. Even unpropertied neighbors took their fish to the stalls of village markethouses. When settlers determined the type of farm they wanted, they also took their places within the commercial society. The Shipp controversy was a struggle between individuals as they defined and molded the agri*culture* arising along the old buffalo trace. The simple fact that the controversy occurred at all speaks to the market orientation of those involved. After all, the year was 1789: farmers may have been cut off from the market at New Orleans and distanced from eastern markets, but they remained determined to protect their commercial interests.

Mills like Shipp's became common along western rivers and streams because the art of farming began with the cultivation of corn. "It is by the culture of Indian corn," noticed François Michaux, "that all those who form establishments commence." Gentlemen farmers with larger properties and greater expectations were not exempt; recall Charles Julian's warning to "always have plenty & some to grind into meal for market." Particularly suited to the Kentucky soil, corn took on a mythic appeal: Patrick Scott claimed "he could hear the corn go tick, tick—it grew so fast." Citizens of Lexington and Washington advertised Kentucky's agricultural output as fifty to sixty bushels of corn per acre, compared to the fifteen bushels per acre in another "best poor man's country" in southeastern Pennsylvania. Corn production preceded agricultural success: "Wheat . . . is fine in quality, and in quantity averaging about 25 bushels an acre; but where the land is fallowed, from 40 to 50 are frequently had. Fallow means corn land, or land planted first with Indian corn, then with oats the second year, and with wheat the next, which is generally more abundant than when sown immediately after, or amongst the corn at the last horse-hoeing."[25]

Not surprisingly, then, the pursuit of profit began with the culture of corn. Its production dominated commercial activity along the road. Between

1799 and 1802, profits earned from the export of Kentucky corn equaled the value of goods imported into the state. By the latter year, so many farmers attempted to enter into the corn trade that regional prices plummeted to unprecedented levels. In 1805, production was on the rebound; merchant Daniel Halstead applauded how the profits from corn and other agricultural productions again "over Balance the in ports." Farmers quickly found it easier to distill corn into whiskey before shipping it out for sale, and the number of distilleries boomed after the turn of the century. But corn also continued to be shipped as ground meal. In Mason County along the Ohio River, merchant-millers not only profited from grinding cornmeal; they positioned themselves as middlemen and provided export services for farmers. The culture of corn produced sustenance early in the settlement process, became the foundation for future farming and farm making, and early evidenced the commercial mind-set of farmers.[26]

Yet, farming was more demanding and complex than just sitting around and listening to the corn "go tick, tick." Most farmers raised myriad vegetables for domestic consumption, and dabbled in corn, wheat, oats, flax, hemp, cotton, and/or tobacco for the market. With the help of laborers, relatives, and neighbors, they harvested crops, shelled corn, slaughtered hogs, cleared and burned brush, cut firewood, tapped maple trees, and plowed the land. They oversaw the labor of slaves and servants, if they had them, and the home productions of wives and children who guarded livestock, churned butter, salted beef, smoked pork, tallowed candles, baked bread, and spun cloth. Even after spring and summer months passed, farmers spent much of the autumn and winter in farm making: grinding meal, building sheds, shoeing horses, repairing wagons, gathering firewood, mending fences, and traveling to market, to neighbors and relatives, and (in light of the extensive litigation over land titles) probably to court. The multitude of demands on time and energy often took a toll. James Flint passed farms in Bourbon and Fayette Counties where wearied workers left tools "to rot in the field" and the scythe "to hang on a tree from one season to another."[27]

Farmers and their laborers, free and slave, performed the multitudinous tasks as from an unwritten script, each one building upon the previous act. The agrarian cycle began in March with the plowing and sowing of the meadows. By mid-April, most farmers sowed corn, oats, hemp, wheat, flax, and peas, using all the available land including apple orchards. May and June saw little agricultural activity; rather, time was used for repairing buildings and wagons. By July, the fields filled with farmers cutting wheat and oats, and reaping hemp, clover, hay, potatoes, and cotton. In late July, members of the farm household spent hours "seting up" the corn, before the harvests of August, September, and October. While farmers occasionally reaped oats "too green to shock," corn usually remained in the fields until ready. In the meantime, the farmer prepared hemp and oats for sale. By November, the fields were bare, but the work continued: corn shuckings occupied the late autumn and winter months. Hog slaughtering took place in December when, too, the collection of firewood became a priority. In January, farmers burned brush on wheat fields to replenish some of the exhausted nutrients, while pork cured in outbuildings for anticipated sale. Farmers and their laborers spent winter hours in orchards where apple trees received a thorough daubing of salt mixture to heal splits. February brought the tapping of sugar maples, setting up of grapevines, and the transport of corn and wheat to mills for grinding and exporting. While the patterns of farming and farm making were traditions passed with the land from one generation to another, local almanacs also scripted the process, spelling out the methods and patterns of "rural economy" from profitable ways to produce butter in winter to making summer beer to preserving cream and milk during calf raising.[28]

This agrarian cycle required coordination and occasionally extra labor. To succeed as a single farmer with wife and children proved incredibly exacting. The moral economy of the neighborhood was meant to provide economic assistance in such instances; neighbors joined together to harvest crops. At corn-husking time on Joseph Hornsby's farm, friends participated in "shelling Corn &c then littering my Stables &c." Seasonal neighborly commitments kept people involved on each other's farms.[29]

The desire to create profit, however, made neighborliness an insufficient form of supplemental labor. Instead, the hiring of slave labor became common along the Maysville Road, even by individuals like Isaac Drake who did not own slaves and opposed the institution. Neighborhoods represented the breadth of slave ownership. In 1792, Fayette County had a substantial slaveholding population (nearly 32 percent of heads of households), but Old Bourbon had only half as many, and fewer than 13 percent of Mason County heads of households were slaveowners. As the 1790s progressed, as commercial agriculture was promoted, and, as the gentry legislated slave codes, slavery became more pervasive. Fayette County's slave ownership changed very little, but Mason County's grew to 21 percent and Bourbon's exploded to 27 percent. More importantly, the average size of slaveholdings increased everywhere, but most notably in Fayette and Bourbon Counties along the southern end of the old buffalo trace.[30]

While slavery was evident, the plantation setting and accompanying ideology requisite to huge slave operations evolved more slowly. Republican farmers cautiously engaged in the institution because slave labor generally contradicted the family orientation of the operation. François Michaux recognized among Kentucky's farmers "so decided a preference to agriculture, that there are very few of them who put their children to any trade, wanting their services in the field." Central to the moral obligation of passing the farm on to one's children, parents encouraged their offspring to engage in the workings of the enterprise early in life.[31]

Within their schemes of the farm, those who did use slaves necessarily committed to a behavior of market participation. Slaveowning farmers needed the market to supply and maintain a labor force that, in turn, helped to produce profitable crops to sell. Farmers invested that profit in more labor or improvements; some money went towards taxes on land, slaves, horses, wagons, and cattle. This pattern applied to large landholders like Charles Julian and others who aspired to the type of farm that he championed, and to small farmers, men who would have seemingly preferred the patterns of Harry Toulmin's more subsistence-oriented farm, who purchased or hired out slaves and, consequently, became participants in the market.

While access to slave labor was crucial to the successful operations of farmsteads along the old buffalo trace, the most visible investments were herds of cattle, hogs, and sheep. Livestock thrived on the abundance of the woodlands. François Michaux discovered hundreds of hogs "kept by all the inhabitants," wandering forested areas. Daniel Drake and his siblings daily herded their family's cattle and sheep from nearby forests and meadows, not returning home until after nightfall on most occasions. In 1802, a good milk cow cost over one hundred dollars, as much as an improved acre of land in Fayette County. Because farmers originally relied on tilled acreage and the woodlands, they paid little attention to making and improving pasturelands for domestic herds. As the loss of cane and other wild herbage created a noticeable difference in the taste of milk and butter, however, settlers faced problems of livestock grazing on open, treeless plains. Thus, farm making also came to include the construction of fences and rock walls to protect cultivated fields from hungry livestock.[32]

In many regards, farmers shared a universal experience along the beaten path. They attended to production for sustenance, as Harry Toulmin encouraged; they acquiesced to communal roles with the construction and operation of support institutions, as Laban Shipp demonstrated; they heeded their moral roles in securing a farm for one's posterity, as John Breckinridge described; and they pursued profit with as simple a crop as corn, as a multitude of Kentucky farmers did at the turn of the century. They also frequently used the yields of the fields—wheat, tobacco, oats, and corn—as "commodity monies" at local stores. As crops with market-driven values, these products directly bound farmers to regional and national market networks; the storeowner and farmer seldom dickered over the value of these commercial crops out of fear that one or the other would be cheated out the fair market price. For that reason, farmers remained aware of market values. "Accounts of an expected fall in the Price of that article [wheat],"Valentine Peers explained to his wife in 1801, "render it prudent to take the benefit of the present Market with my small Crop." The fruits of farm making—firewood, beeswax, whiskey, flour, and salt—also were bartered in stores, and their values in distant markets drew farmers' notice. Given the complexity and intensity of the agricultural cycle, few persons would have created a surplus of any of these items unintentionally. Nor was it easy or thrifty to produce an excess of the products of animal husbandry (beef, pork, lamb, chicken, and turkey) or domestic manufactures (homespun cloth, butter, lard, candles, and sugar). To suggest that farmers participated in the market only to relieve themselves of surplus products is to ignore the demands of agriculture and the frugality necessary to farming success.[33]

More importantly, it disregards the complex economic network that tied farmers together and laid the foundation for a market structure that connected agricultural and commercial life along the Maysville Road. Two crucial objectives of early Kentuckians' moral economy was that the farmer neither stand alone in self-sufficient isolation nor abandon his role as producer and thereby destabilize the cooperative economy. Domestic production of cloth illustrates this point most poignantly. In 1793, the champion of republican-style yeoman farming, Harry Toulmin, wrote how common it was "for all linen which is used in the family to be made at home." Farm women domestically produced quality homespun in great quantities: country linen made from flax, linsey-woolsey woven from flax and wool, and a woolen or more often cotton broadcloth. "Almost every house contained a loom," recalled Maysville resident Mann Butler, "and almost every woman was a weaver."[34]

Between 1792 and 1810, however, purchases of imported cloth at stores in Lexington, Washington, and Maysville grew steadily. If women produced cloth for household consumption, then why did they increasingly purchase imported cloth? As he passed a caravan of Bourbon County women taking their wool to the carding machine, James Flint noticed that "Miss does not wear the produce of her own hands." To complicate the scenario further, purchases of domestic cloth also rose, as did demand for twist, lace, buttons, and needles.[35]

In those years, settlers influenced by genteel displays of refinement cultivated their own taste for finer clothing. One visitor discovered that in rustic log cabins and crude farm cottages "are seen Ladies neatly dressed, who are, as yet, obliged to reside therein for want of better houses." Rural women spun broadcloth or linen for a market where they could trade it for imported cloth and sewing accessories. They found that market in the village West.[36]

The explosion of cloth production led to a demand for private investments that, like the gristmill, became communal enterprises in a cooperative network. Farmers who provided carding machines or bluedying became necessary partners in this burgeoning commercial activity. The use of such craftsmen strengthened the bonds of *communal* self-sufficiency; the collective goal was to incorporate all households into a mutually stable and profitable economy. From field to carding machine to loom to bluedyer to local store to exporter, the domestic production of cloth illustrates an interconnectedness of farm to community to commercial economy. The operator of the carding machine, the bluedyer, and the storeowner did not pose capitalist threats to the stability of farming society; they were indispensable contributors to it. More importantly, they became dependent on local farm and domestic productions to sustain their own enterprises. And because the values of homespun and other home manufactures did not depend upon the market, farmers and merchants arrived at a premarket price through neighborly agreement, underscoring the symbioses of commerce and agriculture, of money and land, of market economy and moral obligation.[37]

At the turn of the century, however, the *Kentucky Gazette* filled with editorials denouncing this intersection of commerce and agriculture. Arguing that society along the road had slipped into moral decline, traditionalists like Adam Rankin and republicans like Harry Toulmin condemned the growing influence of market networks on communities' moral economies. Not coincidentally, these were the same years that Fourth of July orators decried the individualism and self-interest of younger generations, pleading with them to remember the civic virtue of the Founding Fathers. "Aristedes," author of a series of *Kentucky Gazette* articles published in 1803 and 1804 as New Orleans was opening up to western trade, addressed the delicate relationship that existed between agriculture and commerce, one upon which the structures and patterns of agrarian culture formed. According to Aristedes, overemphasis on either side of the equation was dangerous for the individual and the community because one's identity, one's merit within the complex social structures of the Early Republic, correlated with one's relationship to the cooperative networks that emerged in community after community. Some citizens had a responsibility to produce for market; others were to provide services for the public; still others served supplemental roles in production and exchange. Without constant reminder of communal aspirations and individual obligation, the pull of profit could distract a farmer from his responsibilities.[38]

Such fears lay at the heart of the editorialist's "Reflections on Political Economy." He harangued farmers who ignored limitations on their proper economic stations. Their merit within society deteriorated as they pursued their own self-interests. The majority of farmers, migrants from "the most dissapated state in the union [Virginia]," lived above their means, entrapped themselves and their families in debt, and consequently compromised the independence of their farms. "Instead of cultivating the earth to produce materials for domestic manufacture," he criticized, "they have been in the habit of consuming foreign articles altogether." Ambitious to tie their own households to larger market structures, farmers seriously threatened the stability of the region's agrarian culture, its moral economy, their own status as independent men, and the social compact that assured familial and communal development: "The moral reputation became mortgaged for the fanatic productions of foreign countries, and credit ensued."[39]

Aristedes refused to blame the corruption of yeoman farmers on merchants, favorite targets of Americans disenchanted with the workings of a maturing capitalistic economy. Instead, he found the farmer at fault, having diminished the wealth of the soil through his own "indolence and extravegence." Merchants had worked to harness the activities of their agrarian neighbors, opening the *"appropriate market* relative to the *farmer"* in local towns and villages. Yet, as the increased production and exportation of corn demonstrated, farmers had pursued production primarily for regional and national markets, thwarting the role of merchants and merchant-millers by taking on the burden of exporting, and ultimately doing "but little good to the community." Here lay the crux of the argument: Agriculture and commerce had become so interdependent that failure to meet the demands of an agrarian economy naturally undermined commercial development. By exceeding their appropriate economic and social responsibilities, usurping the role of merchants in exportation, purchasing beyond their means and slipping into fiscal dependency, and neglecting responsibilities to provide for the local populace, farmers had subverted moral economy and market economy along the Maysville Road.[40]

In order to salvage the integrity of the agrarian culture, the editorialist encouraged farmers to return to agricultural and domestic productions for the local market, discouraging the production of wheat and corn. The opening of New Orleans freed farmers from relying on such crops and provided an opportunity for redemption. By producing marketable items—in particular, "tobacco, hemp always in its manufactured state, whiskey, brandy, pickled beef, hides, tallow candles, tow linen, cheese, salted pork, nails and nail rods, iron and utensils"—farmers would strengthen the regions' economy.[41]

Aristedes merely reiterated what most people already knew. Charles Julian believed that, in order to succeed, the farmer had to invest in commodities easily sold in the local village, employ his slaves in extra-household manufacturing, and raise the crops and livestock best suited for domestic use and market. His concern over cash prices and investment in bonds communicate a fiscal mentality that stretched beyond barter traditions. Even Harry Toulmin, champion of yeoman self-sufficiency, recognized the opportunities of market connections, although he more cautiously weighed them against the independent nature of the farm. In typical fashion, he initially pushed an idealistic view of republican farming, explaining that in Europe the farmer lived off the land "*indirectly* consuming but a small proportion in his own family," but in Kentucky the settler lived "by his farm *directly* with a view to family consumption." Still, in his calculations as to how much money would be needed, he concluded that £169 (over 42 percent) of the farmer's start-up costs should be retained "for contingencies"; and he did not emphasize that the farmer have "enough" to provide for his family, but instead that he have "abundance." While reiterating the rhetoric of republican agriculture, Toulmin subtly paid tribute to a pattern of economic production beyond the small yeoman farm.[42]

The primary goal for Kentucky's farmers remained to attain a good life in which they could "introduce luxuries . . . in any great plenty," often to the chagrin of Aristedes and others who interpreted such an attitude as extra-moral. The musings of a Mason County resident restated the point in a more lyrical manner: "May the Lord be praised / how I am a mased / to see how things have mended / hot cake and tea / for supper I see / When mush and milk was intended." Subsistence was never the objective; it was just a temporary consequence of relocation. The goal remained comfort and luxury.[43]

As they farmed Kentucky, residents along the beaten path fused the moral underpinnings of a Jeffersonian agrarian republic with the profit seeking of a commercial order. The ideas were not antithetical. Why invest so much of one's time, energy, and money if one had to settle for subsistence? That style of living resembled too closely the circumstances of tenants and the indolent, of Shawnee Indians and Collot's First Settlers. Those who avoided or escaped landlessness adopted a commercial ethos, as Alexander Wilson discovered in 1810: "Every man you meet has either some land to buy or sell, some lawsuit, some coarse hemp or corn to dispose of; and if the conversation do not lead to any of these he will force it." It was not necessarily a positive turn of events. As Harry Toulmin explained about his Kentucky neighbors, the "most unfavorable feature in their character is indolence in some and too eager a desire for growing rich in others, accompanied with no great delicacy as to the means of doing so."[44]

REDEFINING COMMERCE

By the turn of the nineteenth century, the family farm had assumed a dual role along the Maysville Road, grounding farmers in a moral economy while incorporating them in market activities. Even those who failed to secure land were not necessarily excluded from agriculture, accepting roles as tenant farmers until the opportunity arose for ownership of a farm. And farm making reverberated beyond farmsteads into mills, tobacco warehouses, bluedying operations, and even local stores, all supporting the agrarian republic by tenuously serving as both private and public enterprises. It should come as no surprise, then, that settlers expected a similar, dualistic pattern of moral and market, private and public to manifest in the commercial structures of the village West.

For farmers, commerce was bound to communal agricultural life: producing and exchanging domestic goods, and testing the entrepreneurial waters through mills, distilleries, and other private ventures. Such activities stressed personal obligations that they considered crucial to community, even as it allowed individuals to strive towards the primary goal for which most had migrated: economic improvement. In rudimentary ways, stores of the 1780s and early 1790s bound farmers to fragile market networks. Additionally, as the gentry implemented their republican vision along the old buffalo trace, they attempted to replicate eastern commercial networks that would satisfy their own desires to import luxuries and export commercial crops. In Lexington, Paris, Washington, Maysville, and the smaller places along the road, they supported merchants and artisans, men from eastern cities who followed the scent of profit westward. Ironically, financial support for merchants and artisans translated into political clout that would quickly nudge the gentry from leadership roles in local political and economic development.[45]

This was not the story that should be told about early Kentucky. Two-story brick mansions, finely made carriages, and ornamental gardens announced the presence of a well-bred and well-fed Jeffersonian gentry that saw itself as regionally dominant and nationally significant. Yet, comprising only a small portion of the population, their decision to attend to state and national politics as members of an unfolding national ruling class left a vacuum in local affairs that men of lesser wealth but more public station could fill. To succeed, this increasingly powerful commercial sector needed to balance moral obligation with profit seeking as they sought to capitalize on the promise of Kentucky and promote their own visions of the good life.[46]

When editorialist Aristedes chastised farmers for failing to promote the economic health of society, he was thinking of a moral economy that clashed with the individualistic impulse of profit making and affluence that would come to define the men of commerce. As the commercial sector adopted carriages and fine homes for their own use, material possessions became symbols of wealth rather than status, of respectability rather than gentility. By 1800, refinement quickly had become a commodity that people beyond the genteel ranks could acquire, promising distinction and propriety to those who, only a decade earlier, were considered vulgar by the gentry. But with their newfound personal wealth came greater communal responsibility, and many were not up to the challenge.[47]

In 1793, Irwin & Bryson's mercantile opened in Lexington. Their Philadelphia sponsor was William Hunt, who would one day finance his own son's entrepreneurial efforts in Kentucky. Bryson, whose first name has been lost to history, squandered the store's profits almost immediately. In late 1793, local merchants urged Thomas Irwin to dissolve the partnership before his associate dragged the entire enterprise into bankruptcy. Fearful of acting without Hunt's consent, however, Irwin hesitated and left for Maysville where he could oversee the unloading of shipments. Meanwhile, Bryson made daily appearances at the mercantile, spending one to two hours bossing around the employees and "Gambling and Paying his Losses out of the Store." On his return, Irwin demanded that Bryson repair the damage by accompanying a shipment to New Orleans. The latter eagerly agreed, but not for the reasons that Irwin hoped. Upon arriving in the port city, Bryson sold the goods and spent the profit "aplaying at Billiards." Expenses for the trip exceeded the sales. By April 1795, Bryson had foiled Irwin and Hunt's plans for a store in Lexington. "I blame you exceedingly for not destroying the partnership," Patrick Moore wrote to Irwin. As Bryson's uncle and a store creditor, Moore fumed that Irwin "should have done it instantly and secured the property for the payment of y[ou]r debts. ... [When] I get a little more composed I will write you on other matters." Within two years, Hunt granted power of attorney to four Lexington merchants who disposed of Irwin & Bryson's mercantile.[48]

Irwin was not the only merchant to have a difficult time with his partner's gambling. In June 1796, John A. Seitz discovered that his partner, who was in Philadelphia supposedly procuring goods, instead spent his hours "in this infernall Practis of Gaming." *"For Godsake, be more prudent,—more Just—towards our Creditors yourselfe & toward me,"* scolded Seitz; "reflect one minut on the Situation of J. Bryson & you cannot hesitate to quit gambling." Of course, Bryson had brought a whole commercial enterprise crashing to the ground without consideration of the store, its financiers, or its customers.[49]

When conscientious merchants dealt with their prodigal partners, foremost in their minds was how to remain reputable to their eastern creditors and suppliers while establishing and maintaining a satisfied clientele. The shenanigans of their partners, then, disparaged the values—particularly trustworthiness and industriousness—of the consumers that Irwin and Seitz hoped to entice into their stores.

Such episodes did not go unnoticed by the public. Entrepreneurs' moral responsibility became part of the larger debate about values and capitalism taking place throughout Kentucky around the turn of the century. Much of the discussion centered upon merchants and, to a lesser degree, artisans. In tune with the social patterns of revolutionary America, early Kentuckians assumed that men of commerce would act as economic patriarchs over "households" whose members became bound through networks of credit and barter. That they could not restrain their own partners suggested much about their management of the economic households that Irwin and Seitz headed.[50]

In the public mind, merchants and artisans were to act as more than entrepreneurs. As community economic leaders, they were directly responsible for guiding commercial development along the beaten path. "A New and True Friend" wondered in 1798, "Have the merchants in Kentucky arranged and conducted their plan of commerce as to enrich the state; and encreased the quantity of money in the country?" Why had Kentucky's entrepreneurs not created "a system as will combine in one point of view, and evidently promote the interests of the FARMER, the MANUFACTURER, the MECHANIC, the MERCHANT, and the PROPRIETOR OF LANDS"? What were the intentions of merchants in balancing commercial interests with the agrarian orientation of the West?[51]

The beckon of Kentucky drew hundreds of merchants and artisans who, like the multitudes of farmers, expected to profit from the region's economic development. Through consolidation of wealth and power, they became guardians of local interests: social and economic, agrarian and commercial. A yeoman who aspired to rise above his station or a gentleman who sought to protect his own social and economic status certainly disapproved of these haughty entrepreneurs as they elbowed into positions of power. Yet, it may be quite unfair and wholly misrepresentative to agree with New and True Friend that "men of money" promoted only their own selfish needs. Merchants and artisans found themselves between the proverbial rock and a hard place. As economic patriarchs, often officially as town trustees, they sought to protect the moral economy of their constituents. As entrepreneurs, however, they wanted to expand market connections to attract more genteel clients and boost their own profits. The dilemma made it increasingly difficult for men of commerce to contribute to the nascent society of early Kentucky without someone pointing an accusatory finger.[52]

Artisans and merchants provided the economic foundation for the village West. Migrating from eastern and European cities, artisans, in particular, were inclined to settle in Lexington. In 1788, William West, portrait painter, migrated from London via Baltimore. James Ross, shoemaker, arrived from New York City in 1799. The following year, Robert Frazier, watchmaker and silversmith, came from northern Ireland. In 1804, Mrs. White, a milliner recently arrived from London, opened a shop on Main Street. One year later, Joseph Green from London "commenced making patent pianofortes with additional keys, quality, touch and tone"; and Mr. Delisle from Paris, France, advertised his patented metal cutting machines. Luke Usher of Baltimore migrated in 1806 to open an umbrella shop, and became one of the 138 artisans registered in Lexington's town directory.[53]

Yet, when they set up shop, these craftsmen found a very different commercial atmosphere in Lexington, as there was "no overgrown wealthy Capitalist to Screw Down the wages of honest workmen & Cause them to Slight their work, that they might obtain a Scant Living for their Familes," as papermaker Ebenezer Stedman put it. While the gentry-led state assembly had legislated tobacco warehouse rates, millers' fees, black economic participation, slave mobility, and suffrage within villages, it had done nothing to restrict craftsmanship. Artisans' freedom to work inspired their own acquisitiveness. When Harry Toulmin arrived in Lexington in the mid-1790s, he resented that in comparison to his native England where he claimed that skill and honesty characterized craftsmen, in Kentucky such qualities seemed dispensable in the pursuit of cash. Because many an artisan monopolized his craft, he had no "fear of being entirely discarded by his fellow-citizens." Instead, he could expect good profits and be selective in his clientele. As one visitor noted, even a spinner in Paris bragged about the 83 cents per pound he pocketed for making twist, 33 cents more than he would have made in New York City.[54]

Artisans occupied a very unique position in early American society. Their productions *were* the material goods that accentuated the civility process: hats, portraits, carriages, umbrellas, bonnets, stockings, toys, furniture, brick houses or ones of painted clapboards, men's breeches, and women's capes. And the profits to be had from refining the West enabled many to rise to significant economic and political status. As individuals who worked with their hands, however, the stigma of manual labor distanced them from republican gentry who viewed manual labor as contrary to independence and gentility. That was of little matter to craftsmen who understood the attainment of liberty and respectability through a different definition of the good life. Often subordinating their own consumption to the well-being of the community, artisans embraced a corporatist ethic that made it easy, at least initially, to commit to the moral economy of their communities. The artisan world was small and local, and markets seldom extended beyond a dozen miles. Rather than portray manual labor as contrary to gentlemanly republicanism, artisans venerated it as the means to individual and communal independence.[55]

Naturally, apprentices and journeymen often became disillusioned with the rigid training and delayed profits, but few artisans expressed discontent about not achieving republican ideals of citizenship through landownership. Comparisons of Lexington's 1806 and 1818 directories indicate that a core of fifty-one craftsmen remained dedicated to their arts in an era of economic instability. Of the eighty-seven who seemingly fled their professions, most merely shifted to another craft, abandoning neither their roles as artisans nor their residences in Lexington. The practice, according to Fortesque Cuming, was "common throughout the United States, particularly at a distance from the sea coast, for one man to have learned and wrought at two, and even sometimes three or four different mechanical professions, at different periods of his life."[56]

In the artisan world, republican anxieties to prove the regenerative quality of landownership and farming were mute. Their work harbored its own ideological value. Ebenezer Stedman recalled a mechanic's song of the era: "Ye Shoemakers nobiably from ages long Past / Have defended your Rights with your awl to the Last, / and Coblers all merry not ondly Stop Holes, / But work night & Day for the Good of our Soles." Labor, not land, provided regeneration and fulfilled one's moral obligation to the community.[57]

Indenturing one's son to an artisan's shop, therefore, became as much an ideal for many villagers as passing on a tract of good land to one's son was for the farmer. The indenture created a legal compact: the master would teach skills to the youth that he could employ for a lifetime; the apprentice provided cheap labor for a tenure, until he became a journeyman and was replaced by a new apprentice. The system was designed to perpetuate the profession, and artisans taught "the trade and mystery" of a craft to new apprentices every year. The Fayette County Court regulated allotment of apprentices and notarized indentures. In 1805 and 1806, for example, the court approved apprenticeships for a printer, weaver, chairmaker, locksmith, stonecutter, tanner, bricklayer, saw miller, cartwright, fuller, silversmith, and several inn and tavern keepers (whose mastery of entertainment certainly qualified them as artisans). Peter Mason explained the indenture to one parent: "Your son John has set in with me for the term of 4 years To learn the Cabinet business. I have given him two months schooling agreeable to bargain and am to find him in Cloaths." Hence, the 1793 state law that ordered justices to remove children from land-poor families and put them through indenture to "some art, trade, or business" did more than assimilate the children to "responsible" economic participation; it potentially reoriented their *mentalitè* from a land-based, rural world to a labor-oriented, village one.[58]

The village West filled with artisans, aspiring journeymen, and apprentices. As early as 1792, Benjamin Wood recognized the prime conditions for good tailors, "the most of them gits Work plenty . . . from fore Dollars to five for making a Sute of Close and git as much as we can turn our hands tow." By 1807, Lexington's tailors employed forty-seven journeymen and apprentices in their shops; milliners tutored another fifty; sixty worked with cobblers; and nearly thirty students served as saddlers' apprentices. In 1811, David Sayre arrived with $1.75 in his pockets, spent his first three years as a silversmith's apprentice, and then joined Ezra Woodruff's smithy as a journeyman. Most journeymen abandoned Lexington after their training, often relocating to Louisville or Cincinnati to exploit opportunities in the newer West. Those who stayed, like Sayre, garnered a salary between $500 and $700 a year. When Woodruff's business collapsed at the end of Sayre's second year, however, the journeyman's situation looked grim. He had just bought out his indenture from his former craftsmaster in New Jersey, consequently going in debt some $21,000. But Sayre opened his own silver-laying shop, paid off the debt soon thereafter, and went on to establish his own banking house in 1823.[59]

Artisans also joined in the conspicuous consumption of the times. Of the thirty-six carriages owned by Lexingtonians in 1806, nearly a quarter were artisans' conveyances. Lucretia Clay, daughter of merchant Nathaniel Hart and wife to Henry Clay, bragged to a visiting Thomas Hulme that the "decency and affluence of the trades-people and mechanics at Lexington, many of whom ride about in their own carriages" evidenced the refinement of Kentucky. Adopting the manners of a more genteel culture, members of the crafts and service sector (particularly tavernkeepers and innkeepers) targeted a refined and wealthier clientele in hopes of becoming wealthier and more refined themselves. In 1796, Thomas Chapman noted the genteel manners of John McNair, tavernkeeper, who was "a much civiler Landlord than are commonly to be met with in this part of the World." Eleven years later, Fortesque Cuming commented on his dinner at Wilson's Tavern with "our well dressed hostess, who did the honors of it with much ease and propriety."[60]

It is ironic, then, that having established their social place and found profit making in the new western republic, the artisans would be undone by the rhetoric of the Revolutionary moment. Democratic tendencies that swept postwar America questioned the patriarchal role of the master and the servantlike position of the apprentice. The young seemingly imbibed more of the radical rhetoric of 1776 than did their elders, eliciting conservative denunciations about the loss of social values: as the *Lexington Observer and Reporter* admonished, "Woe to the sons that prove unworthy of the virtues of their fathers." In Kentucky, the ease with which individuals could escape the conditions of indenture and disappear into the western wilds exacerbated weakening traditional relationships. Advertisements for runaway apprentices peppered newspapers of the early 1800s, but, as indication of the ease with which craftsmen anticipated replacing their apprentices, the reward for a runaway was usually one penny![61]

Throughout the Early Republic, strains on artisan life, from the challenges of democratic ideas to the legal problem of the breaking of the indenture, raised fears among craftsmen about the future of their trades. In contrast to eastern cities where craftsmen formed associations to retard runaway apprentices and promote occupational interests, no such group arose along the old buffalo trace. Instead, artisans allied with merchants as a political force on village boards.

Like artisans, merchants congregated in Lexington, but their economic networks were never localistic. A corps of merchants—including Andrew McCalla, John W. Hunt, William Leavy, Robert Barr, John Colburn, John A. Seitz, Thomas Irwin, John Jordon, and William Morton—migrated from Philadelphia and Trenton in the 1790s. Others, like James Trotter, arrived as Backcountry merchants who already had credit/debt relationships with many settlers. These Backcountry elite had enjoyed some wealth and had risen to some standing among their eastern neighbors, and like the gentry, they hoped to replicate it in the West. Whether from the Backcountry or from an eastern city, merchants positioned themselves socially to mingle with the gentry and establish political and economic alliances that would, in time, prove beneficial.[62]

Almost immediately upon setting up their economic households, merchants found that the unique conditions that characterized Kentucky settlement placed them in precarious positions. Those who sought success could do so only by expanding beyond local trade patterns and the traditional role of economic patriarch. They had to provide the commercial services that a large numbers of Kentuckians demanded, access to commodities that the gentry had not wanted to leave behind in the East, and exporting services for agricultural and domestic productions. They had to establish market ties as early as possible, in a region physically isolated from market networks.

Upon their father's retirement in 1797, for example, George and Samuel Trotter of Lexington took on these challenges, becoming typical multifaceted merchants who imported eastern goods, exported domestic production, retailed, wholesaled, insured shipments, and offered loans. When the sons assumed control, they expanded the business, setting up a store in Maysville under John Armstrong to facilitate importing and exporting. In time, the store in Lexington served as the hub of an extensive commercial empire with connections to such disparate places as Baltimore, St. Genevieve, Chillicothe, and New Orleans.[63]

The Trotters' success arrived despite the problems of western commerce. Extension of trade networks proved problematic in Kentucky. "Bad men" in the Pennsylvania mountains forced merchants to coordinate caravan shipments. When the Ohio River froze, goods sat uselessly in Pittsburgh. Distanced from the trade centers of the East, western merchants oftentimes received news of national and international economic trends too late to react. When John W. Hunt opened his Lexington store in 1795, his brother Abijah stayed in Philadelphia to purchase and transport merchandise. The siblings expected to clear a 40 percent profit their first year, but as American diplomatic relations with France faltered, Abijah panicked that eastern economic conditions would shatter the business and wrote John: "Trade is much stagnated—produce will not Sell, not a Merchant will Ship at this time, the calls for money are as loud and as numerous as ever." Too late to guard against inflationary prices in eastern markets, he cautioned his brother that "to sell goods without money ought to be out of the question."[64]

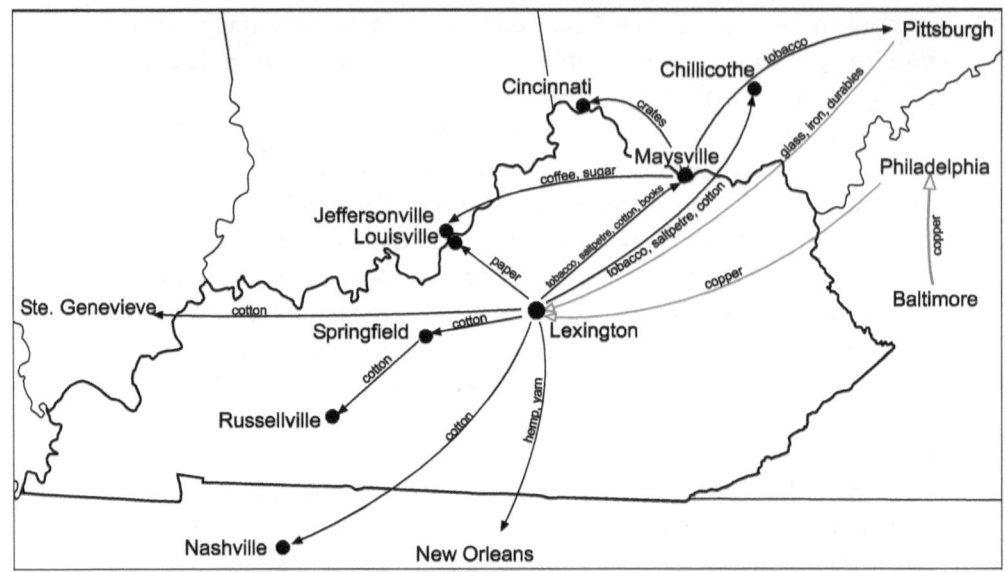

The Mercantile Empire of George and Samuel Trotter, circa 1806. Map by Mapsmith etc.

In the 1790s, a very vocal crowd comprising merchants and farmers cried out for better market connections. In 1793, hoping to push the federal government into securing export privileges through New Orleans, the Lexington Democratic Society pronounced that "the free and undisturbed use and navigation of the river Mississippi is the NATURAL RIGHT of the inhabitants of the countries bordering on the waters communicating with that river, and is unalienable except with the SOIL." In a 1798 congressional debate over taxation, Kentucky representative Thomas T. Davis explained how his constituents would gladly support a direct tax if the federal Congress would aid them in opening the Mississippi River. Kentuckians "had produce of every kind, in abundance," he declared, "but they want a market for it." A Lexington merchant proclaimed that, in America, "five millions of people can raise more wheat than two millions (the number before the revolution) did formerly—and I can neither see, nor hear of any new markets." Yet, the Federalists who dominated the national government in the 1790s showed little interest in aiding western Americans.[65]

In the absence of federal efforts to establish western markets and trade, merchants assumed some responsibility for the process. As the primary

John Wesley Hunt. Courtesy of the Filson Historical Society, Louisville.

problem facing aspiring merchants, the exportation of agricultural produce and domestic productions commanded their monies and energies. They advertised relentlessly for "hands to work my boats" and suffered substantial setbacks when Kentucky-crafted keelboats collapsed in the torrent of the mighty Mississippi. If produce arrived safely in New Orleans, intermediaries normally purchased the goods "in order to Ship." In the 1790s, a bribe to a Spanish official procured access through the port. In 1800, when the French took control of New Orleans, they refused bribes and cut off American trade. Aggressive merchants like Nathaniel Hart sent sons southward to find ways to corrupt French officials. After the United States purchased Louisiana in 1803, these Kentuckians were well positioned to become middlemen. Hart, Bartlett, & Cox became the principal exporter for Kentucky productions by the eve of the War of 1812 and made a hefty profit since the state's agricultural output during the war exceeded that of most states in both quantity and value.[66]

Thomas Jefferson's orchestration of the Louisiana Purchase theoretically should have solved Kentuckians' export problems, in particular those of merchants. Instead, "it will be very difficult to do anything there [New Orleans] for 12 or 18 months," complained Wilson Hunt, "from the quantities of goods and Adventurers that have gone there directly from England." After an influx of entrepreneurs into the lower Mississippi River valley ruined his plans to set up a store in Natchez, the younger Hunt opted for Missouri. Once again, his efforts fell short: "too many goods here [St. Louis], and very little money." The opening of New Orleans exposed the entire Mississippi River valley to commercial expansion, flooding the settlements of an even newer West with people and products. And, in a pattern fairly familiar to many Kentuckians, every westerner that Hunt met "is involved in debt beyond, far beyond their possessions." New Orleans middlemen, a flood of well-supplied entrepreneurs, and the trials of the Mississippi River itself—the Kentucky merchant had to navigate each of these or risk failure.[67]

Along the Maysville Road, expanding commercial networks strained local market relations. Increasingly dependent upon eastern financiers in their eagerness to create mercantile empires, merchants appeared less dedicated to the welfare of the community and more difficult to trust. As early as 1794, "A Farmer" complained to the *Kentucky Gazette* that with "unbounded avarice" and hearts "void of humanity," merchants undermined social stability by sporting "with the distress of those on whom we depend for subsistence." The visiting François Michaux warned that local merchants "are always able to fix in their favour the course of colonial produce, which they take in exchange for their goods; it is only particular articles that are sold for money, or in exchange for produce the sale of which is always certain, such as the linnen of the country, or hemp." The citizenry heaped disdain on any man of commerce who underhandedly profited at the expense of others, and because he supposedly shared the values of the community, his ostracism reminded him of his moral obligations. When citizens interpreted self-interest as greed, when it seemingly interfered with collective goals, commercial men became threats to communal stability.[68]

Yet, "originally too poor to extend their view beyond the simple exchange of their commodities for money," Aristedes explained, merchants had no choice but to send cash to their financiers and to take the "*whole profit* of the commerce to themselves." They offered "glittering commodities" that appealed to the pride and vanity of rich and poor alike who "like Eve, when tempted by the insinuations of the serpent, yields to the delusion, and in time, like Eve, laments his acquiescence to the charm." Still, Aristedes recognized the culpability of those same merchants who, upon deceiving their clientele for profit's sake, risked "fatal detection." Entrepreneurs heard a very mixed message: they should not abandon their profit-making inclinations, but they could not forsake their communal obligations.[69]

While he definitely had the upper hand in local economic relations, a merchant's survival depended upon the support of the community. He needed domestic productions like homespun, butter, and honey to sell alongside imported books, refined cloths, and European goods. Hence, he encouraged barter. Along with account-book credit, the barter system satisfied commercial relations between merchants and consumers even as it symbolized communal obligations between friends and neighbors. Throughout the 1790s and early 1800s, merchants consistently advertised for cash in order to meet the demands of eastern suppliers, but they never abandoned barter as a transaction. Even when the cash supply in Kentucky waned, merchants did not turn immediately to book credit. The continued use of barter suggests that consumers and merchants alike operated with notions of neighborly economic exchange and obligation.[70]

Items made within the farm household but suitable for sale in distant as well as local markets proved central to the merchants' success. Along the old buffalo trace, such productions included pork, beef, corn, flour, country sugar, salt, ginseng, paper, blankets, yarn, saltpeter, cheese, butter, country linen, beeswax, honey, and whiskey. Boats laden with Kentucky domestic productions plied the Ohio and Mississippi Rivers, delivering to frontier settlements in the new American territories. Traveler Zodak Cramer considered the regional market in domestic goods as "the *soul* of our country." Cramer may have been more accurate than he understood. Both consumers and merchants profited from the trade in domestic productions: the former had opportunities to acquire luxuries and necessities through barter; the latter boosted his profits and lessened his debt through the resale of such goods; and both fulfilled their moral contributions to the local economy. In the rural and village West, domestic productions tenuously kept moral economy and market economy in balance.[71]

Barter and book credit (if it could be collected) satisfied the profit orientation of the merchant, promising potentially more return than just a cash sale. In 1800, Thomas Sloo of Washington assured Lexington merchant John W. Hunt "that if I could make a payment of one thousand Dolls you would give me an Indulgence for 12 Months for the Balance of the Debt... perhaps it would Suit you to take flour and Whiskey at a Redused price delivered at Limestone," presumably for immediate shipment out to market where Hunt would profit from having accepted the items at reduced rates. The merchant successful at collecting debts could find equal profit in credit relations with other stores. By 1810, the Lexington mercantile of Barton, Hart & Co. held merchants Benjamin and Michael Gratz in debt for $8,413, and Bartlett & Cox owed an additional $15,000.[72]

Debt collection was a chore, and the merchant's *inability* to manage local credit relations often indebted him further with his own creditors. In their pockets, merchants carried enough money from suppliers in Philadelphia and Baltimore to manage for twelve months, a sufficient length of time to determine whether their stores would be sustainable. Eastern creditors most often dictated terms of trade with western merchants like the Trotters. Their two suppliers—James Adams in Pittsburgh and Bickham, Gellig & Co. of Philadelphia—shipped goods down the Ohio River to Maysville where workers unloaded the deliveries and transported them by wagon to Lexington. Imports ranged from unprocessed items like glass, dyes, and metals (including iron, tin, and copper) to semi-durables and perishables such as Queensware, Bibles, and tea. Kentucky-bound keelboats were scheduled to depart Pittsburgh every four weeks laden with merchandise from Baltimore, Alexandria, Philadelphia, and New York City. Just as merchants dictated the terms of payment from their customers, so too did eastern investors fix methods of payment by merchants. When the Trotters attempted to pay Bickham, Gellig & Co. in personal notes and notes drawn on the Kentucky Insurance Company, the Philadelphians returned the payment and demanded a check drawn on the Bank of Baltimore.[73]

From the first two decades of Kentucky statehood, six ledgers from stores along the road have survived, and they reveal much about patterns of mercantile activity and moral economy. For example, the centrality of cash in the 1790s is unavoidable in reading these ledgers. In the early 1790s, when Kentucky provided supplies for troops fighting the northwestern Indian wars, "money was plenty." A visitor in 1791 recognized that "every thing has two prices, the *trade* and *cash* price." Merchants preferred Pennsylvania notes, but they accepted cash from any eastern state. In Lexington, according to John Moylan's 1792 store ledger, cash exchange constituted over 32 percent of all transactions. Benedict Leonard's store in Washington demonstrated similar commitment to cash payments: 50 percent of all transactions. Articles bartered in the Leonard store—most commonly, country sugar, animal skins and furs, and bacon, but also imported cloth—lend a more rustic flavor to the story, but the coincidence in both stores of bartering and cash transactions, of domestic and imported goods, indicate an orientation toward cash and market that had to be balanced with local moral obligations.[74]

After the 1795 Treaty of Greenville reduced Indian hostilities to the north, the patronage of the U.S. Army dwindled in Kentucky and circulating specie and paper currency declined. Because eastern suppliers demanded cash payments, a siphoning of hard currency from the West made local economies more reliant upon personal notes and account-book credit, as demonstrated in the other four ledgers which date from the mid-1790s and early 1800s. John Wesley Hunt's 1796 store ledger from Lexington recorded 85 percent of exchanges as personal credit, contrary to his brother's warning to take only cash. Although barter was more prominent in Daniel Halstead's 1797 Lexington store, the records also demonstrate a rise in account-book credit usage while cash payments declined. William Tureman's Washington store ledger of 1807 had nearly 80 percent of its exchanges in account-book credit. And Edmund Martin's 1808–1810 store ledger from Maysville contained similar percentages: over 78 percent account-book credit, 9 percent cash, and 12 percent barter.[75]

Hunt, Halstead, Tureman, and Martin operated under social and economic conditions quite different from John Moylan and Benedict Leonard. The decline in cash payments reflected the departure of federal troops and their money. The decrease in barter indicated a newfound stability in a developing society, facilitated in part by the construction of public markethouses, directed by entrepreneur-dominated boards of trustees. Of the sixty-eight Lexingtonians who served as trustees between 1791 and 1812, merchants and artisans held the majority every year. But each had somewhat different reasons for board membership. Merchants found themselves under scrutiny and had to demonstrate a commitment to local moral economies. Artisans were more secure in their pursuit of profit but had to find ways to protect the craftsman tradition.[76]

In the early 1790s, trustees faced a growing population and greater demand for a market at which to sell agricultural and domestic productions. Dominated by merchants in these years, the board initiated construction of a markethouse on the village square and advertised for subscriptions to help build it. In 1793, however, the number of subscriptions was low, and carpenters Benjamin Stout and John Higsbee had not been paid. They quit without finishing the building, forcing the trustees to refund the private subscriptions that had been raised. For the next several years, Lexingtonians met in the basement of the old "state house" to trade goods.[77]

The absence of a markethouse evidently became a topic of local concern by the summer of 1794 when Harry Toulmin dismissed the need; a markethouse "is hardly any object to an American," and in the agrarian republic of Kentucky, most persons "attempt to raise on our own land what we want." Again on his Jeffersonian bandwagon, Toulmin approvingly noted that farmers found sustenance on individual farms and did not have to forage for food in forest or marketplace. Yet, the structures of Kentucky's local trade differed vastly from the market-towns that Toulmin had left in Britain, and his assumption that markethouses meant little to Americans mistook the need for such institutions. A reliable supply of food concerned town governments throughout the Early Republic. By 1795, over one thousand Lexingtonians lived in the largest of trans-Appalachian towns, and most lacked adequate land or incentive to grow "what we want." Contrary to Toulmin's notions, townsfolk needed to shop for sustenance and already had an unofficial marketplace. Furthermore, for residents to re-create any semblance of the moral economies from their eastern homes, the farmers in Lexington's environs desperately needed a market at which to sell produce and domestic productions.[78]

Markethouses were not new institutions on America's economic landscape. In the seventeenth and early eighteenth centuries, individuals in New York City and Boston constructed markethouses with the acquiescence of local governments. In other places like Philadelphia, local governments built and controlled the markethouses. But the republican rhetoric of the 1780s and 1790s associated market charters with special interests and government intervention. Although townsfolk and farmers wanted a venue for commerce, the construction of government-funded markethouses presented an ideological quandary. Voluntary subscription became the practical solution, allowing citizens to reconcile governmental suspicions with economic needs, and still applaud their civic-mindedness. The markethouse offered an economic outlet for agricultural and domestic production; it promoted civic virtue by providing a milieu for political discussion and social interaction; and it reinforced the commercial definition of agriculture by encouraging farmers to produce for market. Because it existed for local trade, few luxuries from the East or from Europe appeared in its stalls. Theoretically, a subscription-financed, republican markethouse seemed the perfect vehicle to discourage indulgence, luxuriance, and large-scale industrial manufacturing in a society that teetered on self-interest.[79]

The halt on construction of a markethouse in 1794, therefore, was a failure on the part of the trustees to meet their moral obligations to their constituents. In 1795, they again collected subscriptions and contracted John Spangter, a local German carpenter, to build a structure sixty feet long by twenty-five feet wide with brick pillars and a joint shingled roof. Spangter's work progressed so slowly that two years later he had completed only the first floor, leaving unfinished the second floor interior when the trustees dismissed him. Still, a markethouse sat on the town's commons. Every market day before dawn, the clerk lit lanterns hanging from the ceiling. Twelve feet from the building, a post and rail fence with turnstiles on each side regulated the throngs of customers, and visitors remarked on the "number of horses belonging to the neighboring farmers" tied to the fence.[80]

Although funded through private subscription, the markethouse came under the authority of the town trustees: they called for the subscription, hired the carpenters, and established its operations. Symbolic of their patriarchal status as leaders of the town's economic household, the trustees relocated their meeting space to the second floor of the building in 1803. Literally under their watch, the markethouse bolstered Lexington's economies, both moral and market. Unforeseen by the trustees, however, the republican markethouse also modified patterns of local commerce. Whereas privately owned stores presented a greater assortment of imported durables like Queensware, cloth, and books, the Lexington markethouse increasingly offered a wider choice of durables and perishables, including meats, vegetables, and dairy products. One visitor to the Lexington market claimed surprise "at the profusion and variety of most of the necessaries and many of the luxuries of life." The market provided an alternative to privately owned stores and eroded much of their appeal. People could find meats, tobacco, whiskey, salt, butter, and vegetables "in great abundance and very cheap." In the mid-1790s, whiskey, shoes, spices, and tobacco dominated customers' shopping lists in John Moylan's store. By the turn of the century, Lexingtonians found such commodities, most produced locally, in the markethouse. In contrast to the Moylan store, then, very different goods commanded customers' attentions four years later in John Wesley Hunt's store as he sought to attract clients with imported cloths, sewing accessories, and country linen. In the years after markethouse construction, Hunt sold more nondomestic goods than Moylan had in premarkethouse days. This most certainly was a change in demand that sparked a change in supply: Society's maturation brought about more refined tastes. But it does not communicate the entire story. Moylan and Hunt stocked the same types of goods, but Hunt's proceeds from domestic products plummeted, undercut by the Lexington markethouse.[81]

William Tureman's store even more forcefully demonstrates this pattern. In 1799, Washington's trustees sought subscribers to build a markethouse, but nearly seven years passed before completion. Not until 1806 were the building ready for occupation, the rules of the market established, and the clerk ordered to remain open from sunrise to 10 p.m. every Wednesday and Saturday. With such regular opportunities to buy from or barter with friends and neighbors, purchases of domestic productions declined substantially in Tureman's store. Customers still purchased imported cloths, sewing accessories, teas, and coffees, composing nearly 35 percent of his sales, but domestic productions—salt, sugar, bacon, butter, and cheese—totaled less than 5 percent. Not surprisingly, they were regular staples in the markethouse stalls.[82]

The commercial consequences of markethouse construction leads to the obvious question: Why would merchant trustees advocate such an institution? The construction of markethouses provided a simple and visible symbol that the merchants respected the moral economy of their constituents. By providing an economic outlet for the productions of local farms, markethouses also released merchants from dealing in perishable domestic productions, either as goods for sale or as payment for purchases. Instead, they could attend to greater importation of refined goods and bolster their own profits. Merchants had discovered a way in which to harmonize entrepreneurial desires with communal obligations.

Yet, only five years after the construction of Lexington's markethouse, the trustees apparently believed that citizens had abandoned mercantiles altogether, purchasing goods at the markethouse and retailing on the streets. In 1799 and 1800, merchants were in overwhelming control of the board of trustees and used their power to forbid purchases "in market during market hours any kind of Provision to sell again." Only licensed grocers could officially operate in markethouse stalls, serving as middlemen to local farmers.[83]

The rise of the markethouses paralleled several trends in economic life along the Maysville Road. First, stores assumed more specific roles in local economies. Harry Toulmin explained that "as to markets to *sell at,* the mills and the stores are the market." Merchants and merchant-millers became the primary overseers of exportation. Second, dependence upon mercantile stores for clothing and luxuries increased as the markethouse became the venue for the sale and purchase of foodstuffs. Third, farmers had to turn to licensed grocers to retail their produce, thereby reducing their direct participation in the market economy.[84]

Trustees' decisions concerning the markethouse also represented lingering moral concerns. In 1798, "A Worn-Out-Merchant" explained how the "farmer, who is the main pillar of the state, complains he has no market for his produce; or if he has, it is so scanty as to afford very little for his labour!" But, the editorialist continued, "hard as the fate of the farmer and manufacturer may be," the merchant "finds that in the present state of things nothing but cash will make his remittances in time, & this he receives so little of, as barely to save his credit with his tradesmen who supply him with his goods!" Cash not only existed at the center of Kentucky's agricultural development, but it circumscribed the merchants' world to such a degree that their political decisions were made with an eye to profit. Thus, the concern of New and True Friend in 1798 that the merchants had not "encreased the quantity of money in the country" was particularly astute. Suppliers needed cash; merchants used whatever would promise a profit; customers wanted barter or account-book credit so as to conserve funds. By the early 1800s, any person who visited the local store or the markethouse without cash contributed to the further entrenchment of merchants in a network of debt stretching from local households and stores to Philadelphia suppliers and New Orleans middlemen.[85]

Pressures to find a market quickly, establish credit with eastern suppliers, secure as much cash as possible from customers, and maintain the personal relationships associated with barter forced merchants into difficult positions. On occasion, the merchant acted apparently without motive to profit. In Washington, William Tureman accepted barrels from a local coopersmith as payment for some cloth and whiskey. He then sold the barrels to another customer without increasing their value to make a profit. As other store ledgers from early Kentucky indicate, however, the laws of supply and demand dictated the degree to which merchants would go to be neighborly. Charging interest on account-book credit transactions, although often uncollected for months or years, promised greater profits than barter. Merchants may have wished to be good economic citizens who bartered for domestic productions and extended unlimited credit, but their eastern creditors never agreed to be so accommodating. Increases in book credit transactions, even after the construction of markethouses, suggest that merchants did not encourage a markethouse economy in order to free themselves from obligations to neighbors and community. Their intentions to fulfill moral obligations remained; their ability to do so eroded.[86]

THE ENVIRONMENTAL REVOLUTION

By 1800, as the gentry sought to impose a republican model of society, as the merchants struggled to meet moral expectations and pursue the capitalist impulse, and as farmers sowed and reaped their way to comfort, the Maysville Road continued a rough, dirt trail. Only settlers' seasonal roadwork kept it from degenerating into a gutted, muddy track. But it was the lone primordial feature on the landscape. The rural environment along the old buffalo trace no longer resembled the wilderness that pioneers had found in the 1780s or even the wooded landscape that gentry came upon in the 1790s. With each strike of the axe, each furrow plowed, and each harvest, farmers modified the landscape around them. Thousands of individual decisions to cut down a tree here or dig up limestone rocks there collectively re-created the natural world in which they lived. The village West contributed by denuding hinterlands for firewood and encouraging greater agricultural production for village populations.[87]

The crusade undertaken by waves of immigrants wrought extensive changes very quickly. "Where the early stations were established [amidst] the wild herbage, consisting of Cane & pea vine," lamented David Meade in 1796, "[the native vegetation] is entirely eat out and the place of it supplied by weeds." A year later, John Breckinridge regretted that "the range & game are totally gone which all the first settlers considered an invaluable." The rapidity of this massive transformation surprised everyone, and it all seemed to culminate around the turn of the nineteenth century.[88]

The environmental revolution began with land. When, in 1793, Moses Austin condemned settlers headed to Kentucky for passing "land almost as good and easy to obtain'd," he betrayed his own ignorance about how individuals sought out superior property. The tracts that migrants disregarded en route were *not* as desirable, at least not from the perspective of late-eighteenth-century Americans intent on finding the best lands. In the Backcountry regions that Austin believed "almost as good," the natural symbols of fertile soil were sparse: limestone rock, luxuriant cane, and specific densities and species of hardwood trees.[89]

Along the Maysville Road, Pioneers early celebrated limestone in naming Limestone Creek, Limestone Hill, and Limestone village. "There is a species of flat, or split Limestone that pervades all the country, laying at unequal depths," observed one traveler in 1791. "In the rich and black-looking soil, it lays near the surface, and in general the nearer the stone lays to the surface the richer the land is found to be." Limestone distinguished soil fertility because its porous and permeable character served as an aquifer, filtering and retaining underground waters that, in turn, helped to push phosphatic nutrients into surrounding top soils. Most effective in the Bluegrass where a gently rolling topography retarded erosion, this process produced phosphate-rich soils. In the shale hills around Blue Licks, the angularity of the hillsides brought different results. Steep slopes, narrow V-shaped valleys, and impervious shale formations forced rains and underground waters full of phosphatic nutrients to wash away fairly rapidly. The region *looked* less cultivable: shale, not limestone, pushed through the topsoil. Along the northern end of the road, limestone protruded even less. Many settlers passed through the region not realizing that the fine, black topsoil and the second strata of red clay concealed a layer of limestone that would have provided the familiar symbol of natural fertility. Those who remained became pleasantly surprised at their good fortune.[90]

A more widely employed indication of fertility was flora. Dense forests featured a variety of hardwood and softwood species. "Such is the umbrage cast over the face of the earth by those vast trees," wrote an impressed David Meade, "that the air circulating in the Woods is always cool which renders traveling here more pleasant than I have found it elsewhere." The prevalence of forests contributed to an environment "which has none of the severity of our winters, but enjoys a climate that is always temperate—a continual Spring and Autumn, as it were." Of course, Kentuckians enjoyed and often endured all four seasons, and the winters usually struck much more harshly than those that Meade had experienced in Tidewater Virginia. Nevertheless, the woodlands created seasonal conditions that astonished travelers, especially in the Bluegrass. "Should you visit the country in the spring, you will be surprised at finding no leaves under the trees," wrote a sojourner in 1791. "The reason, is the ground is so rich and damp, that they always rot and disappear with the winter."[91]

Although the impressiveness of the woodlands drew attention from many, those determined to settle along the beaten path literally could not see the forests for the trees. They looked for specific species as they deduced soil fertility. On first-rate lands, cherry, white walnut, buckeye, elm, hackberry, ash, black jack oak, honey locust, coffee, and paw-paw abounded roughly in descending order. Where the latter three flourished, migrants determined the richest soils. Other varieties, again in descending order, thrived on second-rate lands: chestnut, red and black oak, sassafras, persimmon, and sweet gum. Third-rate lands produced black and red oaks, chestnuts, pines, and Virginia cedars. While the inclusion of conifers in the final category did not prove settlers' bias against softwood trees, the thinly populated shale hills did. Stunted pines were useless as building materials and barely straight enough for rail fences.[92]

Even as many settlers made reference to the decline of bison and cane in the 1780s and early 1790s, they wrote in awe of the perseverant trees. Size excited visitors and residents alike: "Oak and locust on the flat lands are common at five feet diameter. Poplars growing on the beach lands are so common at five and six feet through, as hardly to be noticed. The beech grows to the thickness of four and five feet, and both the last mentioned to the height of one hundred and twenty to one hundred and thirty feet." As late as 1791, the magnificence of the trees and the consequent "advantage of pasturage in the woods, constitute the great excellence of Kentuckey." The region's forests seemed more than capable of supporting throngs of new arrivals.[93]

As populations exploded in the 1790s, however, deforestation along the road developed suddenly and dramatically. The census of 1790 indicated 27,750 black and white residents in the three counties along the old buffalo trace, an area that covered 11,020 acres. Over the following decade, with the population of the state nearly tripling, traffic along the old buffalo trace boomed. County census figures initially do not appear as dramatic, less than doubling to 41,716 by the turn of the century. But as the state grew, the legislature increased the number of counties by reducing the size of existing ones. By 1800, Fayette County, originally 13,170 square miles in 1780, shriveled to 290 square miles. Bourbon County slipped from 10,500 to 310 square miles between 1786 and 1800. Mason County had 5,710 square miles when founded in 1789; by 1800, only 2,060 square miles remained. In 1800, the legislature combined some of the territory from Bourbon and Mason Counties to form Nicholas County, comprising 280 square miles. Taken together, shifting populations and changing county boundaries meant that while the numbers of people living along the old buffalo trace had not multiplied in proportion to overall state growth, the density of settlement became greater than anywhere else in the West. In 1790, human habitation along the road approximated 2.5 persons per square mile; by 1800, it had increased to 14.1 persons per square mile.[94]

Such rapid growth placed overwhelming demands on the natural environment, especially the woodlands. Trees were essential to everyday life: houses, barns, mills, furniture, rail fences, wagons and coaches, and fuel for heating, cooking, and salt manufacturing. Despite the significance of tree species to land identification, increased numbers of white settlers and their slaves ravaged the forests. Red elm, mulberry, paw-paw, sweet crab, hazelnut, redbud, and flowering dogwood disappeared at frenzied paces. Sturdier species remained, but only at settlers' whims. Oaks fell as people scouted honeybees and then chopped down the trees to steal the sweet cache. White poplars endured only as long as the yellow poplars remained in demand: "Before they fell a tree they satisfy themselves by a notch that it is of that [yellow wood] species." When the Drake family acquired new lands "covered with an unbroken forest" outside Mayslick, father and son "charged on the beautiful blue ash and buckeye grove." The destruction of trees became second nature, even a passion, to many who appreciated a tree, as young Drake rationalized, "in proportion to the facility with which I could destroy it."[95]

As deforestation accelerated, settlers' early enthusiasm for specific species proved the undoing of many groves. "When they announce the sale of an estate," remarked François Michaux, "they take care to specify the particular species of trees peculiar to its various parts, which is a sufficient index for the purchaser." In land claims and survey records, migrants chose the most recognizable species and largest trees to identify property boundaries. As yellow poplars, oaks, honey locusts, and other species disappeared, however, so too did boundary markers. The legal impasse created by deforestation forced judicial decisions to be made on assumption and argument rather than fact. In 1801, the Kentucky Court of Appeals conceded that as "the law requires every official surveyor to see the survey plainly marked by trees or natural boundaries, the presumption is, that every survey has been thus marked, or bounded, when made, though the abuttals may not be found."[96]

Alongside the woodlands were open meadows or savannahs. Pioneer accounts of the 1770s and 1780s described large fields with grassy undergrowth and small patches of trees. "You frequently find beds of clover to the horse's knees, sometimes a species of rush-grass commonly called wild rye, from the similarity of it's stalk to the rye so called among us; in other places we meet with tracts of wild cane, very much esteemed by the wild and tame cattle," commented an unknown visitor in 1791 as he traveled the old buffalo trace. "There is also a species of vine called the pea vine, from its producing a small pod, resembling that of the garden pea, of which both horses and cattles are extremely fond."[97]

Some clues suggest that indigenous savannah vegetation, especially the cane, was in danger before the mid-1790s. Contrary to David Meade's pronouncement, winter did come to Kentucky, and the "Hard Winter" of 1779 and 1780 proved a trying time for many. Those who owned livestock learned the perils of wilderness life: "Go through the cane and see cattle lying with their head to their side, as if they were asleep," recalled one pioneer, "just literally froze to death." Near the falls of the Ohio, settler John Floyd noted how "the cane seems to be all dead by the hard frosts." Along the Maysville Road, canebrakes survived the "Hard Winter" only to suffer a decade later when particularly severe winters struck between 1790 and 1793. While settlers knew cane to survive several days of winter weather, the harsh freezes of the early 1790s may have proven devastating to both livestock and to what remained of the canebrakes.[98]

But it was the incessant grazing of domestic livestock that ruined the savannahs. Not surprisingly, many of the earliest stations along the beaten path arose beside meadows where cane or other lush herbage blanketed the earth. As with the woodlands, settlers' dependence on cane, wild rye, and pea vine as markers of fertility did not slow them from rapidly destroying the plants. Unlike migrating herds of bison, cattle and horses became year-round residents of savannahs-turned-pastures. By the turn of the century, their grazing depleted the canebrakes, the wild rye, and the clover, creating conditions for weeds that, in a perverse sense, also came to represent the fecundity of Kentucky. "Such is the fertility of this Country," wrote David Meade, "that wherever air & sun can get to the earth it produces weeds of prodigious size—the most prevailing in open places are the James Town weed and Iron weed or (as it is sometimes call'd) Devil's bit."[99]

The beauty of some weeds deceptively masked their uselessness and occasional deadliness. Jamestown weed produced toxic leaves, stems, and hard, prickly fruits; its poisons seductively tempted practitioners of love potions and other folk medicines. Elderberry and hawthorn bushes that thrived along the edges of the meadows yielded poisonous berries. White snakeroot also flourished, producing a toxic sap that induced milk sickness in cows and humans who drank infected milk. Also known as the trembles, milk sickness manifested as nausea, vomiting, stomachache, intense thirst, low temperature, slow respiration, coma, and eventually death. White snakeroot and other poisonous weeds early symbolized a new imbalance between natural and human life.[100]

More suitable pasturage also proliferated in the open savannahs and ravaged canebrakes. Pennyroyal, a member of the mint family, took root in the rockier soils along the northern end of the trace. Along the southern end, bluegrass symbolized the ecological imperialism under way. Colonial Americans knew bluegrass, when combined with white clover, as English grass. It spread westward across the Appalachians in advance of white settlers, overwhelming the savannahs and seeking out the cool shadows of the woodlands. The speedy invasion of bluegrass into Kentucky meant that by the mid-1790s, as the old buffalo trace swelled with migrants, a familiar welcome mat awaited many of them.[101]

Like the floral transformation of the late eighteenth century, fauna along the old buffalo trace underwent significant change. In the 1770s and 1780s, overzealous pioneers thinned the populations of bison, killing the beasts "just for the sake of saying so." By the mid-1780s, springs at Blue Licks and other salt licks that had in earlier days drawn large mammals were converted into salt manufactories, thereby reducing animal access to the licks. If they escaped the gun, elk, deer, bear, and bison migrated elsewhere in search of dietary needs. Like cane, however, the story of the bison rendered only a more familiar chapter of the re-creation of Eden.[102]

In 1784, when William McConnell constructed a station alongside the road in Bourbon County, bison and other large game still wandered central Kentucky. Archaeologists have uncovered evidence at McConnell's Station, however, that its residents did not typically eat large mammals. Rather, the trash pit contained bones of opossums, cottontail rabbits, squirrels, turkeys, passenger pigeons, snapping turtles, bobwhites, killdeers, and king rails. The bounty of the wilderness certainly provided staples for the McConnell family diet. Yet, with the exception of some elk remains, the evidence indicates that by the mid-1790s, hunters no longer sought larger game, most certainly because of the depletion of herds but also because of changing tastes. Whereas long hunters of the 1770s had developed an appetite for bison meat, many settlers less than a decade later never had opportunity to cultivate a fondness for wild flesh; they consumed domesticated meats with which they were more familiar. Consequently, the pit at McConnell's Station also contained evidence of chickens, swine, cattle, and sheep that had quickly appropriated much of the pasturage previously used by wild animals. Of the larger livestock, only sheep required protection; cattle and swine often roamed the countryside unsupervised for lengthy periods.[103]

Even smaller wild animals and birds felt the effects of the region's environmental transformation. In 1795, David Barrow explained how populations were changing: "Our common kind of birds were very scarce I am told when the country was first settled but they have greatly increased such as Partridges; the gray mocking bird and kill dee are very rarely to be seen. They have no whipoorwills off from the cliffs. They have an abundance of woodpeckers, crows and ravens. Their woodcocks have white or what some call ivory bills. They have plenty of pheasants in places also wild geese and ducks are plentiful in the fall season on the Ohio. Wild turkeys are much reduced on the settlements but plentiful in the borders. It is the same with deer, bear, etc. They have no rats or common mice except in the neighborhood of boat landings. There are but few hares and no fox squirrels but the like of gray and ground squirrels I have never seen before." While animal populations indigenous to trans-Appalachia declined, those familiar to eastern settlers were on the increase. Honeybees that settlers tracked through the woods were migrants of European origin. Old World brown rats were found in Maysville as early as the 1780s and, by century's end, had reached McConnell's Station and other interior homes, prompting an increase in cats and domesticated raccoons. Even hogs contributed to revising the environment, reducing snake and reptile populations through their appetites. Settlers and their supporting cast of livestock and pets remade natural life along the old buffalo trace in the image of eastern environments.[104]

The introduction of large numbers of domestic animals demanded an even more serious revision of the environment. Despite widespread growth of bluegrass, many savannahs remained filled with weeds, and cattle and swine often roamed into cornfields to forage. Thus, settlers had to cultivate pasturage: on farm after farm, "scarcely with a single exception," arose meadows of timothy grass separated by fences—symbols of the privatization of land that accompanied the environmental revolution. Rail fences required trees easily split into lengths: oak, black locust, walnut, or chestnut. Subsequently, small patches of woodland survived on farms as reservoirs of repair materials, as on the Corlises' farm in Bourbon County. In the Bluegrass, settlers used the familiar limestone rocks from their fields to create low stone plantation fences. No longer in need of timber for fence repairs, those farmers who constructed plantation fences increased cultivatable acreage and still restrained their livestock. Because construction of rock walls required the skills of stonemasons hired from Lexington or Washington, they were most commonly found on larger estates where brick Georgian homes sat amid the rural elegance of refined gardens.[105]

As the architectural tastes of the gentry began to spread to middling sorts, abandonment of log cabins also altered the landscape. Frame housing became fashionable, and sawmills opened across the countryside to provide sawn and planed lumber. To contrast civilized abodes from the natural environment, owners painted frame houses with bright colors—"white lead and red lead, Prussian blue, Yellow Ochre, Pattent Yellow, Rose, Pink, putty, Spanish whiting, Spanish Brown and lamp black." The shortage of lumber that resulted from deforestation made the popularity of frame just as fleeting as that of log construction, however. Additionally, in cramped villages, concerns of fire mounted. The availability of limestone, as farmers removed it from their fields, made it a more popular and affordable building material. Still, it was far too porous to withstand heavy flooding, a problem particularly troublesome in Lexington and Maysville. Brick construction proved most alluring: its less porous quality kept floors drier; its fireproof quality provided a sense of security; and its appearance affirmed the architectural refinement of the inhabitants.[106]

The flooding that forced changes in building materials was another consequence of the ecological revolution. As we have seen, rivers and creeks were unreliable water sources, drying up in the summers and flooding in the winters. David Meade suspected such problems upon realizing that "Fame has in most particulars done justice to Kentucky ... but in the article of water it does not appear to me that she has." Rumors of Kentucky's abundance never included the region's waterways. With the environmental transformations of the late 1790s, there was even less to praise. Settlers complained of "an aridity, which keeps pace with the clearing of the country, and completely dissipates the fond illusions of travelers and land speculators." In 1800, members of the American Philosophical

Society met in Philadelphia to discuss how, in Kentucky, the "moisture of the atmosphere, and the copiousness of springs in general, are diminished, especially on the higher grounds, by the destruction of the woods, by which the condensation of the vapours is lessened or prevented." This was a problem far more critical than the unpredictability of streams that eighteenth-century travelers noted as detrimental to gristmill activities. Decreased water levels, like the demise of indigenous flora and fauna, forced a restructuring of everyday life.[107]

As the roots of tree stumps rotted, natural barriers to topsoil erosion disappeared. As cattle, swine, and sheep trampled the earth, they decreased its capacity to absorb waters, thereby accelerating the process. Without the thick forests of the early 1790s, soils became warmer and drier in the summers, and colder and drier in the winters. In contrast to the 1770s when the ground was "so rich and damp" that William Clinkenbeard believed settlers "co'd n't burn this country," increased aridity made uncontrolled fire of greater concern in the summers of the early 1800s. In the winters, snow melted more rapidly and winds cut across the landscape unimpeded, factors that further contributed to the demise of the canebrakes. Without the insularity of snow, soils froze more deeply, striking the roots of vegetation more harshly.[108]

The harder the soil became, the more rapidly waters ran off the land. The limestone strata became oversaturated and unable to carry away deluges of melted winter snows or heavy summer rains. Underground springs and streams swamped farmlands and town streets. Surface flooding became more common, especially along the Northfork near Blue Licks. Drier weather was no better, as summer evaporations left settlers searching for new water sources. "The creeks are getting so low," regretted Bourbon County resident William Barry in 1806, "that people are obliged to resort to horse mills.... [W]ater has grown so scarce that it is carried 6 or 7 miles; many people are compelled to drive their stock that distance for water." Citizens of Washington, whose public wells had sustained village growth since the 1780s, began to travel to Maysville for water by the turn of the century. On Hinkston's Creek near Millersburg, the little water standing among the flat rocks sun-warmed so "that cattle will scarcely drink it."[109]

As waters pushed underground, riparian ecosystems changed. In considering mills and river rights, state legislators did attempt to preserve what remained of river wildlife in order to uphold citizens' rights to kill it. Still, fish populations dwindled, and water fowl, although found along the Ohio

River, became rare in the interior. In their place, mosquitoes bred in the shallow pools among the rocks, and settlers began to complain about the miasmatic odors that arose from stagnant ponds.[110]

Lexington, having originated in a shallow valley formed by a tributary of Elkhorn Creek, had struggled with unpredictable water levels since the early 1780s. Town Creek, situated as the rear boundary for most properties along the south side of Main Street, annually burdened many leading citizens with its "enormous overflows in the rainy seasons." When the rivulet spilled over, the streets became streams. Women traveled from house to house by "Small Cannew on the mud" as the whole town resembled a "Soft meadow." In 1795, Thomas Chapman witnessed how residents' inability to control Town Creek's flooding and the "want of pavement renders it very muddy in winter time, and rainy weather."[111]

By the turn of the century, the water level in "our little branch," with "the opening of the country, the felling of the trees, the plough, the hoe, the harrow, the opening of wells and clearing of springs," succumbed to the same gradual decline that afflicted waterways along the length of the old buffalo trace. As the waters of Town Creek receded, the merchant-controlled board of trustees decided to eliminate the erratic flooding that had been a nuisance to stores along Main Street. Citizens covered the rivulet with an arch and "levelled over it the length of the street." Then, they macadamized Main Street and lined it with brick footpaths.[112]

It was not just the flooding that moved Lexingtonians to pave over Town Creek. Americans understood endemic and occasional epidemic diseases to be the results of miasma—poisonous odors from decomposing organic matter produced by land clearing or released from beneath the earth. In late 1793, a smallpox epidemic struck Lexington; over forty residents died, and miasma was to blame. In November, John Bradford suspended publication of the *Kentucky Gazette* and production of the state's legislative journal because he lost most of his staff. Lexington's trustees, including Bradford, became aggressive. By mid-December, they were pushing for inoculation, a course denounced by citizens at a public meeting. Nevertheless, the Fayette County Court did offer inoculation, and one of every fifteen who agreed to the rudimentary procedure died.[113]

Having failed at public health, the trustees turned their attentions to long-range public sanitation, passing ordinances directing waste disposal and restricting hogs in the streets. By 1796, Lexington had about 1,500 residents residing on 710 acres: a ratio of 2.11 persons per acre, comparable to

Boston during its devastating smallpox epidemic of 1721–22. The population produced more waste than natural means could remove. Regular flooding of Town Creek spread rotting trash, offal, and filth, and when floodwaters receded, an unbearable stench spread across the town. In the late 1790s, a newly dug canal meant to facilitate the flow of waste out of town failed to resolve the problem. By 1801, therefore, as smallpox again threatened the town, Lexingtonians demanded and the trustees agreed that the Town Creek and its receding waters should be paved over.[114]

Their labors were for naught. Despite intentions to eradicate the source of flooding and disease by burying Town Creek, residents instead eliminated the most convenient method for *removing* waters. Overflows became more problematic. Puddles of stagnant water persisted where previously they would have drained off the land, albeit it slowly, and large ponds formed where the creek emptied into underground pools just west of Lexington. Instead of flowing out of town, drainage seeped into the porous limestone strata or remained on the surface. The worst flooding occurred in April 1802 when excessive rains swamped the shallow valley of Town Creek, spread across Main Street, and washed out the gardens and outbuildings of Main Street residents. The waste of domestic livestock, the contents of chamber pots and privy pits, and the refuse of butcher shops, tanneries, brick manufactories, and bluedying operations mixed with underground cisterns and creeks. Nowhere else along the Maysville Road did such a dramatic environmental challenge arise than in turn-of-the-century Lexington. It was a problem that would haunt their children.[115]

Beyond Lexington, fears over miasma and waterways also shaped life. One of the peculiarities along the northern end of the road remained the hillside organization of Mayslick. While reflecting the cultural experiences of its New Jersey settlers, the layout of the neighborhood enabled easier land clearing and provided sanctuary for sheep from wolves and winter weather under the cabins. But in preparing the land for farming, Mayslick's residents pushed uprooted brush into the damp valley, and fear of malaria soon followed. Their hillside homes had been situated to avoid miasma emanating from the spring.[116]

Not surprisingly, the neighborhood of Mayslick—where natural environment, cultural folkways, and practical needs so closely aligned—produced the region's most renowned physician. Coming of age during Kentucky's environmental transformation, Daniel Drake learned early to appreciate the symbiosis of human and natural life and the correlation between environmental imbalance and potential human misery. In 1808, he wrote "Some Accounts of the Epidemic Diseases which prevail at Mays-Lick, in Kentucky," an interpretation of the previous year's local typhoid epidemic. Drake recorded a litany of ecological abnormalities: "summer and autumn were remarkably dry, almost every spring exhausted"; "wheat &tc. ripened two weeks earlier than usual"; "no beautiful colors of leaves before they dried up"; "very little lightning and thunder with showers"; and an uncommon abundance of army worms, green worms on hackberry bushes, and white worms on the beech trees. This early in his career, he was not yet prepared to offer conclusions about the direct relationship between environmental transformation and typhoid, but he did decide that "Evils often seem gregarious" and warned of similar situations in the future when settlers ignored the natural environment.[117]

As an adult, Drake employed his epistemology of environmental causation to reach the vanguard of western medical practice. In 1818, he helped found the Medical College of Ohio at Cincinnati. Four years later, he accepted the professorship of materia medica and medical botany at Transylvania University. Drake taught students to observe keenly the meteorological and geographical origins of disease, and to seek out environmental therapies. Although he left Transylvania in 1827, his emphasis on medical topography continued to influence education along the beaten path. Among the many theses produced by Transylvania's medical students that addressed the relationship of local environments to health problems and medical needs were John Shackleford's "On the Epidemic Fever of 1823" (1824), Samuel McAdow's "On the Epidemic of 1824" (1825), and John English's "An Inaugural Dissertation on the Epidemic Cholera, as it appeared in the City of Lexington in June and July 1833" (1834). Students learned well Drake's conception of environmental causation and therapeutics. His own interest in the topic culminated in 1850 with *A Systematic Treatise, Historical, Etiological, and Practical, on the Principal Diseases of the Interior Valley of North America,* one of the nineteenth century's most comprehensive surveys of topography, climate, disease, and therapy.[118]

Despite Drake's education, his conclusions about the relationship of environment to human misery built upon traditional notions that he had held as a boy in pioneer Kentucky. Drake taught his students that disease demanded distinctive therapies, depending upon the peculiar environmental characteristics of specific locations. Determining those therapeutic practices had long been part of pioneers' medicinal ways. They scanned the skies for omens. They examined the activities of mammals, birds, and insects for hints about the weather and disease. They foraged for flowers, weeds, tree bark, feathers, and even animal excrement to use in medicines and potions. In their views of the land, pioneers recognized a regenerative and sustaining power: Nature provided all needs from medicinal to agro-ecological, if one only knew what to use and where to look.[119]

As they experienced "seasoning"—fluxes, fevers, pleurisies, and other symptoms of their biosocial adaptation to the new environment—pioneers acted on ideas about the interaction and intimacy of the cosmos, and the regenerative qualities of nature. When Fortesque Cuming visited Blue Licks in 1807, he found that the spring's waters had become part of an emerging regional pharmacopoeia. People were "drinking the waters of the salt spring, which are esteemed efficacious in some disorders." The *Kentucky Almanac* regularly printed pioneer recipes for natural remedies. For jaundice, a pound of pigeon's dung boiled in sweet milk would do the trick. Ground black brambleberries mixed with sugar and port wine alleviated dysentery. The flux required tea made with the inner bark of a white oak sapling sweetened with sugar. For cancers, red oak bark burned to ashes, boiled in water, strained, boiled again until creamy, spread on silk or lint, and plastered on the sore every two hours came recommended. What rural settlers understood as part of the natural context of their lives, however, was not so apparent to residents of the village West. Barrels of spring water from Blue Licks were taken to Lexington and other villages, but by the time they were on store shelves, they had lost much of their hydrogen content (and the natural context and regenerative message that rural settlers accorded them). The remaining sulphur begot a more offensive taste. Relying less on nature and more on peddlers and merchants for medicinal supplies, villagers lost the immediate connection of natural source with medicinal need.[120]

In other ways, however, rural and village life shared these folkways. In the *Kentucky Almanac,* readers found monetary conversion rates, lists of government officials, a table of common and feast days, astrological charts, and monthly calendars with reference to planting cycles. While the other items certainly guided a Kentuckian in his relations with neighbors and government, the latter two counseled his relations to the natural world. As in colonial almanacs, the "Man of the Signs"—a drawing of an upright human surrounded by zodiacal symbols indicating which part of the anatomy would be most afflicted during the reign of each sign—was one of the first pages.[121]

Belief in the ability of animals to forecast weather became more widely accepted as well. Unpredictable weather became less so if one only attended to the activities of birds, beasts, and insects. *The Farmers Almanac,* a Lexington-based publication first printed in 1816, provided the requisite details. Evidences of forthcoming fair and warm weather were, among other things, the easy and clear notes of a hooting owl; the flight of bats early in the evening; large numbers of hornets, wasps, and glow worms; and the appearance of spiders' webs on the grass. But if asses brayed more frequently than usual, hogs became playful, frogs moved closer to houses and croaked from ditches, bees stayed near the nest, and one could hear "a growling noise in the belly of hounds," rain approached. Of course, like modern meteorologists, compilers of almanacs disclaimed that "it may not rain or snow at any *particular point,*" regardless of animal activity.[122]

Almanac calendars also instructed farmers on when, according to astrology, to plant their crops. For example, setting melons, squashes, pumpkins, and gourds in mid-April at the cusp of Taurus promised greater yields. Farmers planted radishes "at the decrease of the moon, for they tapered downward," and avoided killing hogs "in the dark or decrease of that luminary, for the pork would shrink and waste away in the barrel." In an era before chemical fertilizers, this agroecological application of the natural and supernatural offered hope for the fertility of one's lands and the quality of one's harvests. In agrarian rituals, farmers sprinkled salt in the splits of apple trees and spread mixtures of gunpowder, brimstone, and grease on sheeps' necks as remedies for disease. Nature offered the cures for all of life's ills.[123]

Salting an apple tree, building a limestone fence, killing a buffalo: the thousands of smaller actions taken to revise the natural environment contributed to one collective consequence. Residents along the old buffalo trace succeeded in remaking the environment to resemble the small worlds they had left behind. The natural environment had been so thoroughly transformed that beliefs shaped in the East could pertain equally to this West. By 1800, settlers had also re-created society along the Maysville Road to resemble the worlds they had left. The gentry secured a slave society where race could be employed to define freedom, revised gender roles so as to strengthen masculinity, restricted political suffrage to those who could be converted into "thinking men," and took control of the religious and educational atmosphere in Lexington so as to promote republican values and religious liberties. Men of commerce established smithies, shops, large artisan enterprises, market networks to facilitate long-distance trade, and markethouses to satisfy their moral obligations even as they pursued profit. Farmers transformed woodlands into fields, built fences to delineate property lines, and worked to be rich.

It may seem that the pioneers as a discernible community were gone. Many had fled as a new society arose along the old buffalo trace. The majority remained, however. Some had joined the ranks of propertied, market-oriented farmers. Others toiled in tenancy or wandered in search of day labor. Small enclaves of pioneers hid in what remained of the wilderness, beyond the gentry's civilizing efforts. Without land, slaves, or wealth, most had been economically and politically disfranchised by the turn of the century. But pioneer culture was about to be dramatically resurrected. Purist Christianity was alive, and in the midst of dramatic environmental and economic transformations, traditionalism had gained new adherents along the Maysville Road.

4 • New Americans?

Revising Identities and Communities

In the early morning of 16 May 1825, five miles west of Lexington at the plantation home of Major John Keene, Fayette County's Revolutionary and War of 1812 veterans jubilantly cheered the arrival of a hero. While many dignitaries had traveled through the Bluegrass region and along the Maysville Road, this foreigner was the most celebrated by far. Everywhere that the Marquis de LaFayette went, he received praise for his pivotal role in the waning years of the Revolutionary War. Few war veterans remained in America, so LaFayette's journey afforded one of the last moments of recognition for the Founding Fathers. General Leslie Combs of War of 1812 fame delivered a rousing address to the assembled. Then the entourage paraded down Lexington's Main Street to Postlethwaite's Tavern where John Bradford recited an official welcome and aging Revolutionary veterans swarmed around LaFayette for a peek. Later that day, Transylvania University hosted a ceremony in which students recited Latin and French poetry to the guest of honor. Immediately following, LaFayette reviewed the county's militias before joining fourteen hundred attendees at an open-air dinner. In the late afternoon, he visited the Lexington Female Academy for a ceremony in which headmaster Josiah Durham rechristened the school LaFayette Female Academy. Then the Frenchman journeyed to Ashland, Henry Clay's estate just east of town, to pay respects to Lucretia Clay (her husband was in Washington, D.C.). A grand dinner and ball with eight hundred guests at the Grand Masonic Lodge Hall capped off the day. He departed the next morning for Cincinnati.[1]

The leitmotif of Lafayette's visit to Lexington and his travels throughout the nation was the Revolutionary moment. Like the multitudes of veterans who clamored to see the aging LaFayette, the tour itself reminded central Kentuckians of their Revolutionary heritage. But in this celebration of liberty and triumphant republicanism, it became obvious to all that they were no longer part of that Revolutionary age. Jeffersonians no longer guided social or political development. In 1818, David Meade regretted how he had outlived "all my seniors in Virginia not only those with Whom I have been on a footing of intimacy, but even those whom I have had any knowledge." By the mid-1820s, the men who had liberated and designed a nation, and those who had fashioned Kentucky, were in the twilight of their years. John Brown, who had served as Kentucky District's representative in the Confederation Congress from 1787 to 1788, lingered, as did John Bradford, revered editor of the *Kentucky Gazette*. But father of the first state constitution, George Nicholas, had died in 1799; and John Breckinridge, father of the second state constitution, followed in 1806. Levi Todd, Fayette County clerk, died in 1807. Merchant George Trotter passed away in 1815, preceding his merchant father James by a decade. Old Captain Peter Durrett died in 1823; purist Presbyterian Adam Rankin followed in 1827. Even the legendary Daniel Boone died in Missouri in 1820. Although LaFayette's visit gave it one last glorious recognition, the Revolutionary moment and the individuals who had created Kentucky in its wake were about to pass.[2]

Between 1800 and 1825, life along the Maysville Road was full of anxiety: environmental transformations, political and economic challenges, religious enthusiasm, generational strains, an emotional rush to war. Not surprisingly, it was also a time of identity making. Older Kentuckians proclaimed the nation in moral decline. Through the Great Revival and the War of 1812, their children awakened to their roles in directing life in an increasingly market-driven world. The Great Revival was much more than religious enthusiasm: it was a catalyst for introspection that empowered individuals to address economy, society, and polity. Likewise, the War of 1812 was much more than a war. It was a rite of manhood that gave new heroes a confidence that they could make a difference; it was a patriotic atonement that assured parents that they had indeed passed their expectations and worldviews on to their sons and daughters; and it was a revival of the sacred legacy, bringing Kentuckians into full communion as Americans. And Lexington, the small town at the southern end of the road, came into full membership among American's burgeoning cities. Its residents, susceptible to revivalism and patriotism, sought ways to incorporate their urbanism into those identities.[3]

BLAST OF THE WILD TORNADO

On the first Friday and Saturday of August 1801, the old buffalo trace through Bourbon County was "literally crowded with wagons, carriages, horsemen, and footmen." Just north of Paris, multitudes turned eastward on the road to Cane Ridge. One participant counted 143 carriages and wagons, over 1,000 other types of vehicles, and, at night, nearly 500 lit candles. Farmer David Purviance delighted that "for more than a half mile I could see people on their knees before God in humble prayer." The numbers were staggering. Estimates of attendance ranged from ten thousand to twenty-five thousand participants. This was the Great Revival.[4]

A decade earlier, Rev. Robert Finley assumed leadership of a small Presbyterian meetinghouse in a former canebrake in eastern Bourbon County. Born in Bucks County, Pennsylvania, and having spent years as pastor to congregations in the backcountries of South Carolina and Virginia, Finley was well versed in the culture of his Irish parents and other pioneers. In a fashion typical of many Backcountry congregations, when Finley chose to move westward, many of his congregants packed up and migrated with him. When, in the mid-1790s, another group from Iredell County in western North Carolina moved into the neighborhood, Backcountry folk peopled the lands east of Paris.[5]

Aided by congregants like John Luckey and James Smith who traveled the farmlands of Bourbon County recruiting new members, Finley worked to organize a congregation. When Luckey and Smith approached Patrick Scott's father, an "old-seceder-Adam Rankin-order" Presbyterian, young Scott remembered that "Col. Smith observed to Luckey, well bro. John What do you think of him. Lucky replied well I reckon if we canna get hewn stone, we must take dorics," the most common and pragmatic of classical Greek architecture.[6]

Like the elder Scott, the meetinghouse that arose at Cane Ridge was plain and practical. Ash logs chinked with mud sat atop a stone foundation. The floor was of planed oak, and a cherry balcony ran along the south wall. In the center of the north wall was a shallow alcove that housed the preacher's platform; a high window lit the pulpit from behind. Even though the congregation soon became well known for "emancipatory" leanings on the issue of slavery, Cane Ridge Meetinghouse was neither architecturally nor theologically unique. It appeared an unlikely place for a religious awakening.[7]

In 1795, Finley left the congregation, embarrassed by his alcoholism and the disciplinary hearing it prompted. For over a year, a rotation of preachers served Cane Ridge and its sister congregation at Concord in Nicholas County. Finally, in 1796, Barton W. Stone accepted the call. Having grown up in Piedmont Virginia, Stone was neither a Backcountryman nor very pious, a self-described "alien in the Presbyterian camp." But through a series of educational and religious experiences, he came to understand his own salvation in the more moderate strain of Presbyterianism. At Cane Ridge, he appealed to his congregants with ideas that they could appreciate: sacramental meetings that resembled frolics, Arminian principles that emphasized individualism, proclamations of God's love that promoted self-worth, and a sense that all were "capable to believe from the evidences given in the gospel" that fostered self-accomplishment.[8]

Many in the congregation were new to church membership, having practiced their faiths in private ways undefined by denominational teachings. "Religion made the hobby of these days," Andrew Todd complained

Cane Ridge Meeting House (1791), Bourbon County. Courtesy of Cane Ridge Preservation Project, Bourbon County, Ky.

from his home in Paris, "every ignorant conceited novice set out for a gospel minister—The blind lead the blind, thro' the intricacies of Error, delusion, passion, pride, party spirit, slander & abuse of all who contradict or stand in their way, or even if they say nothing." Shrugging off liturgical structure, theological debates, and educated clergy, the masses formed spiritual communities that appeared more defined by chaos, and they blazed paths of righteousness that headed in every direction. As Calvinist David Rice witnessed, "They were then prepared to imbibe every new notion, advanced by a popular warm preacher, which he said was agreeable to scripture. They were like a parcel of boys suddenly tumbled out of a boat, who had been unaccustomed to swim, and knew not the way to shore. Some fixed upon one error, and some upon another." Todd corroborated Rice's observations, grumbling that "the root of the prevailing ignorance of the multitude of the deluded thro' our land is laid in the sin of the parents who have not taught their young ones the oracles of God. They are left a prey to every deluded imposter, to be tossed about by every wind of Doctrine." Many a Kentuckian bent with the theological breezes. In Bourbon County, Patrick Scott ribbed his brother about his lack of spiritual roots, evidenced by a recent decision to leave the Presbyterian church: "He had quit fiddling & dancing & joined the Pbyns. . . . when the New-lights came abt., he joined them. . . . It was extremely probable, if he co'd quit the Pbyns. to join the new-lights, he could quit the new-lights to join the Shakers."[9]

The fleeting convictions of the masses, particularly among the younger generations, contrasted greatly with the visions of stability emphasized by purists on one end of the Christian spectrum and Old Lights on the other. Immersed in the individualistic, egalitarian, thinking-for-oneself trends that swept the new nation, younger Kentuckians distrusted anything that seemed too hierarchical or too restrictive. Generally uncomfortable with both purist and Old Light churches, the population of young and largely unpropertied Kentuckians became the most unpredictable and numerous Christians along the Maysville Road.[10]

They were predictable in one way, however. As environmental transformations increasingly threatened what little stability and security they had, many turned to the divine for answers. In 1795, a visiting James Smith rationalized that since the "summer and fall hitherto having been uncommonly dry in this country, has created an alarming scarcity of water, . . . Surely the people of this country as well as the Virginians ought to trace the footsteps of an offended Deity." The God of rewards and punishments had not lost his appeal and was seemingly at work, punishing central Kentuckians for acceding to a society of slavery, gentility, and infidelity. A citizen editorialized in the *Kentucky Gazette* how plagues of "Frost, Flies, Weavels, those feeble, but destructive insects, have, are, and will devour the fine wheat, for which the laborious farmer has wrought and sweat," because "Our sin must be very great . . . tolerating or countenancing of idolatry, or false religion, &c. . . . the persecuting and oppressing the innocent." So too did the increasingly commercial order come under attack. In 1800, John Lyle pleaded to his son, Rev. John Lyle, "What my son are you doing—I hear of your farming and the fruits of it by information—But no fruit from the preach[ing] of the everlasting Gospel. . . . But perhaps you are loading yourself with thick clay, rivaling your wealthy neighbours or striving to hold a place on a board &cc. with any of them." If a godly man like Lyle could not avoid the insinuations of the serpent, how could the multitudes who identified more with the tavern than the church? Herekiah Harriman had the answer: Upon hearing of John Ficklin's abundant harvest in the summer of 1800, he reminded his friend that "GOD is mindful of them that put their trust in him."[11]

In 1797, Mason County Baptists collectively expressed their trust in God with a small revival that temporarily boosted local church membership. But for the next four years, religious enthusiasm blazed only in southern Kentucky. In May 1801, however, camp meetings took place northeast of Lexington on the banks of the Licking River and at Cabin Creek just west of Maysville. In June, revivalists rendezvoused at Concord, at Point Pleasant in Bourbon County, in Lexington, and at Indian Creek just north of Paris. Presbyterians organized most of these meetings for congregants and potential recruits to take Communion. Throughout the summer of 1801, religious enthusiasm grew, but few people anticipated the magnitude of what happened next.[12]

When Barton Stone planned a camp meeting at Cane Ridge in early August, he invited preachers from throughout the region, hoping to inspire a large gathering. John Bradford did not advertise the revival in the *Kentucky Gazette,* but word spread widely, and by August 5, people were on their way to Cane Ridge. Thousands wandered the grounds, listening to one stump preacher and then moving on to another. Most of these men promulgated Arminian ideas that contested the Calvinist views of purity-based Christians, including the notions of election and predestination. Robert Patterson remembered how, as the ministers articulated their visions of heavenly and earthly life, there "was an opposition . . . by some who appeared to be real christians, by nominal professors and by deists." Revivalism attracted throngs of farmers, village laborers, and small artisans who, while attuned to a Judeo-Christian foundation to life, had largely avoided church membership. They "stood astonished, not knowing, and wondering what these things meant; not willing to reprobate it, and many at fact closed in with it." Purist Methodists, Baptists, and Presbyterians who came to watch the revival "call it enthusiasm, hypocrisy, witchcraft, possession of the Devil, sympathy, in fine, everything but what it really is." Adam Rankin did not even travel to Cane Ridge, but still made widely known his conviction that the revival was Satan-inspired, evidenced by the human-produced theologies and songs that echoed through the woods around Cane Ridge.[13]

Deists, Unitarians, and Old Light Episcopalians and Presbyterians had little appreciation for the events at Cane Ridge, as Bradford's refusal to report the revival in the *Kentucky Gazette* confirmed. Of foremost concern was the uncontrolled emotional enthusiasm that seemed to affect everyone in attendance. Watching hundreds of people fall to the ground—some unconscious and still; others shaking and jerking; all exposed to an invisible spirit—frightened many of the gentry who saw little in the proceedings beyond anarchy. This was superstition run amuck. A concern of equal

Adam Rankin's House (1784), Lexington. Photograph by Rod G. Turner.

weight, however, was the egalitarian nature of revivalism. "Of all ages, from 8 years and upwards; male and female; rich and poor; the blacks; and of every denomination" attended the revival, representing an earthly equality that threatened hierarchies of class, gender, and race. "We all engaged in singing the same songs of praise," remembered Barton Stone, "all united in prayer—all preached the same things—free salvation urged upon all by faith and repentance." For the gentry, equal opportunity for salvation was one thing; equal station on earth was another.[14]

One need look no further than the use of space at Cane Ridge to find this egalitarian undercurrent. Traditionally, church buildings were designed with theater in mind. Pulpits were elevated for visibility and, as in the meetinghouse, a window illuminated the preacher from behind. Congregations sat in gradations of hierarchy, and clear distinctions were drawn between races, usually by railings or balconies. Revivalism shattered those patterns. Although preachers pontificated from stumps, the sermons were background noise to the real experience of inner conversion. The chaos that many observers witnessed in revivals reflected a loosening of social constraints on the audience. Faith was an issue of the heart, supernaturally activated within the soul rather than through repetitive prayers and rituals. Artificial hierarchies could not be enforced in the outdoors; even racial and gendered boundaries blurred as people moved from one preacher to another.[15]

Revival preachers communicated a new worldview, one in which greater individualism and egalitarianism restored participants' hopes in the wake of the environmental revolution, the upper sorts' political and social dominance, and the collapse of markethouse economies. This revivalist vision critiqued refinement and the division of society that accompanied change along the Maysville Road. New Lights interpreted conspicuous consumption as an aristocratic vice. Accordingly, they countered republican ideas of civic virtue and individual independence with ideals of personal virtue and neighborly affection. In other words, the revivalist Christian became responsible for self *and* society. Rather than uphold paternal authority, reinforce codes of male honor, and idolize the family, New Light theology affirmed female virtue, questioned racial slavery, contested the idea that youth should always defer to age, and prized communal fellowship over family ties.[16]

But interpreting the Great Revival as a reaction to social disintegration, economic competition, and political partisanship is too simple an explanation. It was not just the failure of individuals' dreams that inspired revivalism; it was the articulation of Christian faith that had no outlet prior to the turn of the century. Purists and Old Light Christians had polarized religion in the 1790s, leaving little room for alternative intellectual and emotional expressions. Cane Ridge provided an opportunity for traditionalists who were nonpurists to formulate a revivalist worldview that allowed for emotional chaos while offering "employment for the mind," as pioneer Robert Patterson explained.[17]

Like other traditionalists, revivalists believed that God and other supernatural beings revealed themselves in the natural world. Revivalism expanded on such theistic notions by situating human conversion and salvation in a supernatural realm attained through visions, mystical experiences, and shamanic encounters.[18]

To observers, even sympathetic ones, mystical experiences were shocking. As an ecstatic state of consciousness overtook participants, they "fell down," testifying to the power of the Spirit as it overwhelmed the fragile earthly bodies of humans. Some who fell remained conscious, relating the conversion under way; others lay in a comalike condition. God reportedly revealed himself to the fallen, triggering a conversion unmatched by other faith manifestations. "I have brought forward the falling down as a certain test of true religion," proclaimed the visiting George Baxter upon his arrival at Cane Ridge some weeks after the revival, "and that I have expressed no doubt respecting the real conversion of those who thus fell down."[19]

Those who fell (and some who did not) attested to shamanic experiences in which they not only envisioned another level of reality, but journeyed within it. John Lyle reported about one man who had fallen, but still awake, "his spirit went out into the earth & saw strange curious caverns & then he thought he would look upwards & he saw a mountain clothed with beautiful trees silver tops or leaves tip'd with silver, he thought this mountain led to God & heaven then above he saw a great light & he pray'd to see a little further and a little to the right he saw still more dazzling light & he sigh'd & sunk before it as the great all in all." The supernatural did not reside only in shamanic journeys. Revivalists had visions of supernatural beings in the fields and woods around Cane Ridge. James Finley, whose father had been the first pastor of Cane Ridge Meetinghouse, sensed a "strange supernatural power seemed to pervade the entire mass of mind there collected." And then there were Adam

Rankin's suspicions that Satan somehow was involved, confirmed by participants whose grunts and hollowed screams were interpreted by observers as barking at the Devil in attempts to repulse temptation. Reports of barking always placed the activity outdoors, in the woods, affirming of a continued association of unconverted forests with darkness and evil. Only by assuming a more animalistic character could participants spar with their Mephistophelean opponent, a reversion to barbarism that republican theorists had never anticipated.[20]

The proximity of both the Holy Spirit and Satan lent an atmosphere of supernatural terror to the proceedings. James Finley likened the spiritual hysteria to "the trees of the forest under the blast of the wild tornado," a metaphor that brings to mind Daniel Drake's memory of his family finding the divine in a tremendous thunderstorm in the late 1780s. The Holy Spirit uprooted the unconverted just as a storm wrought havoc on the natural world. The revival itself intimated a primitivism in which the boundaries between natural and supernatural, order and chaos deteriorated. As one observer described, "Sinners dropping down on every hand, shrieking, groaning, crying for mercy, convoluted; professors praying, agonizing, fainting, falling down in distress, for sinners, or in raptures of joy! Some singing, some shouting, clapping their hands, hugging, and even kissing, laughing; others talking to the distressed, to one another, or to opposers of the work, and all this at once—no spectacle can excite a stronger sensation. And with what is doing, the darkness of the night, the solemnity of the place, and of the occasion, and conscious guilt, all conspire to make terror thrill through every power of the soul, and rouse it to awful attention."[21]

Just as converts, through shamanic and mystic experiences, were made in a new image, so too was God remade. In contrast to Old Light views of God as transcendent and purist views of a theistic God of requirements who wagged his finger and threatened eternal punishment for those who strayed, revivalists proclaimed God as immanent and benevolent. He acted dramatically through the Holy Spirit to convert individuals and to battle Satan. He worked dramatically as well in his creator role, regenerating a deteriorating wilderness and the hopes of the dispossessed, as evidenced by a newly discovered species of fish in Kentucky's ponds and streams that settlers named New Lights.[22]

In this sense, revival-based Christianity employed the preternatural mentality of traditionalism. People understood Satan and God as spirits working surreptitiously within human society. Satan tempted humans through sinful pleasures; God dissuaded humans from the same. For many

Americans of the Early Republic, Jehovah and Beelzebub had become anachronistic. The Enlightenment had redefined sin and lessened the impact of Satan and God. Transgressions remained, but the lines blurred between the talent of entrepreneurship and the sin of avarice, the virtue of gentility and the temptation of vanity, the mastery of individualism and the offense of pride. To traditionalists, the appeal of revivalism was its acknowledgment of the creator God operating in their immediate lives.[23]

By reinforcing the very supernaturalism that John Bradford, William Barry, and other republican gentry condemned as superstition, revivalism appealed to the masses. The omnipresence of God, the Holy Spirit, and Satan in the lives of revival Christians accentuated the urgency of individual salvation, communal accountability, and social action.[24]

The Great Revival, therefore, had an unprecedented effect on life along the beaten path, not because it awakened all of society to an Arminian version of Christian salvation or a theistic interpretation of God, but because it vitalized Christian spirituality in a region that had appeared increasingly deistic and even atheistic. Religion suddenly became important. Revivalists intended to cleanse community and combat both purists and Old Lights to do so. Slavery, gentility, infidelity, and commerce were forbidden apples from which revivalists meant to turn themselves and all of society. After Cane Ridge, they proclaimed a new era for the new West. George Baxter exclaimed how "I found Kentucky the most moral place I had ever been in; a profane expression was hardly heard; a religious awe seemed to pervade the country." Terah Templin noted that there was "considerable difference in the face of families, neighborhoods, and these congregations, compared with what formerly appeared."[25]

Just as the Spirit remade them, revivalists were determined to remake society. Their "new light" provided an ethical position constructed of both Revolutionary ideals—virtue, egalitarianism, and even patriotism—and revival idealism. The political concerns of 1776 resurrected as the spiritual concerns of 1801: corruption and aristocratic privilege, the relationship of church to state, and the need to promote a social compact. In fact, Presbyterian minister John Finley early realized the comparison. "There are many irregularities among us, so it was in 1776 among the whigs in their enthusiasm for liberty," he offered as an explanation for the fanaticism at the Cane Ridge.[26]

Revivalist Christianity upset the religious world along the old buffalo trace, including the definition of community found among purists. Returning to the witch episode momentarily, it is not unreasonable that

Nancy King may have been one of these New Light Baptists, openly suggesting an Arminian view of salvation as widely available and, by challenging their understanding of the heavenly purposes for good deeds and good thoughts, forcing her fellow congregants to expel her from Ray's Fork Baptist Church. For Calvinists who understood themselves as God's chosen, the cries of Arminians in their midst imperiled the sacredness of their exclusive communities. For revivalists, the purpose behind reform epitomized the difference between themselves and purists who were exclusive and relied on scriptural dictates to distinguish between the faithful and the faithless. Revivalists were inclusive, exposing their own sins and those of their neighbors to awaken all of society to salvation.[27]

At first glance, despite its heavy reliance on supernaturalism, revivalism had much more in common with the most liberal of the Old Lights. Both espoused Arminian ideals of redemption, and both appealed to ideologies arising from the Revolution. Yet, beginning with the appearance of revival chaos, there emerged significant differences that rattled Old Light notions of social order. "These meetings exhibited nothing to the spectator unacquainted with them but a scene of confusion, such as is scarcely to be put into human language," explained James Finley as he recalled "seven ministers, all preaching at one time, some on stumps, others in wagons." The disorder of revivalism contrasted markedly with the rituals of denominational religion, especially the Episcopalians who did everything by the *Book of Common Prayer*. The masses seemed as capricious as ever, turning to revival emotionalism for spiritual uplift and moving from Baptist to Methodist to Presbyterian preachers to imbibe different themes and perspectives. John Lyle noted how the pandemonium "disturbed some of the Old Presbytereans verry much." Doctrine seemed to be supplanted by irrational emotionalism.[28]

Physical manifestations at Cane Ridge—falling down, rhythmic dancing, hymn singing, and shouting during sermons—appeared not that different from the festivity of pioneer frolics. Frenzied dances at revivals resembled the rambunctious dances at taverns, quilting bees, and barn raisings. Bodily writhings and convulsions at camp meetings hinted of the sexual pleasures at corn huskings and log rollings. In fact, the passion of revivalism often led to indiscreet intimacy; John Lyle recorded his shock that as a result of revival meetings, "Becca Bell,—who often fell, is now big with child to a wicked trifling school master. . . . Raglin's daughter seems careless. . . . Kitty Cumming got careless. . . . Peggy Moffitt was with child to Petty and died miserably in child bed." The Cane Ridge revival and other camp meetings exhibited characteristics similar to frolics: everyone joined together for a purpose (in this case, preaching and conversion), renewed friendships, competed in contests, enjoyed the interaction between genders (including courtship and sexual intimacy), and drank whiskey, often to the chagrin of revival leaders. Even revivalists' emphasis on individualism and efforts to seek out and expose the sins of neighbors reeked of the vigilantism of Backcountry patterns of justice.[29]

The sacramental moment at camp meetings produced an interdenominational, multicultural free-for-all that clashed with the propriety and ceremony of Episcopal and Old Presbyterian services. Anyone could participate, and worse, Eucharist occurred in the woods or in a field. Elizabeth Corlis questioned, "why can not we worship God rationaly like civilized people in our houses or in meetinghouses instead of praying in the woods sleeping on the ground *all together* &c?" Civility demanded worship within a consecrated church, whether that be the small-frame First Presbyterian Church from which *Watt's Psalms* still echoed into the streets or the new brick Episcopal Church. Ironically, the gentry found advantage in adopting religious enthusiasm. Emotional demonstrations of faith became a fad, supplanting material refinement as the defining quality of the in-crowd, at least for the moment. In 1802, one observer proclaimed that "those that have it shows it, and those that have it not wish to be considered religious for the credit it gives in society."[30]

Revivalists also had a significantly different interpretation of Thomas Paine than did liberal Christians. Paine's radicalism contributed to their definition of virtue, emphasizing individualism, personal morality, and the permissible intrusion of their self-interests into politics in order to create a heavenly kingdom on earth. Let us return to the definition of democracy as "a widespread participation in public affairs, a diffusion of leadership, a widespread sense of personal competence to make a difference." More so than the average pioneer, revivalists embraced this definition; indeed, they counted on it. As life along the road became democratized, revivalists expected it would also become evangelized.[31]

Revivalists' differences with other Christians extended the contentious religious atmosphere of the 1790s into the 1800s. Soon after Cane Ridge, Presbyterian and stalwart Calvinist David Rice cheered when "here for the first tim a Presbyterian minister arose and opposed the [revivalist] work ... labored hard to Calvinize the people, and to regulate them according to his standard of propriety.... The consequence was, the meeting was divided, and the work greatly impeded." In 1802, as he baptized converts, revivalist John Lyle faced similar interruptions and "gangling," the mass exodus of his crowd to the other preacher. A few months later, in June 1803 as a purist rallied a crowd in Paris, revivalist Barton Stone "got on knees began praying and exhorting others to pray. 10 mins. later noise so great as to compel McPheeters [the speaker] to stop. Kept up while rest of cong. left in disgust." Even within families, the strain of revivalism rent relationships. When Joseph Corlis became a "*shouting* Methodist," letters between family members disclosed a growing alarm. His sister, Elizabeth, wrote her father in Providence, Rhode Island, that Joseph "at different times has been absent about several days, attending the camp-meetings; he has now gone to a third, where he has erected a tent and carried a quantity of provisions, more than *I* think *we* can well spare to distribute.... Joseph and all the family, I believe think that I am opposed to the Methodist because they are not *fashionable* ... there is something so indelicate, so disgusting, to me, in shouting in the manner he does.... I do not like such noisy christians, with them religion loses its dignity, its beauty.... I have not much opinion of their camp-meetings either they leave ample room for scandal and good christians ought to set a better example." Another sister, Mary Ann, likewise complained to her father that "Joseph is Methodist mad—goes in the woods at night I hear and hallows enough to frighten one." From Providence, their father responded to yet another sister who defended Joseph, "say what you will of camp meetings I know full well of what they are Composed too well to want any Child of mine ever to be present at one of them. I am the last to think lightly of religion or religious persons," he continued. "On the contrary, they have my highest respect but it must not be merely professional let me see them act the Christian, a hypocrite will always over act because he means to deceive."[32]

Elizabeth and Mary Ann Corlis's sincere panic and John Corlis's condemnation hint of something more than just concern over Joseph's eternal soul. All three related their worries to the here-and-now, suggesting that revivalists proffered an new vision of earthly life that distressed other Christians. Among revivalists, revelation through the Holy Spirit demanded a politics of individualism, the one impulse that purists and Old Lights distrusted. In scripturally pure congregations, anything that eroded community and conformity was heresy. In Episcopal and Old Light Presbyterian churches, anyone who challenged hierarchy and liturgical tradition was frowned upon. The culture of revivalism validated the individualistic impulse in traditional folkways.[33]

Revivalism, then, was a revitalization of pioneer culture, as well as a response to the dramatic changes of the past decade and the birth of a new theology. While the magnitude and number of camp meetings did fade, the Great Revival stimulated a period of church growth that guaranteed greater cultural emphasis on individualism. Baptists enjoyed the greatest boom in church membership. In 1790, Kentucky Baptist congregations numbered 42 churches, 3,105 members, and 4 percent of the population. By 1803, Baptist churches numbered 219 with nearly 15,500 congregants, and 7 percent of the population. Between 1800 and 1805, Methodists in central Kentucky doubled in number, from 1,642 to 3,550. The demand for Methodist ministers forced the Western Conference to increase its clergy ranks from fourteen to thirty in those same years; the clergy doubled again by 1811. But despite growth in numbers, these religious enthusiasts still composed a small minority of the population. Additionally, the increase was not evenly distributed along the Maysville Road. By 1812 in Fayette and Bourbon Counties, there were nine active Presbyterian meetinghouses, twenty Baptist churches, and two unaffiliated New Light congregations. In Mason and Nicholas Counties, there were two Presbyterian meetinghouses, nine Baptist churches, and no New Light congregations.[34]

The density of population in the two southern counties partially contributed to the differences, but the geography of revivalism also suggests that its greatest influence was where politics and economy most clearly favored the gentry. In 1799, a constitutional convention met in Frankfort to revise the 1792 constitution. While granting some democratic changes, particularly direct election of governors and senators, the overall direction of the new document reinforced the dominance of landed slaveowners and the republican government they had created (including its restrictions on suffrage). Revival meetings had engaged residents of the Green River country of southwestern Kentucky for years, but not until the failure of democratic reform in the 1799 constitutional convention did revivalism strike along the old buffalo trace. In comparison to populations south of the Green River who faced lower rates of landlessness, guaranteeing more equal participation in local government, settlers between Maysville and Lexington struggled to acquire property and often became mired in tenant farming or urban labor pools. Whereas revivalism south of the Green River echoed the more egalitarian opportunity of that region, revivalism along the beaten path announced the frustrations of marginal peoples who were deprived of egalitarian society and democratic government.

Ultimately, revivalists' leanings towards egalitarianism and individualism portended a modern, relativistic society. After all, if one believed that each individual could interpret scripture, then there could be many different "truths." In the words of Cane Ridge preacher Robert Marshall, "believing a truth made it truth to the believer. . . . 2 + 2 make four to those who know it," he proclaimed, "but to another 2 + 2 might be 6." The cultural pluralism that characterized pioneer culture along the old buffalo trace made such relativism conceivable to many. Yet, the majority of people, accustomed to claims of absolute truth in discussions of religion, often descended into self-doubt as they heard contradictory versions of salvation. After listening to Richard McNemar preach "a discourse unintelligible to myself & others with whom I conversed about it" at the Cane Ridge meeting, John Lyle fretted that "I humbly hope I understand the gospel of Jesus." But then he overheard Barton Stone who, with McNemar, claimed the sermon "the true gospel which (say they) none preach but ourselves." There were many New Light messages, all bucking the doctrinal rigidity and complicated theologies of established denominations. The audacity of revivalist preachers offended Calvinist David Rice who condemned the ignorance and vanity of men who thought "themselves called of God to preach the gospel, and to go on, relying on their inward call, and neglecting almost every ministerial qualification required in the sacred Scriptures."[35]

If "truth" were contestable, then the structures established on its supposed absoluteness were as well. Purist and Old Light Christians offered visions of community based on patriarchy, biblical ideas of covenant, and classical notions of civic virtue. Religion, like politics and economies, perpetuated their worldviews. Revivalists, in comparison, were iconoclasts, minimizing the significance of ecclesiastic traditions and societal distinctions. Religion was a matter of the heart—salvation in the afterlife, righteousness on earth, and peace of mind—and offered the most common of people individual self-respect and collective self-confidence. One needed merely to be open to the Spirit. But there were obstacles that kept individuals from exposing their souls to the new Christian ethos. Consequently, revivalists took up reform that targeted profanity, drinking, dueling, and slaveownership.[36]

Questioning such behaviors directly confronted the ways in which gentry and pioneers defined masculinity. Revivalism thrived on inducing fear; pioneer masculinity required vanquishing fear. James Finley "prided myself upon my manhood and courage, I had no fear of being overcome by any nervous excitability, or being frightened into religion." Yet, upon attending a revival, "My heart beat tumultuously, my knees trembled, my lip quivered, and I felt as though I must fall to the ground." He fled to the woods "to rally and man up my courage." Revivalism engaged the heart over the mind; republican manhood demanded that the mind control the heart. At an 1801 gathering in Lexington, James Bradford, son of John Bradford, came upon an audience of ladies emotionally charged by a revivalist's message. Because "he was a man," Bradford took it upon himself to protect their sensibilities by threatening to beat the preacher. At the same meeting, George Nicholas's widow, Mary, wandered the crowd with vinegar and bread and, when female congregants appeared ready to succumb to enthusiasm, she "rub'd their faces & noses & put it in their mouths." Through Bradford's eyes, these women demonstrated a willingness, even a mindlessness, to be overcome with religious enthusiasm; as a gentleman, he had to protect them. His gallantry was matched only by Mary Nicholas's tenacity: the response of a genteel woman who emotionally dissociated herself from the message in order to preserve other ladies' reputations.[37]

The women whom Bradford and Nicholas tried to guard from revivalist salvation, however, were quite mindful of the evangelical message presented. The more egalitarian and inclusive social vision of revivalism disputed the patriarchal structure that restricted white women. During and after the Great Revival of 1801, more women than men converted, claiming new levels of spiritual, intellectual, and social agency. Of course, for those like Bradford who were suspicious of revivalism, women succumbed to religious enthusiasm because of their feminine weakness.[38]

For many white women, revivalism proffered a new effeminized culture in which they could flourish. Among New Light Baptists, congregational "courts" charged and punished church members for transgressions from blasphemy to profanity to domestic discord; and, although Baptist women outnumbered men more than two to one, men received more than twice as many admonishments. Although males comprised church tribunals and grand juries, their decisions reflected the revivalist interpretation of personal morality that became pervasive in the early 1800s. White women began to use congregational courts to resolve legal disputes over wills and estates, to force merchants to pay appropriate rates for homemade fabrics, and to compel white men to fulfill their roles as husbands and fathers.[39]

Revivalism also defined femininity, elevating the Christian Mother to a position of cultural significance. Through sermons and essays, preachers appealed to women to actively convert their husbands and children, thereby domesticating faith. The home, where the woman had the greatest role, was to become a refuge where familial religion would be nurtured, away from the temptations of a sinful world. The moral and religious words of a revivalist mother in the privacy of her home became more important than the public pronouncements of revivalist men.[40]

The empowerment of revivalist women, although only a small percentage of the female population along the Maysville Road, buttressed the influence of revivalist men who defended their masculinity as not effeminate but Christianized. In 1813, James Fishback, former professor of medicine at Transylvania University and editor of *The Western Monitor,* argued that the true principles of '76 were "*Christian* bravery, and perseverence," and, that at the end of the American Revolution, "the great patriots, and warriors; with hearts beating high with gratitude to God for their success, laid by their swords, and organized the plan of the American Republic. A dependence upon the great Creator through Jesus Christ, and the light of his word, gave wisdom, union, and decision to their councils, banished far from them the projects of ambition.... Through the revelation's of God's Spirit in the Gospel, they beheld the rights, and the interests of men explained as consisting in, and arising from not only earthly privileges, and enjoyments, but more especially from their invisible relations to their creator, which are established, and supported, by his everlasting truth, and goodness." Fishback proposed the image of a "Christian Republican," a kinder and gentler husband, father, and citizen. "His heart is melted; he is all tenderness," elaborated James McGready, a revivalist in southwestern Kentucky. In contrast stood "Anti-Christian Republicans," as Fishback labeled them, Unitarians and Deists who had increasingly turned their backs on God even as He designed the new nation and set it in motion. Thus, even as the bodies of many churches became more female, by accepting reform of manly characteristics and allowing the use of church tribunals to rectify gender inequalities, white men continued to dominate the minds of revivalist congregations.[41]

For blacks, revivalism's emphasis on individualism became particularly relevant. During the Cane Ridge revival, the crowd broke into four camps: one in the meetinghouse, one 100 yards to the east, another 100 yards to the west, and the last 180 yards to the south. In this last group were African

Americans, "hearing the exhortations of the blacks, some of whom appeared deeply convicted and others converted." This was not an instance of racial segregation; blacks and whites alike roamed between the preaching areas. The distinction between camps appears to have been whether the preacher was Methodist, Presbyterian, Baptist, or, as with blacks, independent. Since African Americans, most of whom were slaves, fully participated in revivals presumably with equal opportunity for salvation, white revivalists came to understand them as individuals, thus raising questions about their treatment within a prematurely designed slave society. Of course, revivalist fellowship did not abandon a system of racial hierarchy. White church members remained vigilant, especially as larger numbers of blacks joined their congregations. But African Kentuckians received different treatment within revivalist churches where whites expected blacks, even enslaved blacks, to uphold the same moral standards as themselves. On some level, this acknowledgment of the equal station of all before God empowered black congregants. Congregations encouraged both blacks and whites to bring disciplinary charges against fellow congregants when they failed to live up to communal standards. At Bryan's Station Baptist Church, Brother Arch charged his owner with "refusing to suffer Sd. Black Bro. To Speak to him and for raising a Stick against him." The congregation reprimanded the slaveholder. Revivalists expected a more equal earthly relationship between the races, for such was the kingdom of heaven.[42]

The emerging relativistic character of religion empowered black preachers like Old Captain because, as revivalist David Rice explained, "there are members of this family [of the Redeemer] who know little more than that they are lost sinners, —that the Lord Jesus Christ is able to save to the uttermost, and—that there is salvation no where else. And they know and believe these truths merely upon the testimony of God, without being able to understand much even of their connection with one another. It is probable that old Captain's understanding of divinity did not extend much beyond these three points. They were enough for his own personal salvation, and they were enough for the salvation of those among whom he laboured."[43]

Ironically, as the swell of religious enthusiasm offered new opportunities to African Americans and white women, it also breathed new life into purist Christianity. As the Arminianism of revivalism threatened to erode the disciplinary force of Calvinism, purists bolstered their missionary and educational programs. They suddenly appeared desperate to control the soul of the American West (and the nation itself). Nowhere was this more evident than at Transylvania University where, by 1804, the departures of John Bradford as chair of the board of trustees and James Moore as president gave purists hope that they could regain control.[44]

In what must have been a shocking valedictory address in 1803, Joshua Wilson suggested that, despite his appreciation for freedom of scholarly inquiry, the "lovers of liberty" and "promoters of learning" whom he labeled "infidels" should "follow the present conduct of their great Exemplar [Thomas Jefferson whom he apparently hoped would lose reelection], viz. Feast upon *Egotism* and retire to *Independence*," and allow the college to become a steward of pure doctrinal truth. The board of trustees, dominated by gentry, had a different idea. They courted the president of the College of William and Mary and convinced President Moore, wearied by two administrations and the incessant struggle with purists on the board of trustees, to retire. But the new president did not come, and through legislative appointments in 1805, Calvinists regained control before another candidate could be identified. Almost immediately, they announced in the *Kentucky Gazette* that under their watch, Transylvania University would indeed become a steward of doctrinal truth: "It had been, and shall be, the care of the Trustees and Professors, to guard against the baneful influence of skeptical principles; and while they carefully prevent the inculcation of the peculiar opinion of any Christian sect, they feel themselves bound to see, that the great leading doctrines of Christianity be warmly inculcated by precept and by example." Purist direction for Transylvania proved a far cry from the "perfect liberty in the formation of their religious creeds" espoused only six years earlier by Moore.[45]

The unanticipated and immediate success of the purists in controlling the board disconcerted trustees placed awkwardly in the middle. By 1806, both William Morton and Levi Todd, men who previously had sided with the purists, transferred their religious affiliations to the Episcopal Church and their pedagogical loyalties to the greatly weakened republican faction which could not stop James Blythe, the renegade, reactionary purist who had challenged the religious doctrines of James Welsh's Wattsite Presbyterian congregation in the mid-1790s, from becoming the new president. Joined by newly appointed professor of moral philosophy Robert Hamilton Bishop, Blythe used his institutional pulpit to further refashion the university and to attack revivalism, culminating in Bishop's *An Apology for Calvinism* that condemned Barton Stone and his Arminian teachings as heresy. By

1808, Transylvania University's new motto—"Pietate et Doctrina tuta Liberatus"—defined liberty as the sum of doctrine and piety, an understandable interpretation given the Calvinist minds who wrote it.⁴⁶

The newfound influence of purists over Transylvania University, the growing presence of African Americans and white women in Baptist and Methodist congregations, the growing tensions between husbands and revivalist wives, the familial and communal strains created by individuals who had gone "revivalist mad": following the Great Revival of 1801, religion became the defining feature of life along the Maysville Road. Even Joseph Charless, the bookseller who profited from sales of *The Age of Reason* in the 1790s, found new profit in religious texts. In 1803 in the midst of revivalism, he opened a new bookstore, boasting that "I could easily sell a hundred Bibles in a month." By the end of the year, he sold out the first run of twelve hundred hymnals and made plans to publish some two thousand copies of a new edition. Over fifteen hundred subscribers financed Charless's publication of Jonathan Edward's *Some Thoughts Concerning the Present Revival on Religion in New England,* which, he alerted supplier Matthew Carey in Philadelphia, would be ready soon for shipment back east. Of the twelve books he published in 1804, eleven had religious themes, including *The Importance of Family Religion* which bolstered the notion of domestic Christianity and the morality of the Christian Mother. Ironically, the very market structures that brought into revivalists' midst the

1806 Seal of Transylvania University. Courtesy of Special Collections, Transylvania University, Lexington.

accouterments of temptation also made possible the extension of revivalist thought.[47]

In determining how long revivalism remained a potent cultural force, we need look no further than Charless's shelves where, in 1809, none of his publications were religious (he had turned his attentions to schoolbooks). A year later, as Alexander Wilson toured Lexington, he commented on deteriorating churches that five years earlier had benefited from the boom in religious enthusiasm. "Though religion here has its zealous votaries," he quipped in his typical sarcastic tone, "yet none can accuse the inhabitants of this flourishing place of bigotry, in shutting out from the pale of church or church-yard any human-being, or animal whatever.... The birds of heaven find a hundred passages through the broken panes; and the cows and hogs a ready access on all sides."[48]

The Great Revival and the religious enthusiasm that spread along the Maysville Road before 1810 was the last gasp of pioneer culture. Thousands of pioneers had already migrated westward and northward in the 1790s, fleeing the gentry's efforts to order society. Those who remained found redemption in a revivalism that transformed their anxieties and failures into a hopeful, joyous celebration of individualism and egalitarianism. Despite their inabilities to secure property titles and guarantee a landed inheritance to their children, despite limited political rights and restrictions on their participation in the markethouse economy, and despite the epidemics and floods and increasingly exhausted soils, God was with them, delivering them, empowering them.

REDEEMING THE SACRED LEGACY

Simultaneous to the wave of revivalism were public debates that dwelled more and more on whether the principles of '76 would die with the Revolutionary generation. Revivalism incorporated more radical elements of revolutionary thought into its theology, but the chaos that it posed alienated older Kentuckians. If members of the younger generation could be so easily drawn into religious enthusiasm, how would they live up to the standard of the Revolutionaries? Coming of age in the 1790s and early 1800s, the first generation of Americans had been molded to patriotic service. They had followed their parents in mourning processions upon George Washington's death; they had been taught the mythic grandeur of the Founding Fathers; they had come to know Kentucky as the future of the nation; and, as we have seen with Mary Ann Corlis's interaction with William Henry Harrison, they did not hesitate to question the political values of their elders. Still, they were derided for lacking virtue, patriotism, and accomplishments. Whenever older residents needed to gain some political advantage, they turned to the Revolutionary moment to promote their position, as when Charles Scott responded to opponents of war veteran pensions: "These men who were dirtying their little clothes when I was fighting for my country ought to have more grace, than to spit lies in my face, because I have purchased the privilege of their doing so." Whether going Methodist mad or challenging a politician's thoughts on democracy, it was a poor time for young white men and women to draw attention to themselves.[49]

But they did: in camp meetings, in challenging professors at Transylvania University, and in militias. In the 1780s and early 1790s, militia membership had largely reflected pioneer ideals. Participation was voluntary. From captains to corporals, the prerequisites for leadership were an understanding of Indian warfare and an example of personal merit, qualities associated foremost with pioneer ideas of masculinity. If he could afford it, every citizen-soldier provided his own uniform, weapon, ammunition, and (if aspiring to cavalry) a horse. If he lacked the funds, then the benevolence of a neighbor might supply him. Muster days occurred six times a year when militiamen gathered for military training, parading, and revelry. On the one hand, unlike other frolics, the primary purpose of the muster was to inculcate civic virtue: duty, firmness, and commitment. On the other hand, musters were like other frolics, uniting participants of common identity: in this case, white men. Even in the amusements, militiamen wanted to demonstrate manliness as these early Kentuckians defined it. From target shooting to drinking to wrestling, militiamen of the pioneer era broadcast their hunting skills, physical strength, and courage.[50]

As the gentry rose to prominence along the Maysville Road in the 1790s, the purpose and character of militias changed. In 1792, the federal government issued the Militia Act, removing voluntariness from volunteer militias. Now required by law to enroll in companies and attend musters, citizen-soldiers between eighteen and forty-five years of age *had* to purchase or borrow guns and uniforms within six months of enrollment. It was a substantial investment: "a good musket or firelock, a sufficient bayonet and belt, two spare flints, and a knapsack, a pouch." Just as reasons for gun ownership—hunting and Indian fighting—became less significant along the old buffalo trace, more and more white men became part of a gun culture designed to replicate the duty, fortitude, and commitment of the Revolutionary generation. In contrast to the men of the 1770s and 1780s who depended on their weapons for protection and food, a man of the 1790s and early 1800s who owned a gun was not necessarily an experienced shooter. Militias were young men's clubs, and chances were that as hunting and Indian fighting subsided, members were less inclined to develop shooting skills. "I carried my gun half my life, for fear of indns," one Kentuckian regretted, "and never saw a wild one."[51]

In the early 1800s, as the younger generation filled their ranks, militias again changed, providing a visible and public demonstration of the character of the Revolutionary generation's progeny. Uniforms and quality horses evoked far more interest than shooting accuracy and physical strength. Style became more important than substance. In Fayette County, where lines of social distinction were becoming deep and contentious, new young leaders of its militia were typically men of middling status who evinced success. Captain Nathaniel G. S. Hart inherited his father's business acumen and ran a mercantile; Captain Stewart Megowan's father owned Megowan's Tavern; Captain John Hamilton operated a hemp rope manufactory; Captain George Trotter was one of the wealthiest merchants in town; Captain John Edmonson was a successful farmer. The 1792 Militia Act had demanded more from officers than regular citizen-soldiers, requiring them to arm themselves beyond the requisite gun and bayonet with a sword and esontoon.[52]

Through investment in guns, uniforms, and military education, militiamen substituted previous definitions of masculinity with the expectation that they could acquire and display masculinity. Appearing a man made one a man. Ephraim P. Ewing's 1810 valediction address to his fellow Transylvania University students laid out the symbiosis of self-control to personal advancement. Dismissing heritage as a prerequisite to success, Ewing explained how "the most brilliant talents spring from the humblest cottages." Self-discipline was central: "Dice, billiards, cards, and many other games. . . . Theatrical exhibitions. . . . Frolics, revels, and gallantry"—successful men staved off such temptations, pursuing honest means to material display of their manliness. These were self-made men, and one's sense of self-significance became important. As Amos Kendall, tutor to Henry Clay's children and future member of Andrew Jackson's "Kitchen Cabinet," wrote in 1814, "Drink whiskey and talk loud, *with the fullest confidence,* and you will hardly fail of being called a clever fellow." One had to fully engage society, work it, eke from it the benefits of business and social relationships. Sociability replaced competition as indicative of manliness. Their contemporaries who had followed the path of revivalism were no different, pursuing behavior reforms by engaging society, not isolating themselves from it as purist Christians were apt to do.[53]

Just as many militia leaders were the sons of merchants and artisans, so too was the worldview of the younger generation borne in the commercial orientation to rural and village life that took shape in the 1790s. Wealth, sociability, conspicuous consumption, self-interest, and self-discipline shaped the perspectives of an entire generation of Kentuckians along the Maysville Road. It was a disturbing trend for critics of market economies: "The man ... who lives upon wealth acquired in dealing, lives upon the advantages he has obtained from other individuals," complained a *Kentucky Gazette* editorialist in 1810, "for what are *profits,* what the *gains* of trade, what the income of *speculators* but *advantages* which cunning men derive in their dealings from men less cunning? ... I allude to the gains which accrue from the *artifices* of shopkeeping, and of dealing, in general, the doctrine of which is not reasonable advance on the article sold, as a compensation for trouble and labor, and risk; ... one man shall pick your pocket in one way, with applause and with impunity; whilst, if he were to do it in another way, he would be whipped, pilloried, or sent to the penitentiary." Their parents had been less obvious, trying at least to find a comfortable balance between individual acquisition and communal obligation. The new men of commerce sacrificed communal responsibility to personal avarice, and supplanted friendly competition with cutthroat rivalry, virtue with ambition, manners with foppishness, and grace with arrogance.[54]

Coming of age with elder republican gentlemen controlling society, economy, and polity, some tried to mask their self-interests by comparing their own aspirations to those of colonial and Revolutionary-era gentry who had operated in a world of deference, patronage, and obligations. Leveling forces unleashed by the Revolution had greatly eroded courtesies and obligatory customs, so that early Kentucky gentlemen like George Nicholas and John Breckinridge more often than not were disappointed by the lack of deference extended them by average citizens. But leveling forces also made it conceivable to young men of middling status that they too could cultivate their own forms of gentlemanly appearances, liberality, public-spiritedness, and honor.[55]

As they aspired to gentlemanly status, however, the terrain became treacherous. Even as they nurtured a gentlemanly comport, the middling sorts overlooked much of the communal significance of gentlemanly obligations. James Russell coached Lexington merchant Thomas Bodley that by doing "everything in your power for my Interest," a gentleman earned the loyalty and deference of patrons. In their efforts to foster an image of gentlemanliness, men of the younger generation put on airs—gentility, appearance, liberality, public-spiritedness, and honor—that made them feel like gentlemen. Lexington merchant William Morton sculpted such a reputation for himself: "From the neatness of his person to the stateliness of his manners he was familiarly spoken of as *Lord Morton*," explained the son of another merchant. But others' use of the nickname "Lord" was certainly used sarcastically to mock Morton's pretensions.[56]

Not all were so shallow. In the autumn of 1779, another future gentleman arrived in Kentucky. John Pope's family migrated from Prince William County, Virginia, to lands just east of Louisville. The family patriarch, William, was one of the many Revolutionary War veterans to flock westward as the military conflict wound down. Financially, he was rather poor compared to other Virginians, but he and his wife had uncommonly good educations and sought out the same for their children, sparing little expense preparing their son for the mantles of respectability and status that Virginia culture valued. In the late 1780s, John Pope studied law with George Nicholas. In 1803, he settled in Lexington to open a law practice, pursue a political career, and aspire to greater wealth and influence. "I have embarked in political life and mean to make a business of it," he announced to a friend in 1808 after his first year in the U.S. Senate. "I occupy a much higher ground here on the scale of talents and republicanism than either you or myself expected," he continued, "except Breckinridge no man from the West ever had more popularity in Congress."[57]

"Talents and republicanism" became the standard for the region's aspiring lawyers and politicians. Law students at Transylvania University learned republican philosophy, but it was a republicanism that few older gentry would have recognized. It had become a means rather than an end. The new goal, made possible by presenting one's self as a republican gentleman, was money. Henry Clay, one of the new generation and a law professor at Transylvania in the 1810s, counseled a student that "Two words will make any man of sound intellect a Lawyer, *Industry* and *Application,* and the same words with a third *Economy* will enable him to make a fortune." William T. Barry complained to his father that his years of education would not pay off since Kentucky "is not a place to make money, and that is the one thing necessary for me." In another letter, he more specifically targeted Lexington as "not a place to make money for a Physician or Lawyer, and indeed I think it a bad place at this time for a Merchant." To protect their professional interests, lawyers and physicians began to regulate their practices. Barry thrilled that the days when "an ignorant attorney could rise to opulence & respectability" were ending. In 1803, the Lexington Medical Society petitioned the state assembly to restrict uneducated practitioners so that citizens might enjoy a "conscious security in the integrity and professional sensibilities of our family physician," thereby protecting the commercial interests of the region's "professional" physicians as well.[58]

Commerce, law, and increasingly medicine were paths to affluence, distinction, and a gentlemanly persona. Pope, Barry, Clay, and others of the younger generation represented a new type of gentlemanly masculinity that combined genteel traits—gallantry, fortitude, honor, and finely honed oratory skills—with more recent capitalistic qualities—self-interest, self-aggrandizement, and self-improvement. It was not enough to display material acquisition and gentlemanly attributes; one had to demonstrate the ambition that buttressed them.[59]

As men defined accomplishment through self-advancement, their wives were to find fulfillment in self-sacrifice with little opportunity for autonomy and self-reliance. Men wished wives to evidence *his* wealth and rank by becoming courteous, moral, cultured, and domestic ladies. As the market economy separated the work of merchants, bankers, manufacturers, and lawyers from their homes, their role as helpmate in the household diminished, placing heavier burdens on women who were supposed to demonstrate more leisured diversions than the average woman. "I have been house-keeping about 4 months & find it a very troublesome business," complained new wife Eliza Todd in 1804; "I have found many difficulties to encounter with, that I never tho't of." Her reward, however, was "the pleasure & satisfaction of an affectionate companion [that] makes up for all of the troubles."[60]

Criticisms of the younger crowd by purists, gentry, and even parents were not merely concerns over the loss of the spirit of '76. They were reactions to the coming of age of a generation of Kentuckians who expected the world to be open before them. When, in 1812, Humphrey Marshall wrote *The History of Kentucky*, he made the younger generation a particular target for his pen. He complained that "demagogues" like Clay and Pope purposely mischaracterized Marshall's own generation of genteel politicians as "suspected, and obnoxious, to the majority," even as their own "avarice and ambition, prompt them, to aspire to the offices, and emoluments, of government." The rise to adulthood and political power by these men demanded vigilance: "Should he [the historian] trace the revolutions of ancient states, and develope their causes; should he single out the ambitious demagogue, who from time, to time, deluded the credulous people, under the mask of *patriotism*, and the name of REPUBLICAN, and thence pursue him to the usurped, and the despot, he would have full scope for moral reflections, and political admonition."[61]

The younger generation just did not face the trials that made men moral and resolute. As war broke out in 1812, James Blythe, the fiery Presbyterian purist and president of Transylvania University, wondered how the United States would win, given the rising leadership of the "free and easy" generation: "the liberties of any people are endangered when they are deposited in the hands of men, in whose breasts the dominant principle is *self-love*." So too did the *Kentucky Gazette*, mouthpiece of gentry republicanism, fill with condemnations.

> When I was a youngster our women were prudent,
> Our spinsters demure, and our matrons sedate.
> In oeconomy's school ev'ry wife was a student,
> And seldom contentious arose, or debate; . . .
> Your quality misses, how different we find them.
> If vice was the fashion, they'd all be profane;
> Our quality wives not a whit are behind them;
> No power can their profligate humor restrain.
> From assemblies to balls, folly beckons and calls;
> Hark! he summons, and see how they pant to obey it,
> While no longer uxorious, their husbands notorious,
> Gods! it drives a plain man nearly mad to survey it. . . .[62]

By the War of 1812, the values and aspirations of the free and easy became identified with the same sense of moral declension that had catalyzed revivalism along the beaten path. At an Independence Day celebration in 1811, Kentucky's Secretary of State Jesse Bledsoe (a member of the younger generation himself) decried "luxury, which bids us to give up our rights and ourselves, before we will forego present gratification; effiminacy, which shrinks from danger and death as the worst of evils; avarice, which places the supreme good in wealth; supineness, which blinds to danger, induces to false security, disarms us and invites the invader." A year later, a contributor to the *Kentucky Gazette* distrusted "the rising generation . . . [that is] too effeminate for republicans."[63]

Fears about effeminacy were nothing new, but Bledsoe's reprobation arose in the midst of a national crisis as the United States advanced toward another war with Britain and a spiritual crisis as residents along the old buffalo trace tried to reconcile the many changes that had bombarded their worldviews. Emphasizing the need for moral renewal, Bledsoe concluded his oration by extolling the Fourth of July as the "sabbath of our political regeneration." [64]

While Independence Day politically regenerated Kentuckians' patriotism, it took an entire war to morally regenerate them. Many historians interpret Kentuckians' madness for war in 1811 and 1812 as symptomatic of their unrestrained nationalism. That will not be disputed here. "The Declaration of War in 1812 seemed to stir every heart in Kentucky," remembered William Leavy, and Lexington glowed as candles in house windows burned in support of the war effort. Nearly 3,800 Kentuckians enlisted as U.S. Regulars; another 21,900 participated in volunteer units that fought in specific campaigns. Immediately, the War of 1812 offered full participation in an event comparable to the Revolutionary moment. But Kentuckians were also determined to get full credit for their participation. Over the previous two decades, easterners had dismissed the Battle of Blue Licks and other western struggles as "Indian Wars." In late 1812, the *Kentucky Reporter* repelled suggestions that the western theater of battle would again be less important, less a conflict against Britain than against Native Americans: "You may as well call the revolutionary war, a Hessian war because the Hessians were hired by the British, as to term this an Indian war."[65]

Fighting to restore national honor presented younger Kentuckians an opportunity to prove themselves worthy of their Revolutionary heritage. In 1810, as war fever escalated, Transylvania University student Ephraim Ewing beseeched free and easy men to "forsake the debilitating pursuits of dissipation and idle sports, and to invigorate their minds with knowledge and their limbs with activity and strength" so that they could prop up the nation's "tottering frame." In that same year, Henry Clay compared his generation and the opportunity before them to the Founding Fathers: "we shall want the presence and living example of a new race of heroes to supply their place, and to animate us to preserve unviolated what they atcheived." After the 1812 Battle of Tippecanoe, one poet lauded "Who to their country's welfare freely give / The sacrifice of life, forever live / As bright examples to the unborn brave, / To shew how virtue rescues from the grave." At war's end, one commentator lauded, "You have done your duty and future generations will say, *you were worthy of your sires, and the sacred legacy which they bequeathed.*"[66]

In the midst of effeminate revivalism and softening commercialism, many hoped that the conflict would revitalize masculinity and restore virtue. The *Kentucky Reporter* described the war's objectives as "something ten thousand times more valuable" than importation rights and diplomatic honor: "The *lives of the women and children on the frontier* are at stake, and the settlement of the western territories is deeply involved." The same motives that had inspired westerners' participation in the Revolutionary War were resurrected as the reasons for Kentuckians' participation in the new conflict. The centrality of militia culture to community, family, and masculinity reemerged. As Richard Johnson, U.S. Congressman from Kentucky, related to President James Madison, Kentuckians "are a spirited people and nothing can satisfy them but some military enterprise to engage at least a part of our men." A meeting of Millersburg men put it best: "we must prove to the Union at large that we have the Good will, the courage & the power to defend our rights, our property our families our Country."[67]

The diversity of masculinities that evolved along the Maysville Road became vivid in wartime. More genteel commanders expressed frustration with the unrestrained entertainments of some Kentucky volunteers. In 1812, Brigadier General James Winchester of Tennessee denounced Kentuckians' "wrastling and drawing tomahawks or knives over each other" as conduct "very unlike a soldier." For some Kentucky soldiers, fighting had less to do with national honor than personal gain. Thomas Bedford of Bourbon County anticipated the choice of an admiring wife after his return. When his sister wrote a letter detailing the latest social events, he reacted: "hearing of the girls getting married so much, I must confess disturbs me a great deal, but, however, I expect there will be some left for me yet." Still, many Kentucky volunteers grasped the seriousness of war for their nation and for themselves. Some were praised for their silent resignation to fatigue and hunger out of a "manly firmness." Even Christian Republicans carved a role for themselves by arranging shipments of domestic goods to the troops. After gathering 899 blankets, 38 homespun shirts, 72 vests, 178 overalls, 455 pairs of socks, 182 pairs of mittens, 145 pairs of shoes, and 37 hunting shirts, John Lyle lauded his congregation's efforts to Kentucky's governor: "I would observe that the benevolent Ladies who furnished the clothing, are very solicitous that it be preserved from rain, and they express great anxiety that, it should be well guarded from robbers & especially from the Indians & British." The war provided an arena for the expression of masculinity, even if it was not the heroic brand that critics had hoped.[68]

Ironically, as the war reinforced masculinity, it also gave women an opportunity to expand their own spheres of activity, calling on them to fuse private life with public need by producing clothing for the soldiers and maintaining the home front. In 1812, Mary Owen Russell received a letter from her nephew applauding "the patriotism of the Kentucky Ladies [to whom] we are indebted for the greater part of the clothing we at present possess." In Washington, at the Fourth of July festivities of 1813, participants toasted "*The Kentucky fair*—They clothed our Republican army. Their patriotism ought to put to shame the *non-combatant states*." Seven months later, during Washington's Birthday celebration in Lexington, the men drank to their patriotic women, "May they imitate those of ancient Rome, and animate our youthful citizens to deeds of valor." The militia band then performed a rendition of "Spartan Mother." Perhaps a series of toasts offered at Lexington's 1814 Independence Day observance best encapsulated the ideological and patriotic continuity that many hoped to draw between the Revolutionary moment, the soldiers of the War of 1812, and the public spirit of the Spartan Mothers:

> The heroes and sages of the revolution. So long as political virtue exists they will be esteemed as the great successful champions of liberty.
>
> The Patriot—He does not spend time in intrigue for office—he meets the enemy in the field of battle—or if forced to remain at home, he stimulates his countrymen and government to energy.
>
> The American Fair—Patriotism with them is inherent—they despise the tory, coward and traitor.

Patriotic women, disinterested citizen-soldiers, public virtue—it was the spirit of '76 resurrected.[69]

For Kentuckians, the association of patriotism to gender roles peaked with the Battle of New Orleans. In early January 1815, American forces under Andrew Jackson defended the port city from British invasion, repulsing the enemy in the most convincing victory of the war. Officially, Jackson's report to the secretary of war criticized one Kentucky regiment that he claimed gave way to the British. His account touched off years of anger, mostly within Kentucky, over Jackson's disparagement of the state's honor. Yet, despite the report, the Hunters of Kentucky soon became symbols of nationalism, masculinity, and the war effort.[70]

A popular song written circa 1820 by Samuel Woodworth, "The Hunters of Kentucky; or, the Battle of New Orleans" portrayed British troops as seeking "Beauty and Booty," the right to pillage New Orleans and rape its women when they captured the city. Drawing upon age-old habits of boasting found throughout America, the song became particularly important to Kentuckians because it bolstered the dual themes of patriotism and manhood before an entire nation. Three specific verses—the second, fourth, and fifth—offer particular insight.

> [2] . . . We are a hardy free-born race,
> Each man to fear a stranger;
> Whate'er the game, we join in chase,
> Despising toil and danger;
> And if a daring foe annoys,
> Whate'er his strength and forces,
> We'll show him that Kentucky boys
> Are alligator-horses. . . .
> [4] You've heard, I s'pose, how New Orleans
> Is famed for wealth and beauty—
> There's girls of every hue, it seems,
> From snowy white to sooty.
> So Pakenham he made his brags,
> If he in fight was lucky,
> He'd have the girls and cotton bags,

In spite of Old Kentucky.

[5] But Jackson, he was wide awake,

And wasn't scared at trifles;

He knew what deadly aim we take

With our Kentucky rifles.

He led us down to Cypress Swamp,

The ground was low and mucky;

There stood John Bull in martial pomp,

But here stood Old Kentucky. . . .

[Stanza:] Oh, Kentucky, the Hunters of Kentucky,

Oh, Kentucky, the Hunters of Kentucky.[71]

At initial glance, the Hunter of Kentucky resembled pioneer masculinity through his actions and attitude. In the second verse, Woodworth alluded to the "alligator-horse," a folkloric creation of the early nineteenth century that embodied the wild, lawless, crude, yet romantically compelling backwoodsmen who would populate American literature for the next half-century. He was larger than life, "hardy and free-born." In 1827, the New York *Courier* related how one Kentuckian, part of a survey crew along Florida's St. John's River, "was seized by a large alligator and dragged underwater. In a short time the Kentuckian and the alligator rose to the surface, the latter having the right leg of the former in his mouth, and the former having his thumbs in the eyes of his antagonist. . . . It is needless to add that the gouger obtained a complete victory. Having taken out one of the eyes of his adversary, the latter, in order to save his other eye, relinquished his hold upon the Kentuckian's leg, and returned to shore in triumph." The Kentuckian became a staple of American tall tales. "I am a man; I am a horse; I am a team. I can whip any man in *all Kentucky*, by G_d," he proclaimed in story after story. "I am a man; have the best horse, best dog, best gun, and the handsomest wife in all Kentucky, by G_d." And then he would proceed to best another boaster. Yet, the Hunter of Kentucky was not a pioneer, for the wilderness context so integral to that ideal was gone, and with it, the opportunity to pass the ideal along. Only where wilderness remained powerful—along the Mississippi River or in the Florida territory, for example—did this protagonist remain a feature of the human landscape. Instead, the Kentuckian became a caricature in traveling shows, perpetuated through song and pioneer costume by actors who reinforced a provincial image of Kentucky even as they sang the praises of Kentuckians' nationalistic fervor.[72]

The Hunter of Kentucky also resembled republican ideals of masculinity through his supposed patriotic principles, but as the song progressed, the analogy faded. Instead of life, liberty, and the pursuit of happiness, the Hunters of Kentucky fought for fame and wealth. Too compromised by the self-interest, individualism, and luxury of a market economy, Hunters of Kentucky failed to recapture the virtues of the Founding Fathers. In contrast to the July 4th toasts given at Lexington's 1814 festivities, those proposed in 1826 carried different messages about masculinity, femininity, and patriotism.

> The first settlers of Kentucky—We this day enjoy the blessings which were gained by their unsurpased enterprise perseverence and valour.
>
> The hunters of Kentucky—always ready to raise the rifle—thrust the bayonet or point the cannon, in defense of "Beauty and Booty."
>
> The fair of Kentucky—Our mothers, sisters, wives and sweet-harts, tender ties that bind us to our country.

Even though Revolutionary veterans and republican gentlemen still lived among them, for these Kentuckians at least, public virtue was but a thing of historical memory. The War of 1812 had transformed patriotism. As the *Kentucky Gazette* reported upon the ceremonial presentation of sword by a local group of women to a hero, it "will, undoubtedly, stimulate him, as well as his competitors in the field of honor, to redouble their efforts in the service of their country, in order to merit the future approbation of their fair country women." The younger generation understood military conflict as foremost the defense of wealth and the admiration of women, consequently reversing women's fortunes, as the toast suggested, by reducing them from Spartan Mothers to sentimental bonds between family and country. At holiday celebrations and militia musters, even as they venerated those women who had contributed to the war effort, men encouraged a domesticity that demanded wives and daughters once again be relegated to the home. How else could the Hunter of Kentucky remain a viable form of manhood unless all women along the beaten path, in their weak and vulnerable states, needed defense?[73]

The young men who went off to war returned with notions of ambition and industry. Artisan Sebastian Derr of Paris cashed in on his veteran status even before the war ended. In 1814, the shoemaker ran for public office on a platform that included tax reform, abolition of the slave trade, ridding the militia of "all rotteness in principle," and establishment of a bank in Paris "so that we can all get money when ever we want it." Just as before the war, Derr, like other men of his generation, believed that he could forge his own political and social rise.[74]

By synthesizing nationalism and economic liberalism, however, the war effort transformed the image of Derr's generation. No longer were men effeminate; they were respectable. When Robert Breckinridge McAfee wrote his history of the war in 1816, his portrait of a regimental gathering west of Paris illuminated the intersection of militias, self-made men, and dignity. "The ranks were filled with the most respectable citizens: the most promising young men of the country, the most intelligent, the most wealthy, had eagerly enrolled themselves for service," McAfee explained. "There was indeed no part of this corps of volunteers, in which citizens of the first respectability were not to be found." Militia culture—parades, dinners, musters, dances—once a pageantry of merely self-promoting conspicuous consumption, became an arena where display was justified by sacrifice. One resident, unhappy with the ostentation, longed for the days when such festivities were the "good old fashioned way at the Republican spring."[75]

Having come of age in the 1790s and early 1800s and passed the test of nationalistic adulthood by fighting and winning the War of 1812, the younger generation that sought to define itself differently from its elders found its identity as an energetic middle class. In the summer of 1819, President James Monroe and General Andrew Jackson arrived in Lexington just in time to celebrate Independence Day. Over the past year, the pages of the *Kentucky Gazette* filled with news of Monroe's presidential tour throughout the East. And many older Kentuckians denounced the fanfare as too pretentious. One reader hoped that Monroe "would be treated on his passage through this state, as a man, and not as a God." But in many ways Monroe was a living member of the national pantheon. As a Virginian, he symbolized the continuity of presidential power from Washington through Jefferson and Madison. With personal and ideological claims to the Revolutionary moment, he wore proudly the patriotic label.[76]

Monroe also identified with entrepreneurs and self-made men who flourished after the War of 1812 as the new political and economic elite. Appropriately, then, Fayette County militia troops of self-made men met and ushered him into Lexington. Two miles west of Lexington, cavalry joined Monroe's entourage; one mile later, light infantry and rifles saluted as he passed; at the edge of town, artillery offered a tribute. For its participants, the procession into Lexington bound commander-in-chief to militias. But, unlike Washington's memorial procession two decades earlier, this celebration was an exclusive affair celebrating the rise of the new American man. Conspicuously absent from the parade were educational and religious leaders who figured prominently in a more republican procession. Similarly missing were artisans, laborers, and other citizens. While a Fourth of July picnic provided opportunity for democratic participation, the procession of 3 July showcased the arrival of the middle class.[77]

Without the traditional threats of Indians, wilderness, and the British, fraternities of armed citizen-soldiers became less essential in the late 1810s. By the early 1820s, militias had become so insignificant to protection and service that the state legislature reduced the number of annual musters to two. Still, as a symbolic institution of self-made men, the militia remained as popular as ever, evidenced at both the southern and northern ends of the Maysville Road by receptions accorded LaFayette in 1825. In these latter days of the Revolutionary generation, the younger generation made clear their claim to the inheritance.[78]

In Lexington, for example, General Leslie Combs delivered the welcoming address. Just thirty-two years of age, Combs attained hero status as a survivor of two British attacks during the War of 1812. After the war, he moved to Lexington, where he studied and practiced law. As a Whig, he became a staunch supporter of commercial development and internal improvements by the late 1820s. Combs very much epitomized the self-made men of his generation, welcoming LaFayette on behalf of the "citizens of Lexington, a *town* called after that, in which the first blood in our revolutionary struggle was shed; and the inhabitants of the *county*, the first in the union called after your name; have confided in me, the honourable office of bidding you WELCOME," Combs extolled from his pulpit on the steps of Keen's Tavern. "This town and county therefore, are, by their associations, recommended to your notice. Your memorable exertions in the glorious struggle, which resulted in the enlightenment of our freedom and independence as a nation, were early impressed on the minds of the first settlers of this western wilderness, in characters never to be forgotten. Enterprize, industry, and perseverence, has wrought an astonishing change among us, in the period of forty years." While acclaiming the

Revolutionary moment as the foundation of freedom and independence, the catchwords of self-made culture—enterprise, industry, and perseverance—thematically eclipsed it. Despite, indeed through, their capitalistic proclivities, the sons had become as morally purposeful as their republican fathers.[79]

After leaving Lexington and briefly visiting Cincinnati, LaFayette arrived by steamboat in Maysville. Once again, a large assembly crowded to watch him step ashore where a walkway of expensive Turkish carpets donated by John Armstrong, once a subordinate in George and Samuel Trotter's mercantile empire but now a prosperous merchant himself, directed the General to Front Street. At that point, LaFayette joined in militia procession to the steps of Maurice Langhorne's hotel where Revolutionary veteran Major Charles Pelham, age seventy-six, delivered the welcome. After a feast, the crowd escorted LaFayette back to his steamboat and watched as it labored up the Ohio River. From imported carpets to hotel steps, the ceremonial landscape pronounced a middle-class society. In particular, Turkish carpets had become a pervasive symbol of consumption and display. Additionally, in later recollections, special note was made of the presence of Julius Levi who served both under LaFayette's command at Brandywine Creek and as the marketmaster of Maysville, a symbol of the merging of Revolutionary heritage with the market economy that made the middling sorts respectable.[80]

ATHENS OF THE WEST

By 1810, Lexington had earned the title of "city" as census officials defined it. With most of its 4,279 residents concentrated in the original in-lots, population density had reached over 6,000 persons per square mile. It continued to resemble the villages along the old buffalo trace, maintaining strong rural ties by providing market outlets for the productions of its hinterlands and serving as a distribution center. Lexington also rapidly developed the unique qualities of early national urbanism: an expanding and dense population, a large artisan sector, increasing numbers of free blacks, mildly better economic opportunities for white women, and the segregation of neighborhoods by economic status. Lexington hardly rivaled the cities of the East, but it was, by far, the most urban of western towns.[81]

The leaders of this urban development were the new generation of self-made men: merchants, lawyers, large artisans, manufacturers, and other

The Marquis de LaFayette in Maysville, 1826. From mural on Maysville Flood Wall, Maysville. Photograph by Craig Thompson Friend.

professionals, many of whom had inherited businesses, wealth, and success from their fathers. It was quite an inheritance. By 1804, the average town lot appraised at a value of $350. Of twenty-eight merchants, nineteen possessed lots exceeding average value. Half of Lexington's merchants had properties valued at more than $1,000 each. At first glance, lawyers fared less well, all claiming lots valued between $350 and $1,000. But in most cases, those were offices. Henry Clay, George M. Bibb, John Pope, and Jesse Bledsoe, for example, all owned substantial retreats on Lexington's periphery. Still, even the values of their professional properties were impressive accomplishments, considering that over 20.6 percent of Lexington's heads of households owned only small parcels of property outside of town, and 38.4 percent owned no land whatsoever.[82]

The out-lots, in contrast, remained so underdeveloped that residents continued to call that neighborhood "the Country." Situated on many out-

Henry Clay's Law Office (1803), Lexington. Photograph by Rod G. Turner.

lots and lands beyond the one-mile town boundary were the rural retreats of an aging gentry, among which were John Bradford's Fairfield, the Breckinridges' Cabell's Dale, and the Todd family's Ellerslie. Beginning in the early 1800s, these homes provided models for the rising entrepreneurial class as its members sought to bring rural refinement into the center of the city. By 1806, only a decade after David Meade's embarrassment over Lexington's architectural vulgarity, Thomas Ashe visited a Lexington "composed of upwards of three hundred houses . . . principally built of brick, in a handsome manner." Among them were the urban retreats of prosperous merchants and lawyers: James G. Trotter's Woodlands (1794), a two-and-a-half-story brick house with gabled ends; William "Lord" Morton's stuccoed home (1810) with brick cornices, dogwood blossoms incised in the marble mantelpieces, and inside shutters on the windows; John Brand's Rose Hill (1811), a smaller replica of Morton's house; Henry Clay's Ashland (1814), a two-story brick mansion with a polygonal vestibule; Thomas Bodley's house (1814) with its elliptical staircase; John Wesley Hunt's Hopemont (1814), with a walled courtyard and carriage house; Lewis Sanders's Placentia (1815), a two-story brick residence with oval and octagonal rooms and connecting "saloons"; John McCalla's Mount Hope (1819), a two-story Federal-style brick house; and John Pope's home (1811–14), a three-story residence with a skylight-illuminated rotunda and twelve-foot-eight-inch ceilings on the first floor. Among the middle class, the demand for refined and unique architecture soared, and architects found a wealthy and eager clientele. Builder-designers like David Sutton (who planned the 1806 courthouse), Matthew Kennedy (designer of Transylvania University's Main Building in 1816), and Mathias Shryock (architect of the Market Street Episcopal Church in 1814) were widely employed. The most celebrated American architect of the times, Benjamin Henry Latrobe, found eager consumers among Lexington's men of commerce, designing the Kentucky Insurance Company building (1802), John Pope's house just south of town (1811), and the two gabled wings added to Clay's Ashland in 1814.[83]

In their youth, the rising middle class had witnessed the gentry's use of grand rural estates to civilize the countryside. Now, in the late 1800s and 1810s, they christened their urban homes in efforts to establish a new type of distinction in the city. No longer living in Lexington, the recently affluent inhabited their own little worlds at Mount Hope and Woodlands. That so many of these urban retreats belonged to merchants did not bypass the notice of less-monied residents who questioned their loyalties to the well-being of the community and resented the further drawing of social differentiations. An editorialist to the *Kentucky Gazette* complained how "the merchant's *dwelling,* his *equipage,* his *apparent amount of his stock in trade,* the *improvements* that surround him, and his *domestic conveniences* bear the aspect of a flourishing opulence." Suddenly, merchants whom the masses had eyed suspiciously for their economic activities in the past became associated with an unrelenting quest to enhance social status, a phenomenon noted by Thomas Ashe as an effort to garnish Lexington "with some pretensions to Europeans elegance." Consequently, many of the newly affluent paid dearly. In 1803, the *Kentucky Gazette* warned that "we cannot recollect the number of houses that have been broke open in this place within a few weeks." The accompanying list of victims reads as a directory of the town's rising economic and political leadership.[84]

The lawyers who joined in this display of wealth and status came to Kentucky for the same reasons that gentlemen like John Breckinridge and George Nicholas had: The complexity of Kentucky's land titles was irresistible and profitable. The contrast between older gentleman lawyers and the new breed is suggestive, however. John Brown, who established his practice in the early 1790s, "was a man of towering and majestic person, very proud, austere & haughty." Some in the new generation did not exude the same type of manly physical presence. Jesse Bledsoe was a lawyer "of the first order of talents, an accomplished classical scholar, a man of exquisite wit—a poet & an Orator." Henry Clay "is rapid in conversation, full of anecdote, and swears most insufferably." Clay purposefully forsook pretension in order to hone a more provincial character. While it certainly worked to his political advantage in later years, such a ploy alienated him from the older-generation lawyers and politicians, forcing Clay to align his fortune and future with the entrepreneurial class which, like him, was a boisterous and self-aggrandizing element.[85]

The younger generation of lawyers and merchants, concerned as they were with profit, used "every exertion in their power to get into their possession the little specie in circulation," according to François Michaux. One method was through legislative incorporation of the Kentucky Insurance Company. In 1803, the state assembly chartered the institution for sixteen years as an insurer of river-trade vessels and granted the company the authority to print its own circulating notes. Such institutions reeked of eastern corruption and favored those with fluid capital over those with land. Kentucky's entrepreneurs had a clear need for insurance against disaster, and the state's commercial growth necessitated some form of accessible currency. Enthusiasm for the Kentucky Insurance Company, not surprisingly, came from Lexington's entrepreneurial sector. In its first year, only seven men held stock in the company: six were merchants in Lexington, and five of them served as town trustees. And the enthusiasm persisted into 1804, when, at a sale of company stock, 143 shares sold at $105 each in less than ten minutes. Cash had always held a critical role in the growth of the region, and its increasing scarcity threatened the stability of local economies. Now, the constant, bordering on inflationary, flow of Kentucky Insurance Company notes facilitated investment by entrepreneurs in the production and exportation of agricultural commodities. The Kentucky Insurance Company provided much needed financial support for commercial ventures.[86]

The institution's initial success encouraged the state assembly to incorporate the Bank of Kentucky in 1806. Because many citizens remained suspicious of the Kentucky Insurance Company and viewed it as an aristocratic institution, when chartering the Bank of Kentucky, the legislature included a provision retaining the right to replace the board of directors if, or when, the bank implemented policies contrary to popular or legislative opinions as to how a bank should operate. Legislators hoped that this one principle, that of moral accountability to the larger society, would differentiate the bank from the increasingly less-than-popular Kentucky Insurance Company. In only four years, however, the Kentucky Insurance Company had become firmly entrenched, and its notes flooded the regional economy just as the Lexington branch of the state bank opened its doors in 1806. By 1811, the bank coffered more insurance notes in its vaults than eastern state notes, Ohio notes, and gold.[87]

The establishment of the Kentucky Insurance Company and the Bank of Kentucky swamped the domestic markets with their respective notes, but the currency of neither commanded respect outside the state. An abundance of Kentucky notes depressed the regional circulation of eastern state and bank notes, the very currency merchants needed to pay suppliers. The Bank of Kentucky recognized the benefit of accumulating eastern notes over its own and offered customers a 1 percent advance in specie when they exchanged eastern notes for Kentucky currency. Consumers more than likely chose to tender eastern notes at the bank than at mercantiles. Faced again with a shortage of valuable currency and inflationary prices, merchants either adopted account-book credit as their primary form of transaction and depended on the extension of further credit by suppliers, or they faced the prospect of losing business. In contrast to Moylan who, in 1792, tolerated only 23 percent of transactions on account-book credit, John Hunt accepted 85 percent on account-book credit, William Tureman and Edmund Martin each permitted 79 percent, and Daniel Halstead allowed 54 percent. As domestic goods found their way to the markethouse and as notes negotiable beyond the state became scarce, the most feasible mode of purchase for both merchant and customer in private stores became account-book credit.[88]

As precarious as their world appeared—with its increasing dependence on account-book credit, its indebtedness to eastern creditors, and its reliance on inflated currencies—entrepreneurs who succeeded became prosperous and too affluent for a republican community. By 1808, sixteen men, most of whom were merchants and lawyers, controlled over one-third of Lexington's total property value of $1 million. One was John Wesley Hunt, who became the West's first millionaire through his mercantile establishment, his investment in industrial development, and his fiscal maneuvers. In that same year, an elite group of thirty-nine men, led by Henry Clay, monopolized the receipt of $80,947 worth of loans at the Lexington branch of the Bank of Kentucky. The same men who received loans acted as second parties for other applicants. Thirty-four of the loan recipients were merchants and mercantile companies, receiving nearly three-fourths of the total funds.[89]

As successful lawyers and merchants coalesced into a political and social force in Lexington, material acquisition parlayed into class consciousness. Most Kentuckians of all economic ranks partook of gentility in some way: refined material goods had seeped into everyday life in the forms of blue-edged Queensware, cashmere and muslin cloth, silver tea services, and hyson and bohea teas. Still, the markers of gentility—brick Georgian homes, delicate carriages, command of large numbers of enslaved black servants—remained beyond the grasp of most. As families of new wealth appropriated refined symbols and sensibilities and as the pioneers found themselves summarily marginalized, elite individuals like Margaretta Brown, who in the 1790s appeared quite the snob even to her husband, became smug in the progress of Kentucky society. Upon a trip to New York City in 1811, she wrote her husband that she would proudly introduce the elite of New York to the high society of Kentucky, "confident that we should lose nothing by the comparison, either in point of beautiful dresses, genteel arrangements, or choice of refreshments."[90]

The rise of the new wealth was not welcomed by the gentry. Their fashions, manners, and dining compared favorably to that of New York City, but merchants and lawyers lacked the heritage and often the classical educations considered imperative to a republican society as the landed gentry had intended it. Whereas genteel dinners, balls, and weddings drew guests into private rural retreats, merchants and lawyers appeared ostentatious with their public picnics, teas, and celebrations. In 1806, Fortesque Cuming described a coffeehouse in the middle of Lexington that catered to this crowd, behind which, amidst the vineyards, "the musick of two or three decent performers sometimes excites parties to dance on a small boarded platform in the middle of the arbor." Tea parties, once exclusive affairs among the gentry, became "a continued festival from the time you enter to the time of your departure," wrote Richard Fowler. "I have known collected at these parties from one to two hundred persons." Whereas the Bluegrass gentry had been rather exclusive, Eliza Todd discovered that, as refinement became available to more and more social aspirants, high society began to fracture: "My acquaintance in this place [Lexington] is not extensive, owing to its being so much selected off in parties."[91]

The gentry's resentment became apparent in a series of verse published in the *Kentucky Gazette,* beginning with this assault on self-made men: "Hunting, shooting, never thinking; / Chat'ring nonsense all day long, / Humming half a stupid song; / Chasing baubles, rings and jewels, / Writing verses, fighting duels, / Mincing words in conversation, / Ridiculing half the nation; / Admiring their own pretty faces, / As if possess'd of all the graces." The author of another piece condemned pretentious men of new wealth, apparently intending to drive a wedge between them and a politically and economically powerful artisan sector. "From his own dung-hill lately sprung, / So buxom, debonnair, and young; / Yet, on his brow sits empty scorn, / He hates mechanics meanly born. / He'll not debase himself to think—/ 'Tis too da———d low! but he will drink, / Politely curse, blaspheme and swear; / Twirl his whip and sing an air, / Dance to show his grace and shape / Brisk and sprightly as . . . an ape." The gentry minced few words in their ridicule in another poem titled "Ole Times and New": "Dash away, splash away, / Squander their cash away. / Visiting twenty score friends in a night, / With the harps, pipes, and tabors, / Distracting their neighbors, / Common sense stands confounded and reason takes flight." Merchant Thomas Hart Jr. did not deny the accusations: "What a pleasure we have in raking in money and spending it with our friends." The "new times," economically and politically orchestrated by commercial entrepreneurs, threatened to rip apart the "ole" republican social order that the gentry had tried to establish.[92]

Throughout the early 1800s, however, with the exceptions of occasional poetic battles, the emerging contest between the gentry and self-made men failed to turn confrontational. As we have seen, with attentions turned toward regional and national affairs, the gentry had relinquished local politics to the entrepreneurs. In Lexington during the 1790s, the major undertaking of this merchant-dominated leadership was construction of a markethouse. Then, around 1802, the composition of the trustees changed. The recent flurry of editorials about the moral obligations and failures of merchants influenced many townsfolk to vote against them. Six of the next nine years, they lost control of the board to artisans.[93]

Years of catering to the gentry and merchants had enabled artisans to appropriate more gentility and establish themselves as a powerful faction in their own right. With control of the board, artisans turned to preserving the crafts tradition. Many shops and smithies sat along the banks of Town Creek, or actually along the pavement that covered the rivulet. Continued flooding problems affected artisans more than any other occupational group, particularly several tanyards whose by-products rotted in the swampy conditions. In early 1808, an artisan-controlled board of trustees ordered the reopening of Town Creek. Laborers dug Water Street Canal down the center of the pavement. By expanding the width of the stream's basin, the board hoped to reduce the flooding problem. A few months later, citizens petitioned the board to address the dogs roaming Lexington's streets: "that they go mad . . . destroy many sheep . . . interrupt the quiet of the night by their howlings . . . infest the market, and are troublesome and disgusting to those persons who, attune there both buyers and sellers." The solution was to limit dog ownership to one per home. Yet, the interests of artisans, in this case butchers and tanners who relied on canines to eat the refuse of their labors, had to be protected. Men of those occupations received permission to keep two dogs.[94]

The artisans and merchants who served on the board after 1803 were men of the new generation. John Springle, a highly successful bricklayer despite his failure to complete the markethouse as a much younger man, began a six-year tenure in 1806. Edward West, son of wealthy silversmith William West, spent three years on the board, beginning in 1806. George Trotter, having inherited his father's store a decade earlier and created a mercantile empire, began a four-year tenure in 1806. His brother Samuel served in 1811 and 1812. Daniel Bradford, John's son, also became a trustee in 1811. To give Lexington the same airs of wealth and affluence that they enjoyed, these trustees decided in 1806 to abandon the old log courthouse and construct a refined, brick courthouse. The three-story Federal-style courthouse doubled in height with the addition of a square clock tower, the weight of which threatened to collapse the entire structure. As Alexander Wilson recalled, "it was found necessary to erect, from the floor, a large number of large, circular, and unplastered pillars, in a new order of architecture, (the thick end uppermost), which, while they serve to impress the spectators with the perpetual dread that they will tumble about their ears, contribute also, by their number and bulk, to shut out the light, and to spread around a reverential gloom, producing a melancholy and chilling effect . . . elevated at the opposite extremity of the building, the judges sitting, like spiders in a window corner, dimly distinguishable through the intermediate gloom." Like their own collective refinement, the commercial class had built a structure ornate but cacophonous. Its splendor faded rapidly. Six years later, Henry Clay pronounced the building "the disgrace of the town and the derision of everybody." In 1818, as he toured Lexington, William Faux noted "the temple of justice, in the best square, which, with its broken windows, rotten window-frames, rotten broken doors, all ruined and spoiled for lack of paint and a nail, looks like an old abandoned bagnio, not fit to be compared with any workhouse in England."[95]

As they abandoned the log courthouse, so too did the trustees decide that what remained of the markethouse economy no longer served a valid purpose. Less concerned with balancing moral obligation with market orientation, they ended use of the building as a community markethouse. To curb wholesaling and the loss of profits, they began renting stalls annually to the highest bidders in 1808, eliminating casual sellers and creating a roster of market participants. Having cleansed the markethouse of the rabble, the trustees thenceforth rented only to select members of commercial

George Trotter. Courtesy of the Filson Historical Society, Louisville.

society, primarily artisans and grocers. Nowhere along the beaten path did the new association of commerce, politics, and refinement prove more evident than upon the public square where a markethouse of specialized artisans and grocers and a courthouse of grotesque proportions loomed over Lexington's townscape.[96]

While the economic liberalism promoted by the younger generation of merchants, artisans, and lawyers was to the advantage of white men, white women occasionally found ways to operate within it. Gender had always determined economic participation in the early West. Store ledgers that listed customers by gender offer a sketchy but suggestive portrait of women purchasers and their economic status. When she entered the local store, a woman functioned as either an independent who established her own account and exerted her own purchasing power, or as a dependent—wife, daughter, sister, mother, slave—who relied on the patriarch's account to purchase goods. The differences between the two categories demonstrated much about the perception of women and the empowerment that the market economy made possible. Widowed and single women most often acted as individuals and composed more than a third of women purchasers. They established accounts in their own names and were directly responsible for eventually paying debts. During the 1790s, merchants extended credit more often to independent women (91.6 percent) than to men (85.6 percent) or dependent women (88.8 percent). By 1807, however, the tide had turned. Independent women received less credit than the other groups, and by 1810, the percentage of individual women purchasing on credit had plummeted to 53.8 percent.[97]

Like other customers, independent women were greatly affected by the consequences of the markethouse economies. Rebecca Green's name appeared twice in John Wesley Hunt's 1796 store ledger: in early summer, she purchased tea, broadcloth, and imported cloth on credit; her slave, Mary, visited the store in late summer to buy trimmings, again on Green's credit. As a housewife, Green should not have warranted the independent financial status that Hunt granted her. Men dominated the household and the market, and the state legislature buttressed male guardianship roles with extensive laws that restricted women's economic activities. But there were no laws that dealt with abandoned women. Rebecca Green's husband had left her in 1791, and in 1798 she won a divorce on grounds of desertion. By 1804, she transformed her home into a cape manufactory where she and Mary produced mantas. As in Rebecca Green's life, a woman's outwork facilitated her participation in the market, and merchants would gladly transform those domestic productions into buying power.[98]

One reason that Rebecca Green was able to create economic security even as her marriage crumbled was the artisan ethos in which independence and virtue emanated from productive labor, self-reliance, and talent. Granted, the more acceptable role for urban women was as symbols of their husbands' accomplishments. But when the husband was a scoundrel who failed to fulfill his proper gender role, women like Rebecca Green could expand beyond housewife to establish their own independence.

Blacks were not as fortunate. Despite the opening of economic opportunities, enslaved and even free blacks continued to be circumscribed by white tradition and laws. Store ledgers again tell the story. No blacks appear in the account books of Benedict Leonard's Washington store (1791) or John Moylan's Lexington mercantile (1792). Not until after the construction of markethouses were African Americans present in store ledgers along the Maysville Road. In 1796, fourteen visited John W. Hunt's store in Lexington; the next year, Daniel Halstead registered two blacks in his account book; in 1807, eleven visited William Tureman's store in Washington; and, in 1810, twenty-two blacks shopped at Edmund Martin's Maysville mercantile.[99]

Although an increased black presence in these stores suggests some advances, how and why they purchased proves equally important and offers a dramatic counterpoint to the notion that stores allowed blacks and whites to operate in relationships of trust. In Hunt's store, each black—eight children, two women, and four men—shopped as a dependent, representing a white master and listed under his name. Of the fourteen, eleven paid through the slaveholders' credit accounts; three girls paid with cash. In contrast, just up the street at Halstead's mercantile, the merchant listed his two black customers, David and Sam, as autonomous individuals in the ledger. Although Halstead did not record Sam's method of payment, David paid with both cash and bartered iron castings. A decade later, Tureman recorded ten slaves (eight men, one boy, one girl) using their masters' credit. Only one black, Philip, seemed to be an independent purchaser, making three visits to the store in 1807 and using credit, cash, and bartered whiskey. Finally, in Martin's store, the pattern found in Hunt's ledgers again appeared: twenty-two African Kentuckians (only one child, a girl, among them) purchasing for their owners, all but two using credit accounts. The others used barter, although what they bartered went unlisted.

With the possible exceptions of David, Sam, and Philip (since we do not know their statuses), the African Americans who shopped along the old buffalo trace acted as slaves fulfilling their masters' needs. No commercial opportunity existed, at least not in stores. Yes, they purchased on credit, but it was the credit of their masters. Only Philip used his own name as a free customer would. The only relationship of trust obvious in these ledgers is the confidence placed in three girls to carry cash and spend it appropriately at Hunt's store, but like the others, these girls had to perform their chore or face the wrath of an unhappy slaveowner. In mercantiles at both ends of the beaten path, white restrictions on black life played out in ways that reinforced the racially restrictive legislation of the 1790s and early 1800s.

While indicating racial motives, these patterns also suggest that black economic disfranchisement was rooted in profit seeking. Why extend credit to a slave when reason dictated that the customer would never be able to pay off the debt? Philip, in William Tureman's mercantile, received complete latitude in forms of payment. But he represents the exception among the many African Kentuckian consumers who frequented these stores. Although the institution of slavery rested upon racism, often race did not prove the motivating factor in relations between blacks and whites, especially when economic self-interest was involved.

Economic autonomy, therefore, became a significant factor in the construction of black/white and even black/black relations in Lexington and along the Maysville Road. Undoubtedly, the inability to act upon economic self-interest distinguished enslaved African Americans from whites. But it also distinguished them from free blacks who, even with their political rights stifled by the 1799 constitution, participated more openly in commerce and society. While his status remains unknown, Philip, the customer at William Tureman's store, was most probably a free black because he enjoyed full commercial participation. In Washington and Lexington, and to a lesser extent in Maysville and Paris, free black populations grew large enough that the cooperation of white Kentuckians became less significant to their lives. As an example, Edward Abdy beheld a scene that confirmed for him the autonomy and even affluence of free urban blacks. In 1833, as his stage left Lexington for Frankfort, it "stopped at one of the best houses to take up two ladies . . . [who] announced to the agent, who had accompanied the coach, that the ladies had declined going, as there were two colored men inside. He offered to please them on the back seats, so that the other passengers, who would then take the middle bench, might form a barrier between them and the hated objects. . . . This little incident shews that the free blacks are too numerous and wealthy in Lexington to be slighted by the stage proprietors." It also shows that overt racial discrimination was a luxury of the wealthy. Euro-Kentuckians who could afford neither another conveyance nor the leisure to wait had no choice but to the share the coach with free African Kentuckians.[100]

Lexington provided a milieu for many free African Americans to gain some economic advantages. In 1804, Rolla Blue, the city's most successful free black, owned and operated a blacksmithie on a small alley perpendicular to Town Creek. By the 1810s, Blue speculated heavily in town lots and provided property for Old Captain's church, contributing to the development of the town's African American community. At his death in the 1840s, he owned enough property to exchange for the freedom of his enslaved relatives. Throughout the early 1800s, another free black, Cyrus Parker Jones, worked as a "huckster" at the Lexington markethouse where he distributed funeral notices for middle-class and gentry whites. As an African American, Jones could not officially reserve a stall in the markethouse, but the trustees did acquiesce to his selling vegetables and fruits from a cart. Considering the trustees' efforts to regulate market activity and black economic participation, their concession to Jones seems extraordinary. Yet, on his visit to the Bluegrass, Fortesque Cuming noted how, outside the markethouse, vegetables were "sold mostly by negro men and women," and that African Americans appeared the more prominent buyers and sellers. The conjecture has to be offered that the majority were slaves, participating in social and economic exchange in a very public place where their activities could be monitored. When, in 1808, the trustees restricted economic participation in the markethouse, their prohibition did not apply to commercially active blacks who continued to sell and purchase on the town square.[101]

As economic opportunities beckoned, Lexington's free black population grew. And as revivalism inspired free and enslaved blacks, many turned to Old Captain's church. Between 1800 and 1810, the congregation grew to some five hundred members, worrying many white Lexingtonians. Race-conscious trustees determined to reverse the trend, and a scheme arose whereby the trustees proposed to bring Peter Durrett's congregation under the control of the white Baptist association.[102]

Black preacher London Ferrell arrived from Virginia with the James Overton family in 1811. Soon afterwards, James died, and although the reasons remain unclear, his son Samuel emancipated Ferrell. A free black in his early twenties, Ferrell wandered the beaten path baptizing converts in the Ohio River at Maysville, in Elkhorn Creek near Bryan's Station, and in Town Creek in Lexington. He quickly established a reputation among African Kentuckians as a talented singer and expert funeral preacher. Durrett invited Ferrell to join the Lexington congregation in 1817; Ferrell declined. Instead, by 1820, backed by the town's trustees, he challenged Old Captain for control of Lexington's black congregation and community.[103]

The timing proved propitious. In 1819 and 1820, Peter Durrett faced two devastating losses: the first was John Maxwell, the town father who had championed and protected Durrett's activities since the 1790s; the second was his wife. Town trustees panicked. Without Maxwell's constant oversight and in grief over his wife, who knew what direction Old Captain would lead the congregation? The church's membership had ballooned, and the potential for black rebellion had been on the minds of most whites since the turn of the century. As Charles Scott complained in 1804, "there is strong reason to believe that we Shall have Some trouble ere long with the Blacks, Since the fate of Sendingo they have given themselves Extraordinary airs and some whites among us are not too good to Join them."[104]

Despite legislative restriction on black political and economic life, Lexington's African Americans had successfully formed a community. Those who participated in the boom in church membership following the Great Revival most often came from larger slaveholdings where a sense of community was strongest. Membership lists from Durrett's church do not survive, but it is probable that most members were slaves. While occasions for large slaveholdings were rare within towns, as early as 1810 Lexington had developed an economic culture suited to concentrations of urban slave laborers. Nearly half of Fayette County's slaves (739 of 1,484) worked in mercantiles, inns, ropewalks, and brick manufactories; and they lived in cramped quarters near the businesses. Some of Durrett's congregants most certainly served, like him, as hired-out slaves from the countryside. Many labored in the hemp and cotton manufactories; most interacted socially with fellow blacks. Thus, revivalism strengthened a sense of community already taking shape among black Lexingtonians.[105]

Rather than dismantle Old Captain's congregation, town leaders plotted to replace Durrett with Ferrell, whom they believed they could manipulate. Surreptitiously, trustees enlisted Ferrell who "consented to become the preacher of the colored people" and to work within the Elkhorn

Baptist Association's dictates. Their actions demonstrated a fundamental problem with the racial repercussions of revivalism. On the one hand, many white revivalists enthusiastically welcomed blacks into their churches and occasionally sponsored black preachers. On the other hand, the larger white population feared granting too much autonomy to blacks since it could inspire dreams of freedom and rebellion. The existence of a large, independent black congregation energized by the rhetoric of revivalism posed a significant threat to racial and religious order along the Maysville Road, and urban order in Lexington. Revealingly, the six men who led the movement against Old Captain were merchant trustees and members of either the Episcopal, First Presbyterian, or recently formed Market Street Presbyterian churches. Rather than condone a black church under the leadership of Durrett, they sponsored Ferrell and the black First Baptist Church under the control of the white First Baptist Church. But the effort to diminish Durrett's popularity and authority was ill-fated. The younger challenger, ordained by the Elkhorn Association upon the urging of the trustees, made a public issue of Old Captain's earlier quasi-ordination, thereby upsetting Durrett's followers. Both slave and free blacks initially refused to attend the First Baptist Church; for years, only seven members sat in its pews. By attempting to take back what had been allowed for three decades, the town trustees unwittingly enhanced Old Captain's leadership within the African Kentuckian community. Not until the late 1820s, after Durrett's death, did Ferrell's congregation explode to nearly three hundred as Old Captain's congregants sought out a church.[106]

In their attempts to control Old Captain, the trustees acted upon a growing fear among white men of new wealth. Free blacks and independent white women were numerous in Lexington, liberated by new attitudes toward free enterprise, individual faith, and personal responsibility that allowed them to live in ways ill-suited to the white, patriarchal structure that the trustees represented. That was one reason that social distinction became so important to the middle class. Too many people of too great a diversity could pursue profit and independence in Lexington. As the entrepreneurial class became politically and socially dominant, neighborhoods arose of peoples identified by their economic, ethnic, and social standings—a sort of "social zoning."[107]

The early stages of Lexington's social zoning is available in the memoirs of William Leavy, who recalled the town in 1804 as it "is vividly the most of it painted on my mind." Leavy divided the town into five neighborhoods distinguishable by class and economic purpose. An artisan neighborhood stretched along Main-Cross and West High Street, and included most of the alleys surrounding Town Creek. Populated primarily by Germans, this southwestern sector of the town was filled with bakeries and butcher shops. It was also the area in which Old Captain's church arose, and free black businesses sat aside Town Creek. Along Mulberry Street, north of Main Street, another neighborhood arose as home to large numbers of Scottish and Irish artisans, many of whom were house joiners, brick masons, and stonemasons. Along Second Street in this neighborhood clustered small residences of free black families. But houses of "disreputable people" sat at the southeastern corner of Church and Mulberry Streets, precursors to the brothels that would plague this neighborhood later in the nineteenth century.[108]

The commercial district along Main Street had the highest lot values in town, averaging $781. All but seventeen of its eighty-two buildings served as stores, taverns, or artisan shops. Leavy's recollections of one block

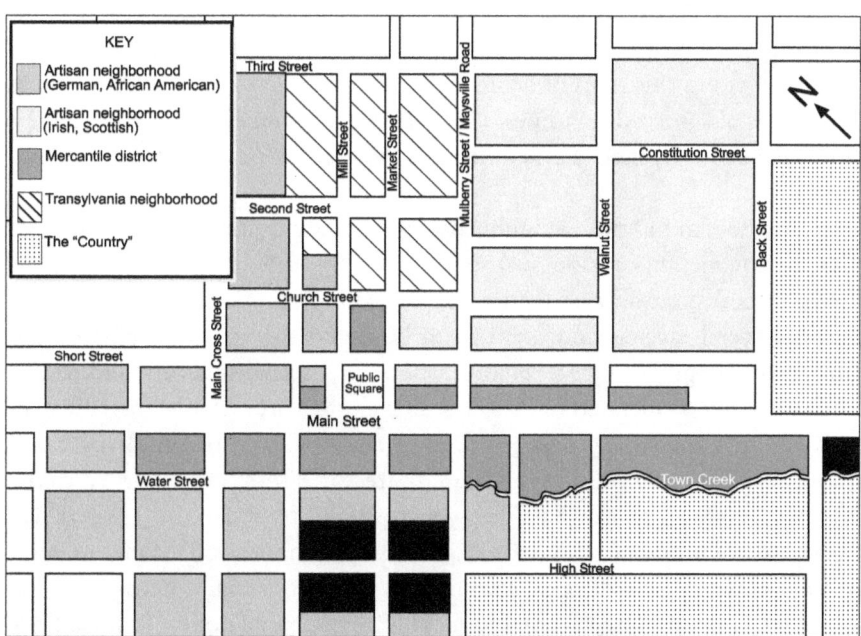

Lexington Neighborhoods, 1810s. Map by Mapsmith etc.

between Mill Street and Main-Cross demonstrates the economic diversity of the "Athens of the West."

Clark & Lowry's hatter's shop	P. Bain's hatter's shop
W. Clark's frame residence	J. Boggs's residence and store
frame residence	L. Young shoe shop
H. Marshall's stone tavern	Fishel & Gallatin log tin shop
law office	J. Bryan sadler shop
J. Bradford's brick and printing office of the Kentucky Gazette	M. Shaugh log residence and doctor's office
Widow Parker's brick residence	L. Sanders's store
Union Fire Co. brick station	J. Weir's brick store
public spring	J. Campbell's store
Cornelius Coyle's frame tailor shop	Widow Plimpton's frame house
W. Leavy's log store	D. White's frame millinary
W. Leavy's brick residence	E. W. Craig's frame store

Construction materials varied from log to brick to frame in an indiscriminate order. Additionally, the eclectic blend of building usage revealed the dynamic nature of central Lexington. Although few identifiable residences appeared along this strip of Main Street, the second stories of many stores and shops also served as homes. This mixture of genteel and rustic, business and residence typified Lexington and repeated block after block along Main Street.[109]

The remaining two neighborhoods had few commercial characteristics. In the neighborhood surrounding Transylvania University, north of Main Street's commercial district, sat the West's first middle-class neighborhood. Several taverns and law offices bounded the southern part of this area, near the public square. Hemp ropewalks, extensive stables, and several carriage makers operated on the west side along north Main-Cross. Elegant town homes owned by merchants and manufacturers surrounded a commons that ran northward from Church Street to Third Street, framed to the east by Market Street and to the west by Mill Street. By the 1810s, there was a clear entrepreneurial definition to the neighborhood. Many residents were known familiarly by their militia ranks: General Thomas Bodley, Colonel Thomas Hart, General John McCalla, Major Thomas Pindell. They took pains to purposefully design the neighborhood. John Wesley Hunt's house at the corner of Church and Mill Streets architecturally mirroring Thomas Bodley's home at the corner of Church and Market Streets. The

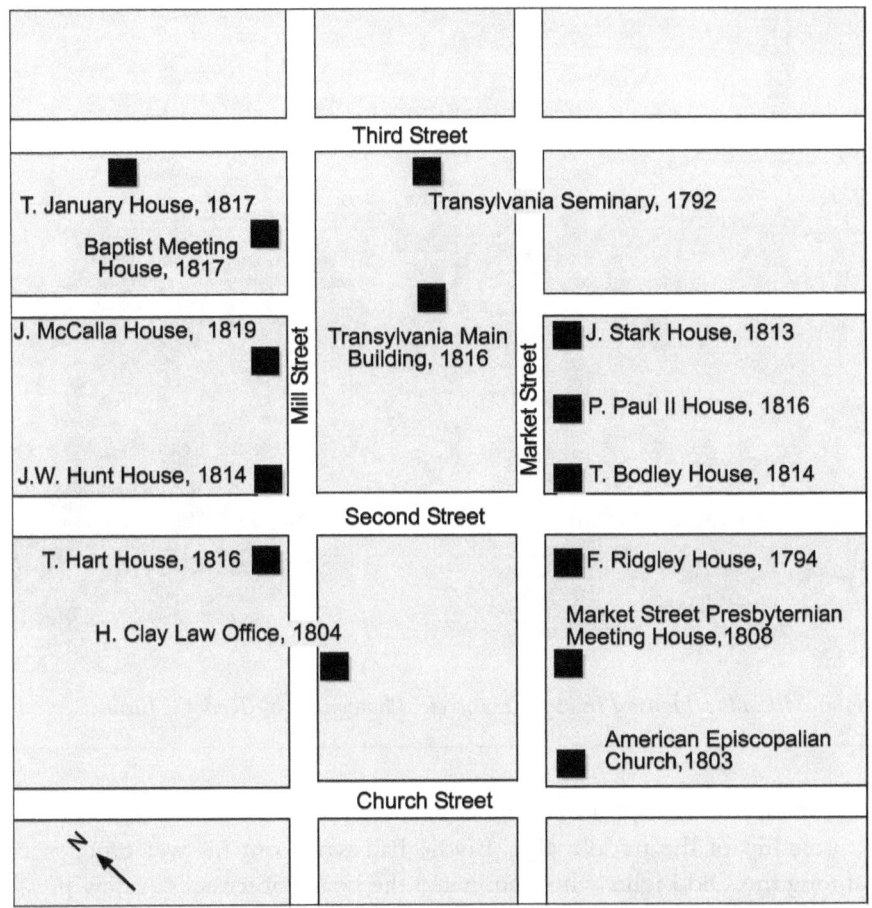

Lexington's Middle-Class Neighborhood, 1810s. Map by Mapsmith etc.

whole quarter, abounding with private gardens and enclosed courtyards, exuded the elegance of Lexington's wealthiest sector. The American Episcopal Church and the Market Street Presbyterian Church, both havens for wealthier Christians, anchored the neighborhood on the southern end. In this neighborhood, as well, were over one hundred slaves who lived in shacks behind the ornate urban retreats of these merchants, lawyers, and manufacturers. When, in 1816, Transylvania University moved from its original two-story brick seminary into a new, larger main building situated in the center of the commons, the middle class also symbolically became the guardians of education.[110]

John McCalla's House (1819), Lexington. Photograph by Rod G. Turner.

In fact, by 1816, Transylvania University was about to come under the trusteeship of the middle class. Blythe had worn out his welcome, even among the Old Lights who dominated the board of trustees. A new president was sought, and Horace Holley received the call. When trustees discovered that Holley was a Unitarian, however, they hastily rescinded the offer, raising eyebrows in the state legislature. An investigation ensued in which the legislature determined that the board was ineffective, to the extent that "the politics taught in the institution have not been pure." In 1818, the legislature named a new board of trustees, and the appointment of entrepreneurs like Henry Clay, Thomas Bodley, Robert Wickliffe, Lewis Sanders, and Edmund Bullock should not be surprising. They offered the presidency again to Holley.[111]

When he arrived in Lexington, Holley wrote that "It has not the pretension without the reality, that so many of the small towns have through which I have passed." Society was not all that impressed Holley. He proceeded to fill Transylvania's faculty vacancies with self-made men: Daniel

Drake in the new medical department, and William T. Barry and Jesse Bledsoe in the law department.[112]

The evolution of Transylvania University in the late 1810s paralleled the rise of the Market Street Presbyterian Church. In 1813, many of Lexington's most prominent citizens formed the congregation and selected the dynamic James McChord as their minister. John Pope, Thomas Bodley, Leslie Combs, Matthew Kennedy, Robert S. Todd, and Joseph Cabell Breckinridge composed the core of the new church. Their allegiance to the culture of a new middle class was unquestionable. In 1819, however, when McChord, growing in his own faith, incorporated a more radical theology of redemption and the commitment to self and society so familiar among the revivalists into his sermons, the congregation revolted. McChord lost favor and, if he had not died in 1820, probably would have lost his job as well.[113]

Demographically, the Transylvania neighborhood had much in common with the final section of town. "The Country" consisted of the lands surrounding the central neighborhoods. Although some dense residential patterns along East High Street could be found in this sector, the majority of lands were larger estates belonging to wealthier and more established

Transylvania University (1819), Lexington. From Charles Caldwell, A Discourse on the Genius and Character of the Rev. Horace Holley *(Boston: Hilliard, Gray, Little, and Wilkins, 1828).*

Lexingtonians like John Pope, George Trotter, and Henry Clay. Twenty-five of Lexington's landowners lived in the Country or beyond. Some, like Joseph C. Breckinridge, used their town lots as office space. Others acted as landlords to the over one hundred householders and their families who rented. In "the Philadelphia of Kentucky," then, a broad chasm between wealth and poverty manifested in the spatial order of the town and flowed over into Lexington's environs. Per capita, Fayette County had fewer landowners, a smaller percentage of propertied householders, and a disproportionately smaller share of yeoman farmers than the other counties along the Maysville Road, and Kentucky in general.[114]

In 1816, Timothy Flint toured central Kentucky and found in Lexington a city on the verge of cultural grandeur. Neighborhoods and demographics were not of interest. Rather, he visited Transylvania University, the Lexington library, and several bookstores and coffeehouses, concluding that "The best modern works have generally been read. The University, which has since become so famous, was even then, taking a higher standard than the other seminaries in the Western Country. There was generally an air of ease and politeness of this town, which evinced the cultivation of taste and good feeling. In effect, Lexington has taken the tone of a literary place, and may be called the Athens of the West." It was a title that Lexingtonians appropriated and promoted.[115]

As Lexingtonians began to understand themselves as distinctive in their urbanism and as Lexington's space began to take on socioeconomic dimensions, observers' comparisons between town life and country life became commonplace. On his 1810 visit, Alexander Wilson noted how "the country people, to their credit be it spoken, are universally clad in plain homespun." Fortesque Cuming, touring in 1808, acknowledged the "polite invitation of general Russell on the road, . . . a specimen of the hospitality of the country gentleman . . . so that I cannot absolutely tax Kentucky with a total want of that virtue." Lexington's *Farmers' Almanac* related country unpretentiousness with the story of a man who visited his sister in "a house jammed in among a great pile of houses, with a door yard about as wide as a carrot bed." He discovered that "not a bit of dinner did I see till after two o'clock"; and at tea time, "judging from our country practice, I hoped I should find some butter-cakes and ham, to make amends for my rough dinner." Instead, a house slave served "red hot" hyson tea, bread and butter.[116]

Portrayed as simple, hospitable, and unpretentious, the mass of country folk became quaint reminders of the 1780s when the ruggedness of an uncleared wilderness and the trials of a continuous war with American Indians had forbade refinement and social distinction, requiring instead simplicity and neighborliness. Just as the folksy image of the Kentucky hunter became part of the national popular culture, so too did men who lived simply and unpretentiously enjoy a type of local celebrity. John Robert Shaw arrived in Lexington in 1793 with little choice but to work "for my victuals." Like many white men of the era, Shaw went to the new West to find a "best poor man's" country. Historians have interpreted that phrase to mean opportunity for unpropertied white men to become independent yeoman farmers. Shaw's life, however, and those of thousands of other whites demonstrates that many poor men found their best opportunities in the towns and villages, beyond agrarianism. Shaw quickly became a successful quarryman, blasting wells across the landscape and eventually accumulating five acres of land just west of Lexington where he constructed a house, stable, spring house, smoke house, wash house, and wagon house. Yet, he refrained from the agrarian vision at the center of republican ideals for the West; his lands were used for a little husbandry but no

Woodcut of John Robert Shaw's Injury. From John Robert Shaw, A Narrative of the Life & Travels of John Robert Shaw *(Lexington, Ky.: Daniel Bradford, 1807).*

farming. Shaw and others like him, as prosperous and propertied as they became, remained beyond respectability, shunned by a new wealth that dictated that society be divided between a respectable, propertied citizenry and unpropertied, urban laborers. Shaw became "not less distinguished for his honesty, industry and usefulness, than for his accumulated evils (if such they may be called) which have pursued him for many years past," as Dr. James Fishback recalled on one of the four occasions when the well-digger blasted off a finger, arm, or leg. When he published his memoirs in 1807, they were devoured on the regional market.[117]

 5 • CHANGING LANDSCAPES

The Triumph of the Middle Class

In mid-April 1818, a mail coach rattled southward along the beaten path. Aboard were several passengers, including James Flint, a British tourist whose visit to Kentucky had thus far been blemished by unaccommodating hosts in Maysville and a coach "stowed full of people, baggage, and letter bags; the jolting over stones, and through miry holes, is excessively disagreeable: and the traveller's head is sometimes knocked against the roof with much violence. A large piece of leather is let down over each side, to keep out the mud thrown up by the wheels. The front was the only opening, but as the driver and two other persons occupied it, those behind them are almost in total darkness." The unpredictability of the weather and the road's inclination to muddy quickly made the coach ride neither comfortable nor convenient, so just north of Millersburg, Flint resolved to walk for the remainder of his journey.[1]

The earliest vehicles to lumber along the Maysville Road were stage-wagons: their makeshift canopies, backless bench seats, and mud flaps made travel uncomfortable; their light chassis made it dangerous. Carrying six to eight persons (including the driver) and a heavy load of merchandise, such vehicles were unstable, as Flint discovered when, earlier in his visit near Mayslick, his stage-wagon overturned. He swore to "go no further with that vehicle in the dark." Unless the wheel base of the wagon matched the deep ruts in the road, passengers and crew alike often found themselves badly bruised as the stage-wagon jostled back and forth. Drivers frequently called on passengers to lean to one side or the other to right a tilting vehicle.[2]

217

A fleet of stagecoaches arose to replace the stage-wagons. By 1800, the first coachmakers set up shop in Lexington and soon thereafter in Maysville. Traveling at about twenty-five miles per day, stagecoaches were enclosed vehicles: the driver sat atop while passengers rode inside, protected from the elements by more than mud flaps and canvas canopies. Wrought-iron springs reduced some of the bumping, and sturdier frames guaranteed safer travel at greater speeds.[3]

By the 1810s, carriages joined the stage-wagons and stagecoaches along the beaten path. In 1815, a profitable carriage-making trade flourished in Lexington. Such vehicles, whether two- or four-wheeled, tended to be smaller than stagecoaches with narrower wheels, upholstered interiors, and fancifully painted exteriors. Serving as mobile parlors, carriages allowed the new wealth to move among the vulgar with minimal interaction. Hotels and taverns, like the Indian Queen of Paris and the Sign of the Green Tree Inn in Lexington, often catered to carriage clientele only.[4]

Alongside these refined conveyances commuted wagoners and drovers. Inns and taverns that hosted middling and upper sorts refused to accommodate these lower sorts. Wagoners hauling goods from Maysville to Paris and Lexington found repose at wagonyards, taverns of poorer quality that provided public scales for weighing wagons and their loads. Drovers had to seek out still lesser accommodations. Towns and villages dissuaded drovers and their herds of swine, sheep, and cattle from lingering within their boundaries. On the outskirts of Lexington and other places arose stockstands where landlords provided watering holes and pasturage.[5]

The evolution of vehicles along the old buffalo trace offers a simple microhistory of early Kentucky. From the unadorned and practical structures of stage-wagons to the more comfortable rides of the stagecoaches to the privacy and refined appearance of carriages, white Kentuckians became increasingly class conscious in the early nineteenth century. Of course, drovers and wagoners also used the road alongside these conveyances, symbolizing the market economy that had already overtaken life along the Maysville Road and the livestock economy that was just beginning to form. And few blacks rode in any vehicle, unless they were free blacks with sufficient wealth to secure a seat in a stagecoach. Riders in carriages, stagecoaches, and even stage-wagons looked down upon others who had less wealth, less property, and less stability, most of whom walked from place to place as peddlers, tenant farmers, day-laborers, hired-outs, drovers, and slaves. By the early 1800s, the beaten path had become an extended social milieu on which individuals of different socioeconomic identities mingled, but it was also where class consciousness became more acute.⁶

LINES OF DISTINCTION

One reason so many people purchased and read *A Narrative of the Life and Travels of John Robert Shaw, the Well-Digger* was their search for a past when society, even if not politically egalitarian, had seemed more simple. By the turn-of-the-century, with the Indian threat diminished and the environment tamed, with land title suits more often than not favoring the propertied and wealthy, Lexington's affluent had begun to assert social superiority in ways that few residents could ignore. A middle class with expendable cash pushed forward its own ideas of respectability: finer sensibilities, taste for artistic and architectural beauty, advanced intellectual and emotional faculties, and distinct class lines. While their business interests compelled them to maintain contact with the less refined lower sorts, the new wealth also fostered a class exclusiveness. As Thomas Ashe recognized in 1806, "A small party of rich citizens are endeavouring to withdraw themselves from the multitude, or to draw a line of distinction between themselves as *gens comme il faut* and the *canaille*. . . . The present better sort of persons . . . live in a handsome manner, keep livery servants, and admit no persons to their table of vulgar manners or suspicious character. As wealth encreases in Kentucky, the *line of distinction* will extend through Lexington to the minor towns, and may possibly pervade the country after a lapse of some centuries."⁷

Attempts to create artificial social distinctions raised the ire of those who had expected more from the political promise of Kentucky. As early as 1793, in the wake of public discourse over the shape of the state constitution, gentry had drawn fire for promoting their social superiority. A disgruntled contributor to the *Kentucky Gazette* sarcastically reproached the masses for holding to notions of equality, what he termed the "obsolete principles of 1776." "In the oeconomy of nature," he explained tongue-in-cheek, "there has been a set of men in every age, born to wealth and invested with innate wisdom to rule, govern and dispose of the rest of mankind to whom an implicit and unreserved obedience ought to be paid." The social revolution that he believed took place alongside the military conflict of the 1770s and 1780s had almost succeeded: aristocracy and deference had been "in a measure lost in the world, and is almost entirely extinguished in the American states, excepting Virginia." But, despite the pioneer legacy of the 1780s, "the seeds of it [aristocratic deference] is got into Kentucky, and from the complexion of our constitution and laws appears to agree well with our soil."[8]

Possibly no other state of the Early American Republic offered so much and ultimately granted so little as Kentucky. As the majority during the 1770s and 1780s, pioneers shaped their collective identity from the struggles of converting a wilderness into a usable environment. Into their midst came the gentry who appeared eager to push aside pioneer peoples, draw wealth from the legal disputes arising from crude land surveys, and establish churches, courthouses, and market networks to transform the nascent urban places along the Maysville Road into bustling centers. After the turn of the century, self-made men built upon the gentry's accomplishments, emphasizing conspicuous gentility and political ascendancy as symbols of their success. Thus, when a social elite attempted to distance itself from the masses, reaction echoed strongly. As Ashe worried, "the public at large conceive this [gentrification] a dangerous innovation; they wish men to continue all vagrants alike."[9]

Fayette County's *gens comme il faut* exuded material refinement, and they were determined that their reputations not be sullied. In 1810, tourist Alexander Wilson reported to an eastern newspaper that he found within Lexington's markethouse "a few cakes of black maple sugar, wrapt up in greasy saddlebags, some cabbage, chewing tobacco, catmint and turnip tops, a few bags of meal, sassafras-roots, and skinned squirrels cut up into quarters—something better than this, I say, in the proper season, certainly covers the stalls of this market place, in the metropolis of the fertile country of Kentucky." Although Wilson qualified his description by noting that his arrival in April "was unfavourable to a display of their productions," Lexingtonians reacted with outrage. Despite their restrictions upon the markethouse economy, it still represented the character of the community. The description of skinned squirrels seemed especially insulting to a crowd increasingly engaged in the pursuit of refinement. "Does Mr. Wilson mean to joke upon us?" ranted one correspondent; "If this is wit we must confess that, however abundant our country may be in good substantial matter-of-fact salt, the attic tart is unknown among us." As if to test his audience again, Wilson admitted in July 1811 (again through the newspapers) that he had indeed misrepresented the truth: "On referring to my notes, taken at the time, I find the word 'halves,' not quarters." Most probably, Wilson accurately portrayed the offerings of the Lexington markethouse. Grocers sold what farmers furnished. In the early spring, farm productions would have been scarce and reliance on the bounty of woodlands greater than at other times; and, by 1810, the ravaging of the rural environment had taken enough of a toll that squirrel would have been more common than venison or even turkey. Whatever truth may have lain in the account, vindication of Lexington's reputation appeared much more important to certain citizens than absolute accuracy.[10]

As Lexington's taverns and inns came under entrepreneurial control, owners actively promoted their city as a resort for America's elite. In 1802, Robert Bradley opened Traveller's Hall, a tavern with stables for *"Genteel Guests Only."* Six years later, Henry Clay purchased the business and renamed it the Kentucky Hotel, building upon Bradley's precedent of serving the wealthy. John Postlethwaite's Tavern became Postlethwaite's Hotel, only to burn in 1820. The establishment arose again as the Phoenix Hotel, home to "wealthy gentlemen from the south, who spend a part of every summer traveling for health and recreation." The reputed healthfulness of the Bluegrass annually drew hundreds of tourists whose contributions to the regional economy made the new wealth "pretty generally like that portion of the people of England who get porridge to eat; stout, fat, and ruddy." Along Lexington's streets dashed "pleasure carriages" in such large numbers that visitors could not help but comment in surprise. "The moral and social influences of the city were felt throughout the country," recalled one observer. "In the suburbs of the town were splendid *villas* and gardens, the houses of handsome architecture, and the gardens arranged in the most tasteful manner. Two of these gardens (Fowler's and Lamphier's) were specially intended as places of resort for the elite in the surrounding country during the summer months. Whatever could gratify the taste and please the eye was furnished here in rich profusion."[11]

Skilled craftsmen and the service sector tapped into this escalating appreciation for refinement and entertainment. By 1803, James Kennedy ran a stagecoach line to the resort at Olympian Springs, some sixty miles northeast of town. Tavernkeepers and merchants capitalized on the popularity of the resort by importing the spring's waters which "are held in great esteem, and the most distinguished personages in the country were drinking them." French silks, including stockings and taffetas, filled local shops. Alexander Wilson, the critic who baited Lexingtonians about the sale of squirrel in the markethouse, complimented the "numerous shops piled with goods, and the many well dressed females I passed; the sounds of social industry, and the gay scenery of 'the busy haunts of men,' had a most exhilarating effect on my spirits." William Essex and Joseph Charless profited from the printing and sale of books. By 1810, caterers of refined crafts and services included a stone-quarrier, a glazier, three stonemasons, a portrait painter, three carriage makers, several house painters, an umbrella and bonnet manufacturer, two stocking makers, a breeches maker, a toy-maker, an apothecary, six bricklayers, and a whitsmith making "all sorts of plain and ornamental Railings."[12]

Entrepreneurs flourished, and much of their profits was invested in manufacturing. A merchant-manufacturer system grew to replace the domestic household and independent artisan forms of manufacturing along the southern end of the road. In ropewalks and other manufactories, owners adapted machinery to bolster production. Carding machines, spinning throstles, spinning mules—the hum and pace of manufacturing increased as the 1810s wore on. At one cotton factory, the owner aspired to engage one thousand spindles "on the principle of the New-England factories." In a hemp-bagging manufactory, the interaction between hired-out laborers, overseer, and machinery "was going like clock-work."[13]

As the nineteenth century proceeded, refinement, capitalistic investment, and lines of distinction spread from Lexington outward. By the late 1810s, carriages and mansions still displayed the wealth of the gentry, major merchants, and some artisans, but lesser symbols of refinement—parlors, tea services, imported clothing—filtered into the homes of even lower-ranking citizens. While display of material possessions was at the heart of consumption, there was also a degree of social integration as people of lesser wealth found some common ground with the middle class. Still, that common ground comprised shared values that entrepreneurs had formulated to promote consumption and distinction. Less wealthy Kentuckians may have purchased refined goods, but, as William Faux remarked, acquisition ultimately strengthened social hierarchies: "storekeepers and clerks rank above farmers, who are never seen in genteel parties and circles."[14]

As refinement spread along the beaten path, so did the temptation towards class differentiation. "Trade and communication between the town and the country, resembles the circulation of the blood," explained an editorialist to the *Kentucky Gazette*, "it is first impelled to the heart, and from thence driven to warm and invigorate the most extreme parts." Thomas Ashe's prophesy that "*lines of distinction* will extend through Lexington to the minor towns" began to actualize as genteel tastes moved northward along the road.[15]

But the drawing of lines of distinction was much more difficult the farther one moved beyond Lexington's environs. Smaller populations did not lend well to class distinctions. "When the number of inhabitants is not too great they may all enter into society," explained lawyer William T. Barry about social life in Paris, "but when they are numerous it is impossible for all of them to associate; they must consequently split into parties." In contrast to Fayette County's heavily urban population, 70 percent of Old Bourbon's landowners continued to live rural lives. Paris residents with ideas of respectability returned from their excursions to Lexington hoping not only to bring refinement into their own lives, but to transform their community into a respectable society. The many fine brick, stone, and frame houses that Thomas Chapman came upon indicated some architectural refinement, and the number of carriages on Paris's streets between 1792 and 1812 increased from three to forty-three. By the late 1810s, the town's merchants were shaping their own entrepreneurial world, and investments in commercial and industrial development were beginning to show profits. Houses in the hinterlands evidenced the architectural pretensions of what remained of the landed gentry and the wealthy entrepreneurs who were finding success in Paris: Jacob Spears's Stonecastle, Jefferson Scott's New Forest, John Hildreth's Mt. Airy, William Thomas Buckner's Xalapa. Middle-class residences in the town became known more by their painted clapboards than their occupants: the "Yellow House" of merchant William Caldwell; the "Blue House" where the brewer, Mr. Ellerbeck, lived; the "Red House" of tavernkeeper Nathan Standeford. Brick and stone row houses lined Main Street where residents like Thomas Arnold, clerk of the Bourbon County Court, laid claim to the town's first piano and carriage. Joseph Duncan's Tavern towered over the village, serving as a stagecoach stop, tavern, and hotel. Upon his visit to Paris in 1819, Richard Lee Mason found theatrical performances opened to the general public; a decade earlier, such exhibitions were offered only in Lexington.[16]

Lexington's rage for manufacturing also spilled over into Paris. On the eve of the War of 1812, the only cotton manufactories along the road were the six in Lexington and the two in Paris. Calculating as well the domestic production of cotton and blended cloths, Bourbon County produced more cloth and of better quality than any other county in the state; Fayette followed a close fourth. Increased attention to hemp bagging accompanied

Joseph Duncan's Tavern (c. 1790), Paris. Photograph by Rod G. Turner.

the boom in cotton cloth production, and hemp factories in the two counties accounted for over 94 percent of bagging produced along the Maysville Road and 73 percent of the state's production. Tapping into the cash crop agriculture that flourished in southern Kentucky, merchant-manufacturers of Lexington and Paris found great profit in hemp and cotton productions.[17]

Industrial development had spread as well into Millersburg. A fanning-mill factory arose in 1810, followed by William F. Baker's fulling mill in 1817, Joseph Miller's 1818 woolen factory, and an extensive spinning operation owned by Robert Batson in 1828. The road ran through a small village of frame and stone rowhouses. But Millersburg's heydays were still ahead.[18]

In Washington, some lines of distinction appeared early. As the primary mercantile center along the northern part of the beaten path, Washington's stores offered a broad spectrum of imported goods. Archaeological research uncovered shards of English-made ceramics at numerous sites, and estate inventories indicate brass kettles, glass decanters, Queensware, silver

James McKee's House (1798), Millersburg. Photograph by Rod G. Turner.

spoons, tea services, and silk handkerchiefs. John Johnston, a transplanted New England physician, built a two-story frame house around 1794. Lawyer Thomas Marshall built "The Hill," a brick Georgian residence, in 1800. Seven years later, another lawyer, John Chambers, constructed "Cedar Hill," a two-story frame house atop a knoll overlooking the road, protected by a stone wall behind which was a row of cedars. Unlike Lexington, however, Washington's neighborhood did not abound with brick homes. In 1802, François Michaux noted the majority of houses built of wood. Even as late as 1810, the only construction artisans living in Washington were three carpenters, three housejoiners, a nailor, and a stonemason; there were no bricklayers. Washingtonians, many of whom were northern bred, did not find in brick the cultural significance given it by the Virginia gentry. In the village, this cultural distinction manifested as Johnston and Chambers, men from Connecticut and New Jersey respectively, chose frame construction while Marshall, a Virginian, selected brick. As refinement sprawled northward, its expression changed with the regional backgrounds of settlers along the northern end of the Maysville Road. Refinement took hold in Washington, but not in the same fashion as in Lexington. "I have been surprized at the state of Society here," commented Valentine Peers. "I was at a Tannery Ball in Washington on Monday last and found things, except in the waists of the Ladies which yet retain a decent length, very much Ball Fashion." And as refinement took hold, it marginalized many of Washington's residents who sought out new venues for their cruder passions. Daniel Drake remembered Sunday visits to Mayslick by "young gentlemen from Washington, many of whom sought our community for amusement, and to be among those whose lower rank would allow them wide latitude of manners and conduct."[19]

Despite this initial burst of refinement in Washington, along the northern end of the road, lines of distinction were muted well into the 1810s. The societies of Nicholas and Mason Counties centered around "small freeholders." Nicholas County, "in purity and refinement," conceded Daniel Drake, "did not rank very high." Both counties had smaller slaveholding averages and larger numbers of small farms than the two southern counties. In 1807, Nicholas and Mason Counties had fewer carriages *combined* than did Lexington five years earlier.[20]

In contrast to the genteel crafts of Fayette's and, increasingly, Bourbon's artisans, the more dispersed, farm-oriented population at the other end of the road needed practitioners of less fanciful arts and services. Few artisans

John Johnston's House (c. 1797), Washington. Photograph by Rod G. Turner.

Thomas Marshall's House (c. 1800), Washington. Photograph by Rod G. Turner.

in Mason County worked out of an urban area: 29 craftsmen lived in Mayslick, Washington, and Maysville; 110 lived in the countryside. By far, the majority provided construction-related skills. Sixty-three carpenters, housejoiners, and masons lived in the countryside in contrast to sixteen in the villages and hamlet. No less than in the Bluegrass, a population and construction boom was under way to facilitate the crowds of new arrivals. Yet, the craftsmen of Nicholas and Mason Counties provided only very basic construction services. For ornamental artisans like a whitsmith or house painter, one had to send to Lexington.[21]

The greatest line of distinction, then, was drawn between the southern and northern ends of the Maysville Road, not based on an urban-rural dichotomy but rather on the extent of genteel consumption and capitalist investment. The rural gentry of Fayette and Bourbon Counties had inspired merchants and artisans to cater to their needs, thereby creating a market of goods and services that elevated middling sorts and enticed poorer sorts. The fairly small gentry class in Nicholas and Mason Counties delayed similar development, allowing the material and occupational structures of pioneer culture to continue predominant. By the late 1810s, the beaten path had become a continuum originating in genteel Lexington, passing through mildly refined Paris, and ending in the less sophisticated villages and towns of Nicholas and Mason Counties.

As lines of distinction were drawn, racial lines were also solidified. The genteel heritage of the southern end of the Maysville Road required that an aspirant to social status not only live in a brick home, distanced from less wealthy neighbors by gardens or walls, but that there be inhabited slave cabins behind that home. Those who could afford large numbers of slaves used them to embellish their own social status. James Flint observed how in the Bluegrass home of one genteel matron the "sable domestics with whom she is constantly surrounded, and who obey her every nod, serve as a foil, or background, which, by drawing a contrast, greatly enhances her charms."[22]

For over a century, historians have struggled with the notion that slavery in Kentucky was milder than elsewhere in the United States. While recent historians have condemned the question as an argument over degrees of evil, why historians believed it could be true has much to do with the sources. When Edward Abdy traveled through Kentucky and Tennessee in 1833, he wrote, "the people of Tennessee were much more kind to their slaves than the Kentuckians,—while the latter are superior, in this respect, to the Virginians." Like Abdy's comparison, individuals' descriptions of Kentucky slavery (and consequently historians' sources for understanding it) depended on which slaveowners they observed. Along the northern end of the road, where gentry populations were thinner, typical slaveowners could ill afford overly abusive treatment of their slaves. Small Backcountry farmers "who brought slaves with them ... often only one" and nonslaveholders who hired slaves on occasion invested proportionately more of their money in slave labor. Loss of a slave or being sued for abuse of another person's slave could be financially devastating. Of course, slavery was often just as harsh in Mason and Nicholas Counties as elsewhere along the beaten path; recall Daniel Drake's anger over Hickman, the Backcountry Marylander who mercilessly whipped his two elderly slaves. Slave auctions were as active and profitable in Maysville and Washington. But, because the institution dispersed among small farms across the hills and dales of the countryside, the cruelty of slavery remained largely hidden from public view.[23]

Along the southern end of the road, slavery and its cruelties were much more evident. Slaveowners regularly auctioned off slaves on the public squares, advertised them in the pages of the *Kentucky Gazette,* and even sold them along town streets. The rage for manufacturing that appeared in Lexington and Paris in the 1810s coincided with an emphasis on large-scale agriculture. Fayette County's first agricultural fair was held in 1814, and residents of Bourbon County attended their first in 1818. Edward Stone found opportunity in the demands for slave labor that accompanied these economic developments. In July 1816, he advertised "to purchase twenty negroes, boys and girls from 10 to 25 years of age. A liberal price will be given for those answering the description." Stone's occupation as a slave trader may have been frowned upon by neighbors, but many of them were guilty of benefiting from his activities. In the cellar of Stone's The Grange (1816), possibly the most impressive of Bourbon County's new middle-class homes, he kept his merchandise in the cellar, shackled to the walls with iron chains. Individuals brought unruly slaves to Stone's estate, which sat along the Maysville Road just north of Paris. Regularly, coffles with dozens of slaves marched northward from The Grange to Maysville, where they were exported to New Orleans for sale.[24]

Increased demand of slave labor raised fears of black rebellion, particularly in Fayette and Bourbon Counties. When, in September 1800, the *Kentucky Gazette* reported about Gabriel Prosser's slave revolt in Virginia, the news compounded already tense relations between black and white Kentuckians. Only two months earlier, Lexington was abuzz with rumors of recent attacks by two blacks on travelers northwest of town. Soon thereafter, the trustees issued restrictions on Saturday and Sunday slave gatherings. By 1802, they rejected state legislation approving the continued slave trade in Kentucky and prohibited the importation of slaves into Lexington from other states because "the Slaves in the South are strongly bent on insurrection." Then the board authorized the construction of watch houses to serve as reminders to whites that a menace was out there, waiting for the opportunity to bring chaos to their world. To blacks, watch houses embodied the persistent scrutiny of their lives.[25]

Still, ambiguous legal status and greater economic opportunities for blacks continued. Although the 1798 slave code and the 1799 state constitution had attempted to resolve the legal status of African Kentuckians, neither fully succeeded. In court case after court case of the early nineteenth century, the intricacies of the institution continued to be called into question. When a slaveholder, upon his death, freed his slaves, were the children to remain legally enslaved until they came of age? Could a slave found wandering *with* permission of his master be arrested and sold by the county court? How could a slave obtain freedom? The new wealth desired to establish strict legal definitions of status for blacks as they pushed these cases through the courts, but the enforcement of codes remained difficult, leaving the legal status of slaves (and free blacks) always blurred.[26]

As Presbyterian minister David Rice recognized, however, "the rich hold Slaves, and the rich make the laws." By 1810, white efforts to strengthen slavery in Lexington and increasingly in Paris had restricted opportunities for enslaved and free blacks to develop community. In Lexington, where watchmen scrutinized black activities, 122 of 133 free blacks lived as independent heads of households, appearing "free" but under constant surveillance. Town trustees often directed watchmen to disperse assemblies of free blacks. In Bourbon County households and southward into northern Fayette County, 193 of 218 free blacks also lived under constant white surveillance, so that a strong black community was difficult to fashion. Where free black/enslaved black interaction became possible because of numbers, therefore, white vigilance worked to limit it. In contrast, in the slaveholding societies of Nicholas and Mason Counties, black populations remained small, and 41 of 61 free blacks lived in independent households. Whites in these counties were no less paranoid over rebellion, but less emphasis was placed on strengthening the institution along the northern end of the road. Black populations, both free and slave, remained too negligible to form community.[27]

As critics like Edward Abdy commented on Kentucky culture, making slavery along the Maysville Road seem capricious, decadent, and morally bereft, refined Kentuckians attempted to soften the harshness of the institution. They responded, as did most Southern slaveholders, with an elaborate set of subterfuges in which planters—whether gentry or new wealth—tried to convince themselves, others, and apparently posterity that their peculiar system differed greatly from other American slave societies. They seldom referred to slaves as slaves. Rather, labels articulated more intimate relations: John Brand and John W. Hunt spoke of their hemp factory slaves as "our boys"; Joseph Hornsby referred to his farm slaves as "my people." Incorporating slaves into a manufacturing or agrarian "family" supposedly moderated the degradation of slavery. Others characterized the institution as a variation on indentured servitude: "The man who objects to slavery," explained Henry Toulmin, "may give freedom to his slave when he has served long enough to indemnify him for the price he paid for him." Still others attempted to ameliorate the cruelties of slavery by appealing to ideals of the Revolutionary moment. In 1796, Henry Clay, writing under the pseudonym "Scaevola" in the *Kentucky Gazette,* insisted that whites, "enthusiasts as they are in the cause of liberty," despaired at the institution, and to insist that they enjoyed the enslavement of humans insulted "their good sense."[28]

These attempts failed to convince anyone, except possibly the speakers, of the mildness of Kentucky slavery. Even Toulmin, the ceaseless defender of Kentucky republicanism, admitted that "no laws can protect a slave against the thousand provocations which it is in the power of a petulant master or mistress to offer." When Alexis de Tocqueville journeyed down the Ohio River in 1833, he observed how "the traveler who lets the current carry him down the Ohio till it joins the Mississippi sails, so to say, between freedom and slavery; and he had only to glance around him to see instantly which is best for mankind. . . . On the left bank of the Ohio work is connected with the idea of slavery, but on the right with well-being and progress; on the one side it is degrading, but on the other honorable; on the left bank no white laborers are to be found for they would be afraid of being like the slaves; for work people must rely on the Negroes; but one will never see a man of leisure on the right bank: the white man's intelligent activity is used for work of every sort." While the description was certainly a metaphor for the emerging cultural distinctions between North and South, Tocqueville found no reason to differentiate Kentucky from other slaveholding states. He described a listless society that, compared to the energy along the northern banks, depreciated the value of humans both black and white: "one might say that society had gone to sleep."[29]

Along the Maysville road, slavery was a tradition rather than a necessity. As inheritors of that legacy, the entrepreneurial class sought to impose standards with which they were familiar on a slaveholding society where, given the minor numbers and dispersed black population, such regulations were not necessary. Under the new order of subtle subterfuges and heavy restrictions, blacks felt compelled to escape white surveillance. In contrast to the most public of black frolics like funerals and weddings, clandestine affairs—dances, boxing matches, and races—occurred in the woods or in abandoned buildings. When whites discovered such celebrations the consequences were dire. In 1831, some fifty slaves gathered in an old distillery just west of Lexington for a dance. A slave patrol happened upon the revelry and, when the dancers attempted to hide, opened fire. One man died; others nursed serious wounds.[30]

By the mid-1810s, the new wealth had accomplished much—the construction of a commercial economy and a refined culture, the appropriation of patriotism and Revolutionary rhetoric to substantiate their political aspirations, the promotion of Lexington as the "Athens of the West," the establishment of lines of distinction between themselves and poorer whites, and the solidifying of white superiority over blacks. With manufacturing, commerce, and refinement spreading northward along the road, their social and economic dominance seemed secure.

As the War of 1812 ended, however, economic troubles began. By early 1814, the initial economic boom that accompanied the war faltered. In July of the next year, all of Lexington's ropewalks ceased operations. Along the length of the beaten path, but particularly in Lexington, mercantile prices and profits plummeted with the reopening of trade with Britain. Hundreds of laborers no longer found work in manufactories. The market for domestic productions dried up, as did regional and national markets for agricultural productions. Lexington, in particular, was hit hard, receiving "a very great check from the evils of the *paper system*." Entrepreneurs, whose fortunes were grounded in the commercial republic, faced financial chaos.[31]

Yet, the men of commerce not only weathered the economic crisis, they benefited from it. Although some suffered as much as farmers and laborers, like Lewis Sanders who lost his cotton and woolen factory to bankruptcy and James and David Maccoun whose wholesale business had been the second largest in Lexington, in general, the middle class found opportunities to purchase land and businesses. The result was "an unwholesome or mushroom growth," remembered William Leavy. "Men were tempted to engage in business with slender capital and false or chimereal hopes, and were compelled to give up, or give place to others."[32]

Although the economy slightly improved between 1815 and 1819, it was a recovery based on speculation and weak currency. "Everyone is afraid of bursting the bubble," reported one visitor. Besides the Bank of Kentucky whose notes had acquired little value since the institution's inception in 1805, forty-six independent banks cropped up across the state, each issuing its own notes. Whereas the *Kentucky Almanac* once ended with agricultural advice, by 1819, its final pages offered a roster of banks. Along the Maysville Road, branches of the Bank of Kentucky, located in Washington, Paris, and Lexington, competed with a branch of the Bank of the United States operating in Lexington and five independent banks: the Bank of Limestone, Paris Bank, Sanders' Manufacturing Company of Lexington, Farmers and Mechanics Bank of Lexington, and Hinkston Exporting Company of Millersburg. By 1820, over $26 million in independent notes flooded the state's economy. "Go on then to incorporate all these various sorts of *Rag Companies* and I will again predict that in five years we will be in rags from one end of the system to the other," prognosticated a contributor to the *Kentucky Reporter* in 1816.[33]

It only took three years. The Panic of 1819 ended whatever optimism inspired the banking boom. Purchasing power diminished. James Flint commented on citizens' handling of money: "Small bills are in circulation of a half, a fourth, an eighth, and even a sixteenth part of a dollar. These small rags are not current at a great distance from the places of their nativity. A considerable proportion of the little specie to be seen is what is called cut money—Dollars cut into two, four, eight, or sixteen pieces." Kentuckians continued to use credit, as encouraged by merchants after the collapse of the markethouse economies. "The immutable maxim, that productive labor is the true source of wealth," he concluded, "has been lost sight of." Critics of the new wealthy agreed: Too much dependence on luxury, debt, and banks had diverted citizens from the simplicity, familial and communal obligation, and independence characteristic of true republican virtue.[34]

The rage for speculation "with slender capital and false or chimereal hopes" was infectious. Just as the people vernacularized gentility beginning in the 1790s and popularized religion in the early 1800s, in the 1810s, they enthusiastically democratized the patterns of entrepreneurial speculation, as one politician complained, "with the hope of acquiring extravagant wealth almost without labor or exertion." Thousands lost farms, homes, and businesses as the market economy faltered in 1819. Economic difficulties converted into social ills as the formerly well-to-do became obsessed with threats of arson, turning on "houses of bad character [which] have been assailed at night by a party of men in disguise (who are supposed to be of the most respectable of our townsmen) and the owners driven out of town and several men caught at them tarred and feathered."[35]

The panic devastated Lexington's artisan ranks. Mercantile imports had gradually eroded local craftsmanship "by importing from Philadelphia, every article which the Mechanics can make here. . . . Scarcely a stock of goods is now opened, which does not contain among them other *dry goods,* such articles as saddlery, boots, and shoes, hats, collars, iron chains, axes, brushes and ready made clothes & co, & co and household furniture brought out to order." When the bubble burst, many artisans were in no position to survive the troubles and left Lexington to find new lives elsewhere. That craftsmen comprised nearly 40 percent of Lexington's 1818 workforce suggests the impact of an artisan exodus upon the local economy. Those who remained faced an increasingly poorer and discriminating clientele. As one critic of importation sarcastically described, "the blacksmith shops with their black and sooty recesses ... looked like the shops of Vulcan in the lower regions ... [and] hatters frequently have their reeking hats lying on double rows on each side walk, as if on purpose to annoy you. All of this you as a gentleman, with the aid of gloves, perfumed handkerchief and quick steps may possibly run by unhurt." By 1819, prices of artisans' goods were too exorbitant for average farmers who, to purchase a locally produced coat, would have paid 144 bushels of corn, equivalent to an entire year's worth of bread.[36]

The historical development of an agricultural economy along the Maysville Road contributed to the widespread depression. Drought struck central Kentucky at the least propitious moment, as banks called in loan after loan. "The earth is parched, vegetation has perished, people and stock are suffering for water," depicted a local paper. Early commitment to commercial participation suddenly jeopardized small and large farmers alike. Even tenant farmers, drawn into the market by the dictates of their indentures, found times trying. Merchants, always pressured by eastern creditors, turned to the courts to coerce their clientele into paying. An Ohio merchant wrote George and Samuel Trotter about their payment of Kentucky notes: "the trash we are obliged to receive in payment for goods is such that it would cost too much to convert it into good paper." Urban laborers fled the towns and villages to become "cultivators in the back woods." Farming, which had once held tremendous ideological significance for republicanism, depreciated into an activity without ideological purpose. Instead, people turned to agriculture for survival. Of course, land was not easily found, and hundreds migrated from village to town to hamlet to farm searching for work and sustenance. By 1821, the state assembly authorized the construction of poor houses to accommodate the uprooted.[37]

As a result of this economic collapse, a second crisis arose of constitutional proportions. "Long ago it was said, when a man left other States, he is gone to hell, or Kentucky," explained William Faux in 1818. "The people are none the better for a free, good government. The oldest settlers are all gone or ruined." The epidemic of debt that swept along the road and throughout the state made evident to all Kentuckians the harsh, impersonal temperament of the market economy that had finally rooted out the pioneers. People who had never experienced heavy debt now wanted extensions on indentures so they would not lose businesses or lands. Bankers, merchants, and others who expected returns on their loans opposed such relief, knowing that foreclosures on property promised still greater profit. Antagonisms heightened between poor and rich; anger swelled against "*all those who do not live by the sweat of their brow:* the Lawyer and Mercantile juntos, which mutually assist each other in preying on the laborious throng." In the rhetoric of masculinity that followed the War of 1812, relief proponents caricatured the "sunshine effeminacy" of entrepreneurial usurers, contrasting it to those who had carried "the honorable scars he received in his country's service" and portraying merchants, lawyers, and manufacturers as men who "live by our wits, and upon the sweat of other men's brows."[38]

While many men of commerce profited from the difficult economic circumstances facing bank and mercantile customers, some were concerned about the communal repercussions. Authors of relief broadsides like "Soldiers to the Field!!!" pined for the days when "No longer will one man drive in his *gilded chariot,* whilst another humbly crawls in the dust, covered with rags. No longer will a chosen few stride stately over their *Turkey Carpets,* with their thousand lights glittering through their *chandeliers,* controlling not man, but nature itself; converting night into day, whilst others are stealing along a dark alley, *meeting at some grocery,* reposing under some outer shade; or sheltering themselves from the inclemencies of the weather under the broad canopy of heaven. Then will true republican simplicity prevail. Then in the language of our forefathers, all men will be upon a perfect equality." George Trotter, the youngest of Lexington's successful mercantile family, signed the broadside. Comparing readers to "brave soldiers," Trotter encouraged them to "enlist" pro-reliefers like himself to state office.[39]

By 1820, pro-relief legislators dominated the state assembly and passed measures to aid indebted farmers which the state Court of Appeals subsequently declared unconstitutional. For four years, pro-reliefers worked to overturn the decision. In 1824, the state senate abolished the court, replacing it with a new court of relief advocates. The Old Court–New Court struggle continued until 1827 when the legislature reinstalled the original court and the economy returned to stable levels for most Kentuckians.[40]

But the Panic of 1819, the Relief War, and the Old Court–New Court controversy had long-term consequences. Since the 1780s, Kentuckians had been looking backward to the Revolution for their moral compass. The Founding Fathers and their rhetoric and idealism had imbued western development with national meaning. As blind patriotism, rampant commercialism, and growing individualism superseded the republicanism so central to the Revolutionary moment, society along the Maysville Road had become increasingly atomized. The cooperative *communitas* of log rollings, gristmills, militia musters, markethouses, and camp meetings disappeared from the landscape. When economic troubles and constitutional crisis struck, the self-interest demonstrated by entrepreneurs left many Kentuckians adrift. Citizens had choices in the elections of the 1820s—relief or bankruptcy, a court that represented the people or a court that represented the monied interests, a government of republican leaders who acted on behalf of the people or a government of democrats who were the people—but they hesitated to choose. Caught between the republicanism of the Revolutionary generation and a future of unknown principles and rules, most Kentuckians were indecisive. Many self-made men, including Fielding Bradford Jr., John Bradford's nephew and editor of the *Kentucky Gazette* in the late 1810s and early 1820s, were ready to jettison their historical baggage for something new, denouncing the continued influence of their Revolutionary ancestors as antiquated: "Those people drowned and burned *witches* and stood with their hats under their arms at the doors of great men."[41]

The middle class emerged in the late 1820s more thoroughly in control of economic and political life along the beaten path. That George Trotter could fashion himself as a friend to the common man, when the rampant investment and networks of credit so integral to his business contributed to the economic crisis, suggests how quickly men of commerce rebounded. No longer was virtue, talent, or republicanism prerequisite, as John Pope had assumed in the late 1790s. Politicians could set aside the notion of the disinterested public servant. Instead, they could openly pursue their self-interests and run for public office. Their ability to survive and even benefit from the hard times contrasted with the suffering of farmers and others who longed for the good old days. Many people were not happy. As one "Lover of Truth and Justice" put it, "Our forefathers have always sent lawyers and merchants to make laws for us, and we their children are doing the same; and God says he will visit the sins of the fathers upon their children even to the third and fourth generation—and it is by such that our laws are made, to justify thieves and robbers, to oppress the poor and crush the needy."[42]

Still, the middle class was not oblivious to the economic needs of their communities. In 1827, Lexington's trustees undertook the construction of two new markethouses. Nearly ten thousand square feet in size, the buildings straddled Town Creek at the western end of Water Street. As one observer remembered, "Each of these houses extended across one entire square.... At the same time, the people of the city—men, women, and children—poured in from every street and alley, all bearing in their hands bags, baskets, dishes, jugs, or plates, in which to receive whatever they might be able to buy.... The throng was now increased to thousands.... Many went to the markets to meet and converse with old friends from the country, and inquire for news from the different localities of the sellers; many went as spectators to catch the manners of *rustics* 'living as they rise.' All seemed to be interested and edified." The new markethouses served the same public purpose as their predecessors: to bring citizens together. Rather than trying to address the needs of a burgeoning urban population, however, the trustees of the late 1820s sought to heal the wounds created by a decade of economic and political rivalry. The markethouses provided venues for restoring a sense of community across lines of distinction. At the same time, they announced the continuing hegemony of the middle class.[43]

Renewed attentions to the refinement of Lexington impressed visitors. "In reality it is very beautiful," explained a visitor in 1837, "but in a positive, numerical way, a money-beauty, in short an American beauty." Susan Yandell, frustrated by the lines of distinction that impeded her own participation in certain circles, criticized the town's social elite for cultivating Lexington's reputation as a place "refined in manners and conversation." By the late 1820s, Lexingtonians proudly wore a mantle of elegance that blended the romantic gentility of the old gentry with their own acquisitiveness. As early as 1817, Samuel Brown remarked on the "fifty and sixty *villas,* or handsome residences in 'the country' neighborhood of Lexington" where older gentlemen, like David Meade, continued to "dress in the costume of the olden time.... He wears the square coat and great cuffs, the long court vest, knee breeches, and white silk stockings at all times: the buttons of his coat and vest are of silver with the Meade crest on them. Mrs. Meade had the long waist, the stays, the ruffles at the elbow and the cap of the last century." The dissimilarity between the "American beauty" of Lexington and the popular image of Kentuckians found in American culture perplexed a *Niles' Weekly Register* correspondent who detailed in 1814, "They have a theatre; and their balls and assemblies are conducted with as much grace and ease as they are anywhere else, and the dresses of the parties tasty and elegant. Strange things these in the 'backwoods!'—The houses are mostly of brick, and some of them are splendid edifices—one or two of the inns yield to none in America for extensiveness, convenience, and good living." Town residences, "commodious and comfortable, and the most of them far superior to those usually inhabited by farmers," successfully mimicked the grandeur of rural estates: "Many of them are surrounded by gardens and pleasure-grounds, adorned with trees and shrubs in the most tasteful manner; and the eye is continually regaled with a beautiful variety of rural embellishment." Five piano craftsmen thrived in Lexington's market; song and dance instructors prospered from greater interest in refined entertainments. Accompanying these private initiatives toward gentility were trustees' efforts to pave streets, lay sidewalks, and erect street lamps. When John Melish visited Lexington in 1811, he prophesied how the refinement of the town would one day be accomplished "as two-thirds of the inhabitants can compel the remaining third to agree to it." By the mid-1820s, his prediction had come true.[44]

The Revolutionary imagery that once elicited patriotism devolved into nostalgia; its heroes became less ideological icons than commercial gimmicks. Papermaker Ebenezer Stedman remembered how, after the Frenchman's visit to the Bluegrass in 1826, "Evry thing was *Lafayette*. All the new fashions were *Lafayette*. It Ran to such Extrems that you Could not By a Hat nor anny thing out of A Store But it was *Lafayette*." Kentuckians' sense of their Revolutionary heritage ceased being ideological and became commercial. The refinement of Lexington that culminated in the 1820s had come at great expense. With intensified class consciousness and with the mediating artisan ranks greatly diminished, society seemed split between the haves and have nots. "A deeper gloom hangs over us than was ever witnessed by the oldest man," reported one Lexingtonian.[45]

As the economic and political turmoil of the late 1810s and 1820s abated, society along the Maysville Road resembled an hourglass. When Mary Ann Corlis arrived at her family's new home outside Paris, the tea parties began and she thrilled at the "polite socialite women" with whom she rubbed elbows. The list she compiled, however, demonstrated that what remained of the gentry had acquiesced to the rising influence of the new wealth, merging into one upper class; the Breckinridges, Hunts, Popes, Clays, Harts, Mortons, Postlethwaites, and many others mingled with each other without distinction. Thus, in the upper globe of the hourglass positioned the genteel: gentry clinging to their social and political stations, a cohort of merchants, lawyers, and prosperous artisans desperate to claim a share of that elite status. In the lower globe were the bloated ranks of rural and urban working classes: less successful artisans, journeymen, grocers, farmers, laborers, and slaves. The funnel of the hourglass allowed grains of refinement and political opportunity to filter downward, but never in great quantity. In the stores along the Maysville Road, purchases of wallpaper and matching china, silk taffetas and upholstered furniture demonstrated that poorer customers identified with the merchant behind the counter or the wealthier customer who just left aboard a carriage. Yet, the sprawl of refinement exacerbated rather than blurred lines of distinction. Especially in Fayette County and increasingly in Old Bourbon, where Lexington and Paris grew more prominent with each passing year, class differences became most explicit. As David Meade explained, "you may live not only well but cheap—if you live in the Country & cultivate a Farm—but on the contrary to live well in a Town your expences will be as great as in the Eastern Towns." The goal of the refined good life was to demonstrate that one could live well in town or country, and to distinguish more clearly the unequal stations and dependent relations that characterized a respectable society.[46]

As the genteel appraised their success along the beaten path, one thing remained awkwardly and conspicuously wrong with the world they had created. The road itself as a lingering primordial feature of the region inhibited their abilities to import, export, civilize, and profit. The twists and turns of an old buffalo trade had to be erased and replaced.

THE VETO TURNPIKE

In the late 1790s, gentry-controlled county courts required select residents to maintain lengths of roads. In Fayette County, Hugh McIlvain was to keep the beaten path clear from Lexington's town limits to Owing's Pond; James Rogers oversaw the length from the pond to Bryan's Station; and

Elisha Warfield Sr. took care of the section to the boundary of Bourbon County. Along the road, overseers recruited citizens, through barter or cash, to clear and smooth the road to a width of at least thirty feet. Maintenance parties placed stone or wooden markers designating directions and mileage at crossroads. Bridges and causeways across milldams likewise became the responsibility of overseers who, with the power of the state assembly behind them, could "cut and take from the lands of persons adjoining [the road] . . . so much timber, earth and stone, as may be necessary" to preserve the thoroughfare.[47]

Overseers and other citizens made little effort to remove completely the impediments to travel; tree stumps and rocks continued to deface the trace. Regulations required that stumps be cut to approximately six inches in height and rounded so as not to scrape the axles of vehicles. Not until after the stumps rotted and disappeared was the surface leveled, turning workers' attentions to the holes and erosion precipitated by the loss of root systems and to the deep ruts incised by the wheels of carriages and wagons. Using rakes and hoes, road crews replaced lost dirt and smoothed the thoroughfare. But the proclivity of the old buffalo trace to muddy and erode quickly meant constant supervision and repair. In 1822, Horace Holley and his family traveled southward along a road "made muddy and uncomfortable. . . . The rich soil of Kentucky is the worst of all for roads in a wet season." The constantly poor condition of the road affected people in all walks of life from papermakers like Ebenezer Stedman who recalled how "thare was nothing but mud roads, it was a day's ride to travel twelve miles," to Holley who groaned that "Kentuckians . . . are the worst road makers in the world."[48]

A flurry of road-oriented activity by the state assembly beginning in the 1810s signified a meaningful cultural shift along the beaten path. The residents of Maysville were the first to push for aid from the state. The road up the bluffs behind the village was a constant nuisance, as the Kentucky House of Representatives recognized in 1809: "the water continually running down from the side of the hill, passing over the road, and there freezing—it becomes in cold weather so slippery, that it is both dangerous and difficult to pass it, either with Teams or on horse-back. And when the weather is sufficiently moderate to break up the roads, it is almost equal to an impossibility for loaded Waggons to pass at all; and is even extremely difficult to be passed by travellers on horse-back." Pressure from Maysville resulted in an 1811 act authorizing a lottery for improvement from the Ohio River to the southern boundary of Washington. The plan designated half of the profits to reducing the steep grade of the road where it climbed the bluffs. The assembly placed the project in the hands of legislatively appointed lottery managers, further diminishing citizens' direct participation, mirroring the decreased sense of communal obligation that accompanied individualism and the market revolution.[49]

Even as efforts began to address the road's condition, demands for internal improvement of the trail became more frequent. The region's middle class determined not to let the economies of Louisville and Cincinnati (both boosted by a new steamboat economy that churned along the Ohio River) surpass them. Displeasure with the condition of the dirt track and Kentucky's thoroughfares in general reached a crescendo in 1817 as the state assembly debated a bill to incorporate the Lexington and Maysville Turnpike Road Company. "Every avenue to seat of justice," complained the editor of the *Kentucky Gazette,* "is, for six months in the year, almost impassable." The assembly approved incorporation "for the purpose of forming an artificial road from Maysville through Washington and Paris, and thence to Lexington."[50]

The 1817 legislation stripped road clearing of its communal purpose, granting charters to private companies "for the purpose of forming artificial roads" and removing citizens from their investment in regional development. Although not acted upon for another decade, clearly a new attitude about the relationship of citizens to roads had arisen. In contrast to the natural track, an artificial road would be unnatural: its course dictated by politics, use, and economics. More people could acquire refinement; village residents had grown distant from nature; entrepreneurs were eager to profit in the construction of toll gates: all of these reasons probably played into the assembly's decision. By the early 1830s, demands for improved condition of the Maysville Road culminated in the Maysville Turnpike.[51]

"An artificial road by the best and nearest route" read the directions of the state assembly. By 1817, there was something anathema about the meandering course carved by buffalo and Indians, the inconveniences of mud and ruts, the conspicuousness of such a primitive road in the midst of a refined region. As an aboriginal trace, the Maysville Road had fulfilled its purpose. In order to keep pace with American development, a human-made road was necessary. The incorporation of the Maysville and Lexington Turnpike Road Company was a declaration against the most conspicuous vestige of natural influence in Kentuckians' lives. As one Bourbonite put it, "they must turn their attention to what nature has left undone." Little was accomplished, however. As Adlard Welby traveled along the beaten path in 1820, he noted how the "roads at present are altogether in a state of nature, the trees only just chopped off about a foot from the ground, and rocks, and stones, and gullied left to be got over as we can; no wonder then, that you see a blacksmith's shop every two or three miles, and tavern by the side of it put up and spend your money while the repairs are doing."[52]

Given the economic and constitutional problems of the 1820s, it was not until 1827 that the state assembly inspired action anew by incorporating the Maysville and Lexington Turnpike Road Company for a second time, requesting federal funds and guaranteeing additional capital from the state. Eager to make good with the new opportunity, in late spring, company officials hired James Darnaby and William Ellis to survey the entire length of the old buffalo trace and suggest a more direct course. The season was unusually wet as the two men began their task. Mary Ann Corlis wrote to her father how the road was so bad from "frequent and heavy rains" that she could not travel from Paris to Lexington. Through the muck and mud, the two surveyors pushed forward, authorized by the state assembly to "run out and mark said route." And so they did, creating a plat on which the actual road was drawn as a dotted line and the proposed route became a continuous solid line directing travelers along the Maysville and Lexington Turnpike and into the future.[53]

"A Plat and Survey of the Road from Lexington to Maysville" views the beaten path from Lexington in the center of the Bluegrass, unlike Victor Collot's map which made Maysville the point of origination. As citizens of Lexington, Darnaby and Ellis naturally began their survey in the "Athens of the West." Since the days when Collot decided that Lexington would never account for much, the town had become the largest and most commercially active in the state—the "*Bath* of America," as one British visitor proclaimed it in 1819.[54]

The manufacturing economy that had permeated Lexington and increasingly Paris made the stretch of the road between the two towns of utmost importance. On Darnaby and Ellis's map, the most notable dissimilarity between the road-as-it-was and the road-as-they-imagined-it-could-be is along that first length where they suggested a direct northeastward route that would eliminate the twists and bends of the old buffalo

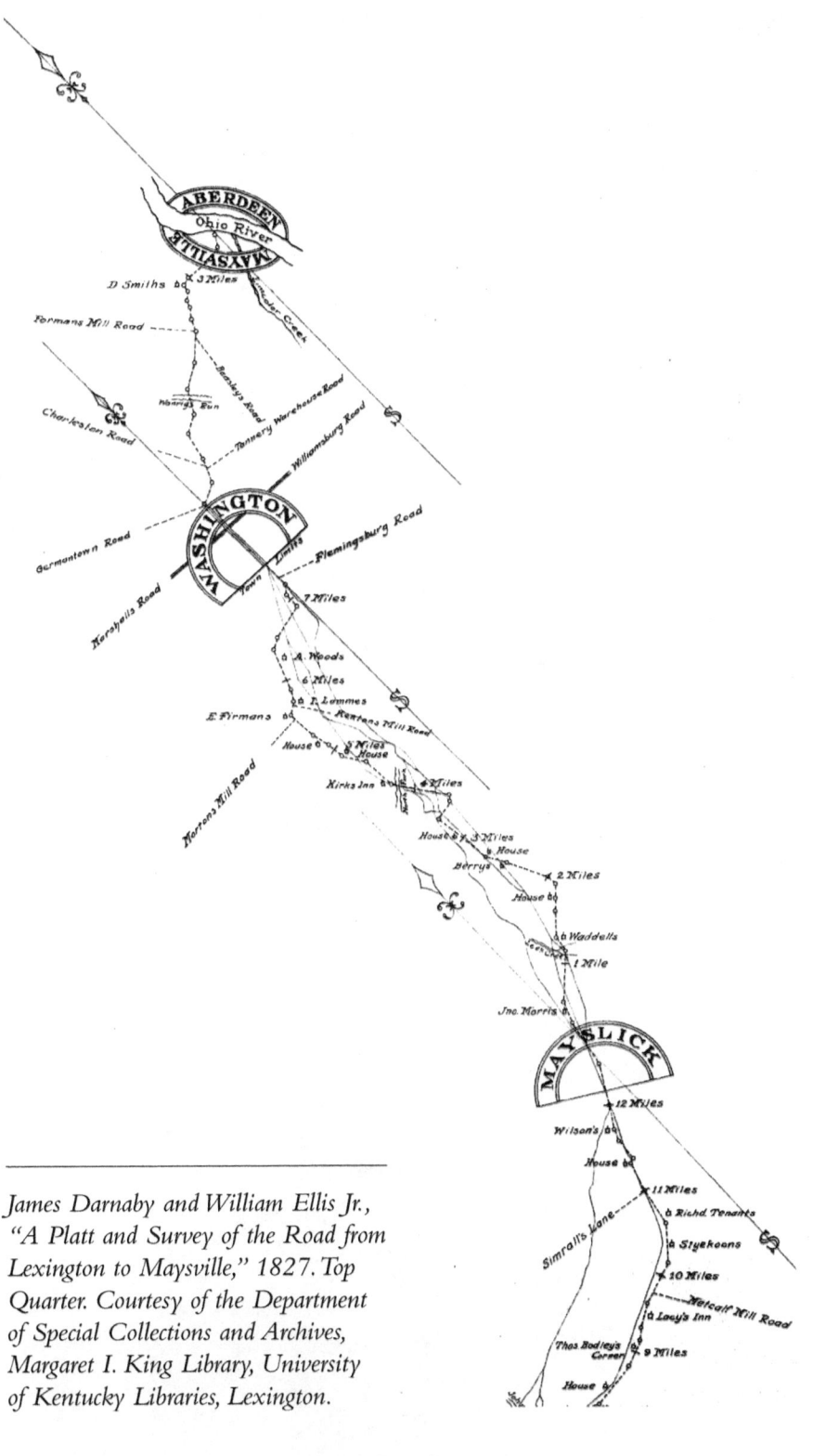

James Darnaby and William Ellis Jr., "A Platt and Survey of the Road from Lexington to Maysville," 1827. Top Quarter. Courtesy of the Department of Special Collections and Archives, Margaret I. King Library, University of Kentucky Libraries, Lexington.

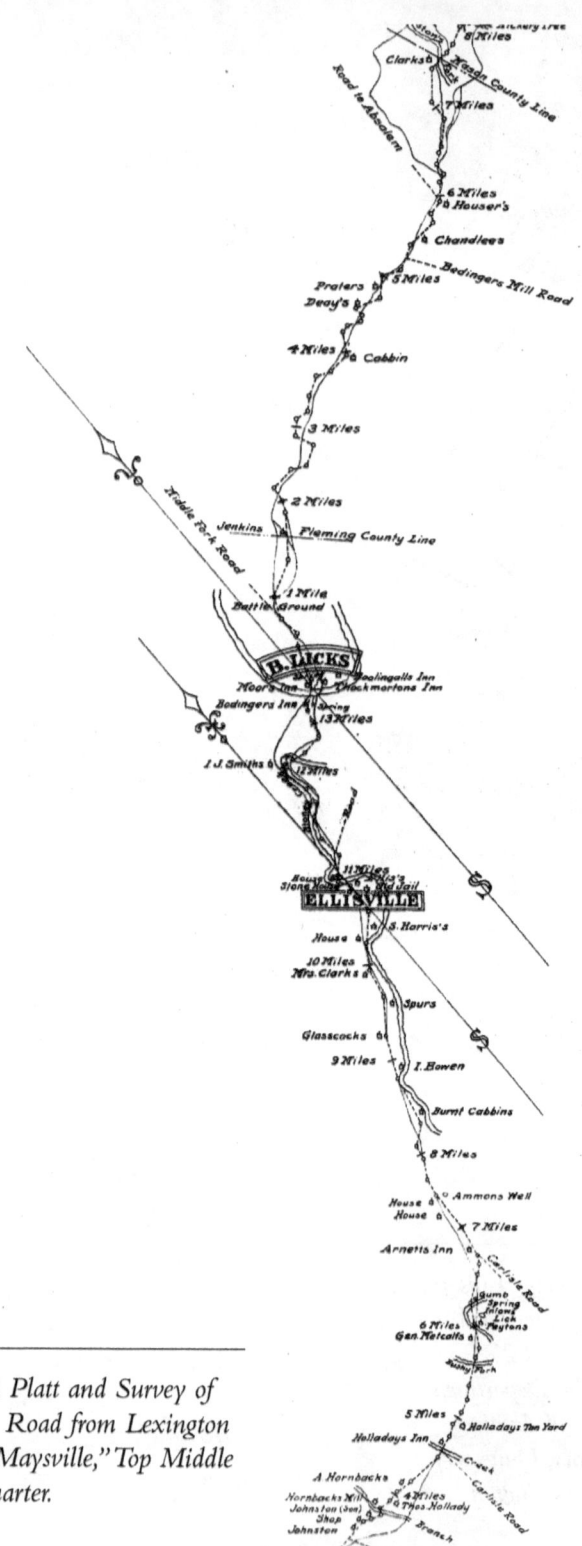

"A Platt and Survey of the Road from Lexington to Maysville," Top Middle Quarter.

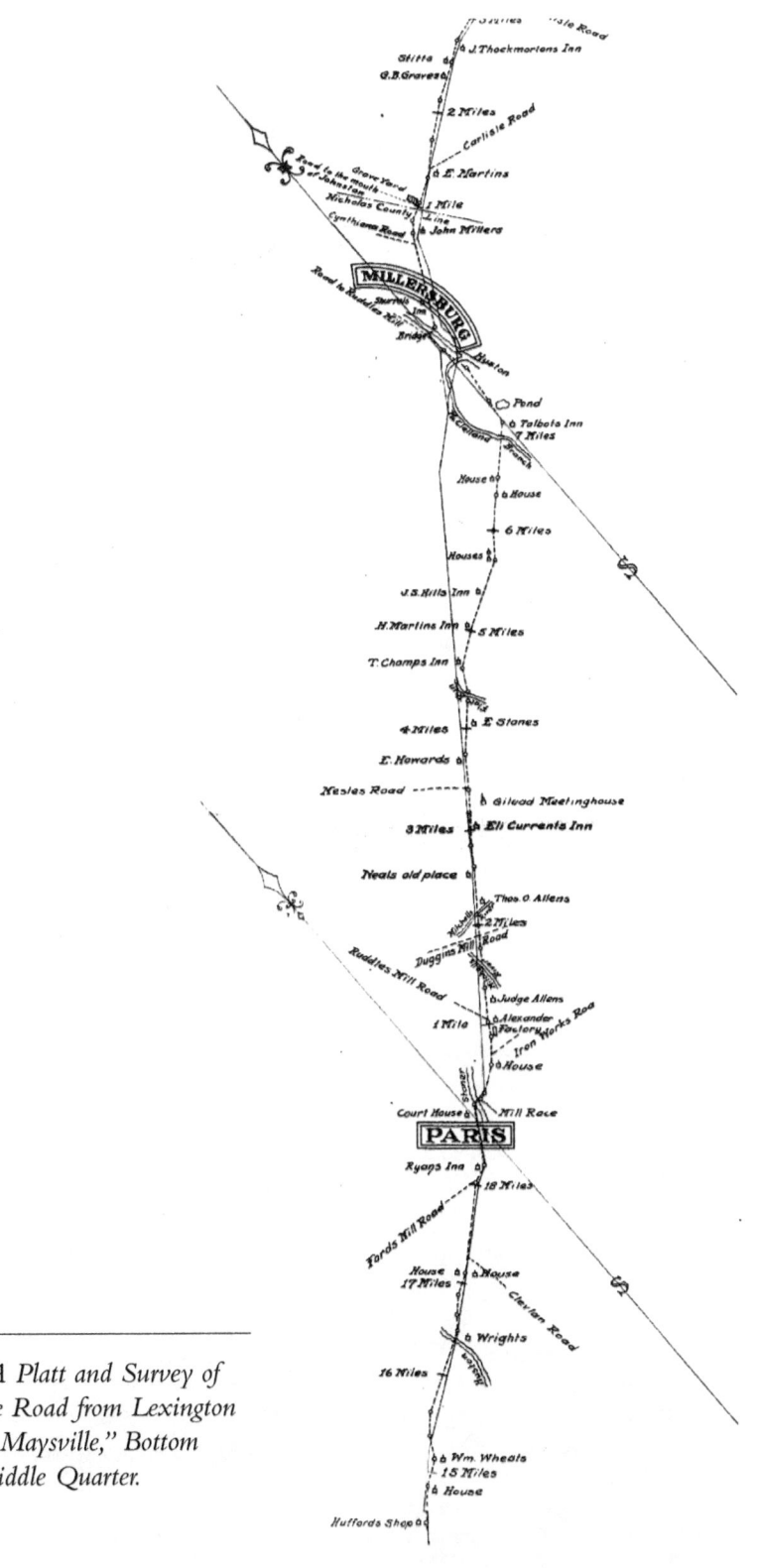

"A Platt and Survey of the Road from Lexington to Maysville," Bottom Middle Quarter.

"A Platt and Survey of the Road from Lexington to Maysville," Bottom Quarter.

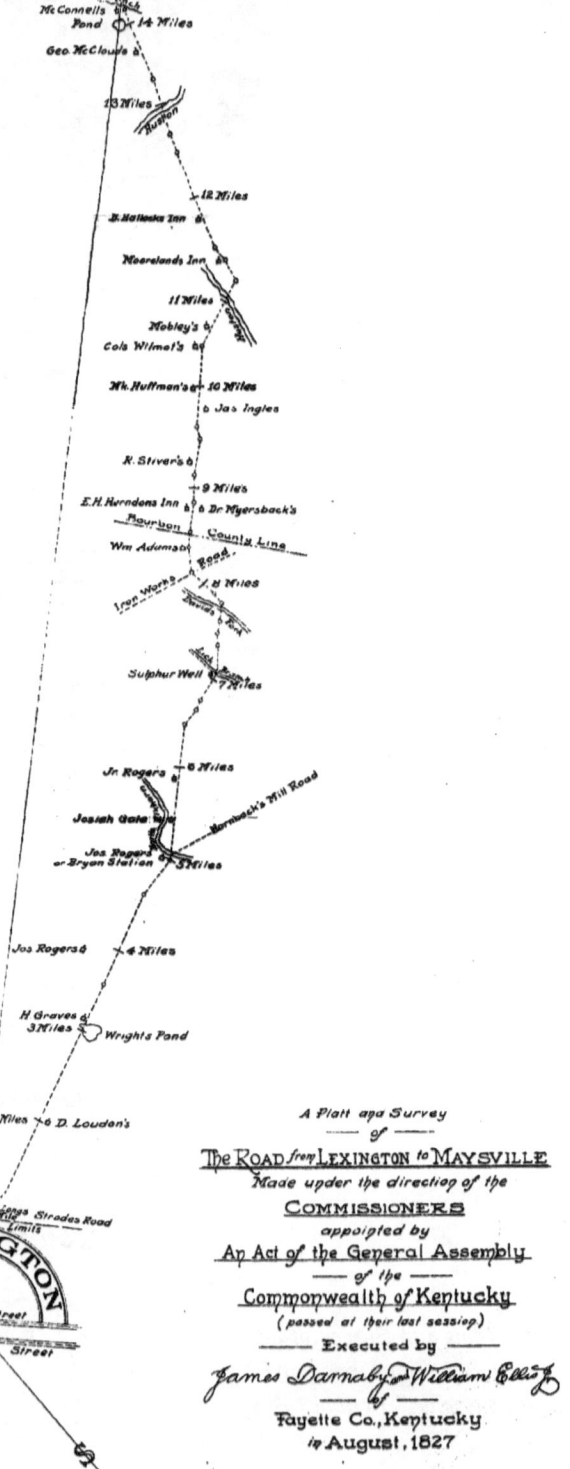

trace. Certainly, the topography of the Bluegrass allowed for a straighter course, but some credence must be given the notion that in its role as an economic conduit, the road between Lexington and Paris needed to be the "best and nearest route." Additionally, the landscape had changed: denser populations, more taverns and inns, greater refinement. The old buffalo trace no longer aesthetically fit. Susan Yandell compared the "fine houses and commodious barns" to "our journey through Virginia." The older Virginia gentry of Fayette and Bourbon Counties, aided by an eager middle class, had indeed replicated Tidewater and Piedmont landscapes, replete with a distinct hierarchy, large rural estates, slave labor, stately urban homes, and conspicuous consumption. The bustle of genteel life was apparent across the Bluegrass. "The finest equipages are seen dashing along the highways and the by-ways," noted one tourist as he moved northward from Lexington.[55]

The busyness of the road cannot be ignored on Darnaby and Ellis's map. Almost nine miles from Lexington, just across the Bourbon County line, travelers came upon Herndon's Inn. In the eleventh mile, Mooreland and Hallack's Inns provided repose and sustenance. On the outskirts of Paris, Ryan's Inn accommodated visitors. Like Postlethwaite's Tavern in Lexington, these enterprises provided infrastructure for road travel. Once upon a time, according to Mann Butler, "every farmer's house was a home for all," and indeed, every settler served as a landlord for sojourners and neighbors. By the late 1820s, however, an explosion of taverns, inns, hotels, wagon yards, and other support institutions lined the length of the old buffalo trace. Cordiality had become a commodity rather than a neighborly obligation, and the state assembly established prices on hospitality.[56]

Three miles north of Lexington was Wright's Pond which, over the years, had enlarged across the Maysville Road. Why residents had not rerouted the path around the pond is unclear, but in 1827, the Fayette County Court commanded William Burkley and William Smith to "show cause, if any they can, why the Maysville Road shall not be altered so as to pass around Wright's Pond and over their land, said pond being impassable, and said road having to pass through it." The two men must have successfully demonstrated reasons for, six months later, the court again took up the issue, this time directing the construction of a bridge. Workers completed it in late 1829.[57]

Upon arrival in Paris, Ellis and Darnaby came upon a replica of Lexington. Its manufacturing surpassed that of its southern neighbor in some categories; its lines of distinction were approximating the genteel "Athens of the West" as well. By the 1830s, Paris, "where the spirit of dictation prevails the most," had become a nest of "blue stocking gentry" of moral earnestness and reform sympathies (both by-products of the revivalism that racked Bourbon County). In their literary and conversational gatherings, bluestocking ladies exhibited a more political and participatory femininity than had female gentry. Mary Ann Corlis, whose political wit had once brought shame to her hosts and herself, would have felt at home. As in Lexington, class differences had disintegrated into rich and poor with the decrease of the artisan ranks. One citizen complained how the "merchants of Paris . . . import from Europe, from the Eastern Cities, and other countries, various articles which could, and should be manufactured in our own country. . . . [L]ook back but a few years, and compare your situations of the past, with the present time. Your villages were once crowded with journeymen, of almost every branch of mechanism, whose employers were animated with the lively hope of providing well for their families. But how is it now? Your journeymen have been compelled to leave the country." Alongside Old Bourbon's artisan exodus marched a cadre of small farmers. As "pasturage is extending its limits," large herds of cattle, horses, and mules wandered where yeoman farms once sat. By the mid-1830s, a grazier system would come to dominate the county and much of Fayette as well. "The rich men of the county are buying out the poorer classes of our citizens," reported the local newspaper in 1831, "and they are going further to the West." Wealthier farmers opened larger expanses of land to cattle, hogs, horse, and mules. By 1836, livestock sales exceeded $2 million. As a visiting New Yorker noted, "This placing their chief dependence upon flocks and herd gives something patriarchal to some districts of Kentucky." In the late 1820s, as Ellis and Darnaby passed through Bourbon County, a genteel, landed elite was laying the foundation for the economic transformation that would bring them, Bourbon County, and Paris full membership in the market economy.[58]

Both Fayette and Bourbon Counties also evinced a change in the mentality of their citizens. Family households became more atomized, distanced from community and economy by heightened notions that public and work space was distinct from domestic life. As early as 1810, Henry Clay's remarks hinted of the desire of self-made men to separate family from community: "Their manufactories ought to bear the same proportion, and effect the same object in relation to the whole community, that the part of his household employed in domestic manufacturing does to the whole family." In contrast to the days when mills, stores, and other family operations served both private and public purpose, reminding individuals of their moral obligations to community, by the 1820s, communal purpose and familial purpose were divorced, freeing entrepreneurs from moral responsibilities. Little could thwart agricultural and commercial development in the Bluegrass. "The farms are already so close together," complained a French visitor in 1820 about the lands watered by Elkhorn Creek, "that one would take it for an immense village." Even immediately north of Paris, the environs of the old buffalo trace became more cluttered with inns, taverns, mills, farmsteads, and meetinghouses before reaching Millersburg. Once again, in this northern stretch of Bourbon County, Darnaby and Ellis considered expedience most crucial to the vibrant commercialism of the region and proposed a more direct route.[59]

In Millersburg were the last accommodations of the Bluegrass: Edmund Martin's Tavern and William Samuel's Eagle Tavern. Then, the refined landscape and architecture of Fayette and Bourbon Counties faded. Once past Millersburg, the surveyors entered the shale hills where the solid line of Darnaby and Ellis's proposed route, with minor exceptions, coincides with the dotted line of the Maysville Road. But, of course, across the region's rough topography, there was little room for deviation. Taverns continued to line the road. Sitting on a hillside six miles from Millersburg was Forest Retreat, home to Thomas Metcalfe. In 1828, Metcalfe, a veteran of the War of 1812 and a progressive National Republican who wholly supported internal improvements and the programs for economic development devised by Henry Clay, won a narrow victory to become governor of Kentucky. Adjacent to Forest Retreat, he operated a stagecoach stop and tavern that, over the years, catered to Clay, Andrew Jackson, and William Henry Harrison among others.[60]

Some five miles farther, Darnaby and Ellis came upon Ellisville, the only new hamlet to have arisen along the road in thirty years. Situated along the banks of Stoney Creek, Ellisville received its name from the original owner of the lands, James Ellis. In 1804, he successfully petitioned the state to move Nicholas County's seat to his farm where commissioners laid out a village the following year, including a courthouse and a jail. In 1817, however, the legislature removed the county seat to Carlisle, a larger town some eight miles to the east, and Ellisville's fortunes deteriorated. By the time Darnaby and Ellis wandered through the hamlet, the old jail and Ellis's tavern remained the only extant public buildings.[61]

James Ellis's Tavern (c. 1807), Ellisville. Photograph by Rod G. Turner.

When comparing Collot's map with that of Darnaby and Ellis, one of the most insightful differences concerns Blue Licks. Collot found only the bleakness of the shale hills and a few salt works. The surveyors of 1827 came upon one of the most economically dynamic locales along the old buffalo trace. In the early 1800s, William Bartlett, a co-owner of Blue Licks springs, began to bottle and sell medicinal waters. A quarter century later, four inns surrounded the spa, accommodating guests from throughout the nation, but particularly from the southern states. The lines of distinction manifest in Lexington and Paris ventured regularly to Blue Licks where a craving for entertainment and relaxation transformed the hamlet into a resort. "There is a good deal of company here . . . [including] Mrs. [John] Pope," wrote William Barry on a visit to the licks in 1817. "We have three houses of entertainment here. I am at the most private, but not the best." The ranking of hotels, the acknowledgment of a senator's wife: both relate the sense of class distinction that mingled with the briny odors at the spas.[62]

Just north of the licks, Darnaby and Ellis acknowledged another particularly relevant place—the "Battle Ground." Their nod to the Blue Licks Battlefield reinforces the success of the gentry in transforming the disaster into a nostalgic commemoration of history. Aware of the rapid changes in the natural landscape, Kentuckians began preserving their past through histories, historical fiction, poems, and memorialization of historic sites. While no monument arose at the battlefield, citizens had protected the cedar grove through which their ancestors had fought and fled in the 1780s. The old buffalo trace ran through the center of the field, cartographically suggested by the road literally splitting "Battle Ground."[63]

Toward Washington, the two surveyors again proposed alternate routes for the new turnpike. Fewer inns lined this stretch of road, possibly making Darnaby and Ellis feel more at liberty to draw traffic away from establishments. The vitality of economic institutions along the road was a foremost consideration to those supporting a new turnpike. In 1829, before any work began on the Maysville and Lexington Turnpike Road, the state legislature officially restricted the corporation by dictating that "no such new way shall be established, if it shall pass over a different side of, or at a greater distance than the present road does, from any house now occupied as a tavern on the present road, unless with the concurrence of two-thirds of the justices present in Court, or with the consent of the proprietor of such house." By offering various routes for the turnpike, Darnaby and Ellis assured that the company could meet that obligation.[64]

While the environs along the southern end of the road had experienced tremendous revision since the 1790s, the landscape along the northern end of the road remained "barren and hilly." The influences of Backcountry Virginians, Marylanders, and Pennsylvanians in particular continued to define culture along the northern end of the road. Remarked one English visitor to Mason County in the late 1810s, "swearing, gouging, and whisky drinking is carried to a greater extent than in any State of the Union."[65]

Only Maysville had truly changed, particularly as the era of the steamboat took off. Residents had learned to accommodate nature. In 1819, James Flint commented on how "the houses stand above the level of the highest floods." The town evidenced the bustle of a port: warehouses lined Front Street; merchant-manufacturer John Armstrong operated an extensive ropewalk; a team boat, powered by eight horses on the south shore of the Ohio, ferried people across the river without their "quitting the stage"; the commercial district had a ropewalk, a glass manufactory, several stores and taverns, and a bank. In 1829, residents built a new markethouse, the second floor of which became the meeting room for the town's trustees.[66]

The steamboat economy most demonstrably affected Maysville as boat construction became a major industry. Like carriages along the beaten path, steamboats represented a new order for which many people were unprepared. When Ebenezer Stedman arrived in the port aboard a broadhorn, a steamboat churned past and "frightened all hands except the pilot For we had never Sean one nor heard one Before. The Splashing of the wheels on the watter, the Roar of Steam as it Escaped from the pipe, was enough to frighten anny one on the Rivver. The First appearance of a steam Boat on the Ohio River produced as you may Supose not a little Excitement and admiration. A steam Boat at that Day was to common observers almost as great a wonder as a Flying angel woold Be at presant." The emergence of the steamboat trade turned the attention of Maysville's residents and merchants from Lexington to Cincinnati. In early 1830, the *Phoebus* began a regular route between the two river ports. In May of that year, 155 steamboats docked in Maysville. Indicative of a new sense of regional importance, by 1833, residents appealed for and received from the state assembly incorporation as the "City of Maysville."[67]

In 1827, Darnaby and Ellis found no alternative to the steep, winding course of the road down the bluffs behind the river town; their solid line overlays the dotted line of the old buffalo trace. Their arrival in Maysville marked the end of their surveying duties. They prepared their map for submission to the company's commissioners and headed home. The new cut, the route by which Darnaby and Ellis proposed to redirect traffic between Maysville and Lexington, immediately became known as the Maysville Turnpike. Of course, minor deviations were incorporated into the design, but no longer would the meandering track of an ancient dirt trail direct human travel. Instead, a grand thoroughfare promised ease and convenience to travelers.

Even with a blueprint, constructing a road in the Early American Republic was no small task. Expectations were that, based upon the survey, Maysville and Lexington Turnpike Road Company's other employees—engineers and chain-bearers—would lay out a thoroughfare fifty feet wide, twenty feet of which would be "of firm, compact and substantial materials, composed of gravel, pounded with stone or other small, hard substances."

The laying of crushed rock to pave roadbeds was the invention of English engineer John MacAdam, and hence the process became known as macadamization. In order to set the twenty-foot swath of pavement, workers had to dig a ten-inch channel in which to pour the crushed rock and then tamp the entire road. The other thirty feet were to remain earthen surface for use during dry weather when traffic would presumably be heavier. Additionally, the company was responsible for four new bridges, hiring Lewis V. Wernwag to design and build the one-arch spans. Every five miles, the company could construct a tollhouse for the collection of fees. Milestones were to be erected at every crossroad; printed lists of rates were to be posted at each turnpike. Thomas Metcalfe predicted that completion of the project would "help us out of the mud & mire into which we hence to long have been most 'ingloriously' stuck."[68]

The construction of the Maysville Turnpike was at the heart of a shifting political culture in central Kentucky. In the 1790s, national affairs were the domain of the gentry as the coalescing commercial interests used boards of trustees to direct local and state developments. In the first decades of the nineteenth century, economic concerns became more national in scope, and the rising class of merchants, lawyers, and manufacturers began to demand greater roles in influencing regional and national growth. As the 1803 purchase of Louisiana guaranteed exporting through New Orleans and Jefferson's 1807 Embargo Act stifled international trade, Lexington's middling ranks sought out someone to champion their interests. They found him in Henry Clay. His outspoken support of internal improvements and commercial development rankled many who considered him solely motivated by the capitalistic impulses of self-made men. Clay's American System championed internal improvements, and the Maysville Turnpike early received his approval.

Not all residents were pleased with the plans for a new turnpike. After reincorporation of the Turnpike Road Company in 1827, rhetoric became particularly acerbic. A "few monied capitalists in the town of Lexington" were behind the plan, warned a Paris citizen. With the memory of the Relief Crisis still resonating, others questioned the morality of a turnpike corporation: "It has heretofore been often urged by old fashioned Democrats, that chartered monopolies are deadly to the interest of the public, and the people have been a thousand times told that they were, in creating such soulless mercenary institutions, forging chains for themselves." By June 1830, a discontent grumbled in the *Kentucky Gazette* that the improvements program for the beaten path was but a "petty scheme for squandering the people's money away." Suspicions of Clay and his motivations echoed along the beaten path; as one critic sarcastically described,

> Take notice . . . we, the truest, most able, wise national republicans, who think a national debt is a national blessing, residing in and about Lexington, with our great father, Henry Clay, whose fame . . . we fondly hope, will rival the fame of the George's and the Louis' and even the Czar of all the Russias . . . that a great road 60 feet wide and 60 miles long . . . shall be constructed from the great, most enlightened, & most refined city of Lexington, to the great stream Ohio, and to our well beloved town, known by the name of Maysville . . . in order that we may ride and roll along in our silver carriages, seeing we are very fat and rich, and the common roads of the country are rough, uneven, and jostle our bodies so much that we often become so fatigued in our rides for recreation and health, that we cannot take our wine and coffee, with as much pleasure as we could desire.[69]

Beginning in 1829, as central Kentuckians meandered along the old Maysville Road, they occasionally came upon construction crews working on a leviathan, the cost and benefit of which they could only imagine.

Clay reacted to such complaints, proudly positioned himself as leader of the proponents of internal improvements. In 1829, he griped that while Ohio and other neighboring states "are obeying the spirit of the age, and nobly marching forward in the improvement of their respective territories, we are absolutely standing still, or rather going backwards." In that same year, as he traveled from Lexington to Paris, Transylvania University professor John Roche commented upon how "the roads are intolerably bad. It seems to me that not only in order that the towns on the road, from Maysville to Louisville, by Lexington, should prosper, but even that they should hold what improvement they have gained, the road must be *turnpiked*." For Kentucky's commercial elite, Clay's ambitions to harmonize local needs with national goals, as epitomized by his ambitious American System, addressed their own aspirations. He expressed support for improvements to the Cumberland Road and its extension into a National Road in order to promote a "National scheme" that would include the Maysville Turnpike. The generation of self-made men who came of age with the War of 1812 could imagine that their needs were America's needs, that their dreams were America's dreams. As Clay phrased it, residents along the

Henry Clay. Courtesy of the Filson Historical Society, Louisville.

beaten path had the "double character of Americans and Kentuckians to support." Consequently, proponents had no doubt that the Maysville Turnpike was a national project, as both a mail route and an economic thoroughfare.[70]

Clay was only the most visible and vocal advocate of the project. While Darnaby and Ellis plodded through the mud, Thomas Metcalfe worked in the U.S. House of Representatives to obtain the funds requested by the state assembly in the 1827 incorporation act. In February, Metcalfe proposed an appropriations bill, but the House's winter session ended before debate began. By May, Secretary of War John C. Calhoun's commission of a survey for a western postal route that ran from Zanesville, Ohio, to Florence, Alabama, satisfied Metcalfe. The old buffalo trace was part of it, and a team of surveyors from the U.S. Engineering Department moved southward along the beaten path as Darnaby and Ellis plodded northward.[71]

Additionally, there was a cadre of self-made men who advocated the project in the village West. Many, in regular contact with Clay, organized town meetings where citizens debated the project and usually produced a common statement of support, as in late February 1831, when residents of Paris agreed "that the completion of the Maysville, Washington, Paris, and Lexington Turnpike Road, is an object to which the citizens of Paris look forward with great interest, and is one for the success of which they will devote as much of their energies and means as they can." This was not a disinterested bunch that pushed forward the Maysville Turnpike; the chairman and secretary of the Paris meeting respectively were John Martin and Hugh Brent, men who promoted livestock breeding as a new direction for Old Bourbon's economic development. In the 1827 reincorporation legislation, the state assembly had appointed these men and other local leaders as company commissioners in charge of subscriptions. The average Kentuckian could not afford one share of stock, priced at one hundred dollars. Rather, these men targeted wealthy acquaintances, like Samuel Williams of Paris, to finance the project.[72]

Sam Williams began as an artisan, completing an apprenticeship for a tailor in Philadelphia, fulfilling his journeyman years in Winchester, Virginia, and arriving in Paris in 1796 to open his own business. A Welshman, he found a wife within the Scots-Irish population that had settled in Bourbon County and, throughout the years, engaged his family in the house-raisings, quiltings, log rollings, weddings, and country church meetings of Backcountry culture. By 1806, he became involved in livestock breeding and helped to establish the Paris branch of the Bank of Kentucky. Two years later, Williams had prospered enough to purchase a 167-acre farm with a 10-acre orchard outside Paris where he constructed a two-story, blue ash house with "every nail . . . made of wrought iron at a blacksmith shop." Within the next few years, he also owned the first two-story brick building in Paris in which he housed his tailor shop and a store, another 93-acre farm adjoining the town, and a ropewalk and bagging factory with Charles Metcalfe, brother to Thomas. In 1812, as the hemp industry began to show profits, Williams built a two-story rope, bagging, and hemp fiber warehouse, a two-story ropewalk, a loom house, and a block of small brick buildings in which he housed his hired-out laborers. In 1816, Sam Williams had become so affluent that he contributed the sixty-five-foot steeple and bell for the town's new courthouse.[73]

When the relief conflict arose, Williams became a familiar face among Bourbon County's antirelief crowd. It was he who sued for invalidation of the relief laws, winning at both the county court level in *Williams v.*

Blair (1822) and at the Kentucky Court of Appeals in *Blair v. Williams* (1823). His investments and his son's recollection that his father loved George Washington, but had a "disciple's veneration" for Benjamin Franklin, certainly attest to Sam Williams's self-made identity. Embracing an entrepreneurial work ethic, Williams's "spartan self-discipline gave a rooted contempt for all fripperies, frivolities, gands, shams, and ostentations," recalled his son. "He found offensive all effeminate self-coddling men, fiddlers, aimless idlers, slaves of vice, and all occupations catering to them." He proudly labeled himself a "Henry Clay Whig," and when the Maysville and Lexington Turnpike Road Company formed in the late 1820s, he quickly became a stockholder.[74]

With the enthusiasm of men like Samuel Williams, commissioners had little trouble raising money. Yet, even though political and financial support appeared unquestionable, the Turnpike Road Company again hesitated. Commissioners made no effort in 1828 to begin converting the Maysville Road into the Maysville Turnpike. Of particular concern remained the bluffs that trapped Maysville against the Ohio River and that made travel into the interior difficult. In 1809, the Kentucky House of Representatives had debated a citizens' petition for improvements. Because the road climbed the north side of Limestone Hill, "in the winter season, the rays of the sun scarcely ever fall upon the road." The condition of the beaten path had changed little since the 1780s, but the urgency of improvement had grown stronger. "When it is recollected that this is the great inlet to this commonwealth from the Atlantic States, and that an immense quantity of goods for the consumption of a large portion of this state, and numerous articles of domestic production for exportation, are received and sent by this route," the state legislature noted in 1809, "a good road can never be expected to be made up the said hill, in any other way than by levelling, and firmly paving the same—having thereunder a sufficient number of arches, to carry off the water from above." Despite this early recognition of the problem, another two decades passed before officers of a turnpike corporation broke ground. Frustrated entrepreneurs of Mason County took matters into their own hands, petitioning for and receiving from the state legislature in January 1829 a charter for the Maysville and Washington Turnpike Road Company. Workers began macadamizing the old buffalo trace up the hill behind Maysville. It took nearly two years to complete the four miles to Washington, but physical construction of the turnpike had begun.[75]

Something more symbolic simultaneously occurred: the project began with a Fourth of July ceremony, and improvement of the Maysville Turnpike became associated with national patriotism. Coincident to their identification as important Americans in the wake of the War of the 1812 was Kentuckians' conceptualization of their region as the crossroads of a nation. One argument for public investment in Kentucky's internal improvements was to draw "citizens of the East and West, the North and South, nearer together, thereby identifying more and more their interests." As Thomas Metcalfe spun the debate, the Maysville Turnpike, "national in its character is calculated to benefit in an eminent degree, not only the citizens of Kentucky, but of many other States belonging to the confederacy." Improvement proponents elicited opposition from Kentucky Democrats who distrusted arguments based on a claim to the sacred legacy. According to Congressman Thomas Moore, projects like the Maysville Turnpike profited "a set of men who assume a superiority over their fellow citizens.... *The people ought never consider themselves safe.*" The collusion between Metcalfe and Clay was reprehensible: "A Governor devoted to him [Clay] and acting under his influence, in gross violation of public opinion, has nominated the bitterest of Mr. Clay's partisans for almost every office of honor, profit, or influence."[76]

On the face of it, internal improvements appeared an economic issue, but at heart it struck at concerns over political economy, particularly lingering worries about corruption and the consolidation of power. Even though their political and social status had begun to deteriorate with the Constitution of 1792, most lower-class Kentuckians viewed the economic and political crises of the 1820s as the final, debilitating blow. Buying into the commercial culture of their era, they had speculated, invested, and consumed conspicuously, only to suffer in ways that merchants and lawyers seemingly evaded. Now, in the late 1820s, self-made men attempted to convince poorer Kentuckians that, even though they might not be able to invest and speculate in the project, they should politically support it. A Maysville circular of 1826 called upon "farmers and citizens" either to purchase shares in a future turnpike company or to "pay the amount in labour on the road." For what purpose? To open a "new era to western commerce" was the response from Maysville's merchants. In Fayette County, according to the chairman of Lexington's committee on internal improvements, "the example of subscribing will be set by some of the Capitalists of the town." For Kentuckians disenchanted by the relief crisis, however, entrepreneurs demonstrating the proper forms of investment appeared more likely to hoodwink rather than benefit them. At a turnpike meeting in 1826, an observer noted that "sneering and squinting at the motives of the movers of this undertaking were visible out of doors." For Jacksonian Democrats, the toll gates of a federally sponsored turnpike exemplified the encroachment of capitalism upon local life and the dependence of local life upon a market economy. For most residents along the Maysville Road, Metcalfe's portrait of the turnpike as integral to the national road system was incomprehensible; they used it for short-distance travel—it was, foremost, a local road.[77]

Work on the four miles from Maysville to Washington proceeded amid this ideological turmoil. Clay successfully pressed Lexington's trustees to initiate a similar project from their end, and in October 1829, macadamization began on Mulberry Street (which became the Maysville Road at the town limits). A committee organized to plan future road improvement to the Bourbon County line; and citizens in Old Bourbon began work as well. In the absence of centralized planning, both counties improved the road along their own designs. When, in 1829, Bourbon County road crews, having followed Darnaby and Ellis's suggestions, completed a new cut from Paris to the Fayette County line, they were dismayed upon finding no road with which to connect![78]

Since the Maysville and Washington Turnpike Road Company appeared the only unifying agency, in January 1830, the state legislature authorized a subscription of $25,000 to finance the company's work. At the same time, the assembly chartered a company to construct a railroad from Lexington to a point on the Ohio River. As Thomas Bodley, wealthy Lexington merchant and proud proponent of internal improvements, elaborated, it was "the most liberal charter ever granted, and many of our most wealthy citizens are sanguine they can carry it into effect. Indeed, I think they are *mad* on the subject—for my own part I would prefer their capital should be employd in *McAdamising*." As the railroad corporation began its work, Bodley and other commissioners for the turnpike company proceeded with their plans. By the end of April, they had raised $67,000 at stockholder meetings in Paris, Lexington, Maysville, Millersburg, and Nicholas County. In early May, Bodley joined commissioners in Paris "to locate the Turnpike road from Maysville to Lexington. Our meeting is to decide whether the turnpike is to be on a straight line, or with the old road, from Paris to Lexington."[79]

By May as well, expectations were high that a federal bill, allocating $150,000 for stock in the company, would pass the U.S. Congress. When combined with the $200,000 of stock already purchased, as Bodley anticipated, the federal money would ensure that "the work will go on rapidly & its calculated to complete it two years." As the commissioners finished their work on the southern end of the road, accepting Ellis and Darnaby's plans for a direct route between Paris and Lexington, news arrived that President Jackson had vetoed the bill with a lengthy condemnation of the project and its sponsors. Tracing his position to precedents set by Thomas Jefferson and James Madison, the president of the common man rejected the bill because of its unconstitutionality and lack of national benefit. Certainly, the acrimonious relationship between Clay and Jackson did little to entreat the president to the bill, but Jackson had a legitimate constitutional concern as well.[80]

For Clay and other National Republicans, the Maysville Turnpike veto had a particular sting because, in the years following the War of 1812, internal improvements had become a gauge of the type of political economy they worked to create. In an earlier era, between the Revolutionary War and the second war with Great Britain, Federalists and Jeffersonian Republicans alike agreed that roads and canals, lighthouses and cleared rivers were necessary for liberty and union. Republicanism was strong, and it inspired men like Jefferson and Madison, who were typically suspicious of consolidated power to promote internal improvements as part of the nation-building process. In fact, only through such proposals could republican fears of reversion to primitivism or advancement to manufacturing be assuaged. The two men wanted internal improvements, but they believed that the Constitution did not explicitly grant such power to the federal government. Both encouraged Congress to pass an amendment establishing authority over internal improvements. Expecting eventual acquiescence, in 1808, Jefferson instructed his secretary of the treasury, Albert Gallatin, to devise a comprehensive, federal internal improvements program. Gallatin proposed a ten-year, $20 million project that, upon completion, would have economically united the nation's diverse regional economies. His plan faltered as the Embargo of 1807 and the War of 1812 drew attention, energy, and money away from the domestic program.[81]

When Madison became president, he still championed a virtuous government sustained by a republican economy in which the citizenry (made independent by a yeoman, agrarian order) placed their faith in their representatives (who set aside their own self-interests out of gentlemanly honor).

After the war, however, Clay put forth a new plan for national development: a proactive government that sustained a capitalistic economy, in which the citizenry (eager to profit from the opportunity of manufactures and markets) depended upon their representatives (who advanced their own class and local preferences to provide that opportunity). Internal improvements, protective tariffs, banking policies, and manufacturing incentives—although they could be framed as nationalistic projects, the planks of Clay's American System rewarded the entrepreneurs. If a loose interpretation of the Constitution were necessary to ensure that reward, so be it. Through Madison's eyes, this younger generation abandoned principle for expediency.[82]

The principles had changed, however. No less than Madison, Clay and his colleagues sincerely hoped to preserve the union and secure liberty. Through regional and local economic development, the American System would unite the states in mutual betterment. The two most influential events of their generation—the War of 1812 and the Panic of 1819—had taught the National Republicans valuable lessons about economic independence. As he conceptualized his American System in the early 1820s, Clay sought to redress unemployment and foreign market closures, the problems inherent to a capitalist economy, by cultivating regional economies that complemented each other. In 1824, in debate over a bill financing a survey of roads and canals, he pronounced, "All the powers of this Government should be interpreted in reference to its first, its best, its greatest object, the Union of these States. And is not that Union best invigorated by an intimate, social, and commercial connexion between all the parts of the confederacy?" The answer, of course, was yes.[83]

The fading of republicanism did not leave the self-assertive program of the National Republicans unchallenged. Another vision of political economy emerged, premised on the notion that only seldom did those without wealth enjoy the profits of investment and development. The Jacksonians arose in response to the inequities of the market system. "We are the friends of the farmers and mechanics," proclaimed Kentucky Democrats in 1828, "and it shall be our constant business to watch, protect, and defend their interests, against a heartless aristocracy." Claiming the sacred legacy, particularly the egalitarian forces unleashed in the wake of the Revolution, Jacksonian Democrats promoted a government sustained by a laissez-faire economy in which the citizenry (made independent by their own citizenship) could represent themselves (since those in positions of power were usually self-serving and corrupt). Accordingly, in his veto of the Maysville

Turnpike, Jackson argued that a national government should pursue projects for *all* Americans, and that state governments should finance those improvements that served their own citizens: "is it national, and conducive to the benefit of the whole—or local, and operating only to the advantage of a portion of the Union?" Although he expressed support for internal improvements and called for a constitutional amendment to empower Congress to pass such legislation, under the current Constitution, he believed he had no choice but to declare the Maysville Road wholly within Kentucky's boundaries as undeserving of federal assistance.[84]

From his editor's office at the *Kentucky Gazette,* George Trotter served as the regional advocate for Jacksonian democratic principles. Before the president's veto, Trotter had denounced national funding for the Maysville Turnpike as fiscal irresponsibility: "There was before Congress at the present session bills of this sort involving an expenditure of TWENTY SIX MILLION OF DOLLARS! Yes TWENTY SIX MILLION OF DOLLARS asked for to construct roads in different parts of the country. For the passage of the Maysville Turnpike bill the friends of Mr Clay exerted all their energies, and for what purpose? It was to involve the nation in bankruptcy: for if Gen Jackson should sign the 150,000 for the Kentucky Turnpike it would follow of course that he must sign appropriations for *twenty six million more,* for roads in other states; roads to which *we have no sort of interest.*" Suspecting that Kentuckians would receive proportionately few of these internal improvement funds in the long run, Trotter promoted the road as a local project and encouraged fellow residents to "put our shoulders to the wheel before we call on Hercules to help. We all feel deeply interested in completion, and our farmers should promptly come forward and subscribe what they can conveniently pay."[85]

Reactions were not so clearly cut, however. In early June 1830, an effigy of George M. Bibb hung from a makeshift gallows in Maysville. Bibb, a former ideological colleague of Clay on issues like war with England in 1811 and restitution to dispossessed landowners in 1823, had voted against the improvements bill. A crowd of over two hundred cheered the hanging of Bibb's effigy before placing it in a coffin and interring it "with great solemnity and profound silence in the middle of the Turnpike." A similar scenario played out in Millersburg. What was most fascinating to observers, however, was that "Jackson men took the lead in this disgraceful business. All this is for violating previous pledges and instructions." The mobs consisted of Jacksonian Democrats *and* Clay National Republicans, cooperatively expressing their anger over events surrounding the veto.[86]

When news of Jackson's veto reached Lexington, emotions ran high. Although he printed none of the criticisms in the *Kentucky Gazette,* there were enough made in public and private circles to push Trotter to defend Jackson's veto and his own earlier comments. On June 21, National Republicans rallied in support of Clay, penning resolutions denouncing Jackson and those who supported him in Kentucky's congressional delegation. Trotter, in reporting the affair, took the opportunity to mock the partisan leadership, especially when the speaker

> contended with all the suavity of manner and elegance of gesture so peculiar to himself that they [the Jacksonians] were "traitors—traitors—traitors." The indignation of all was aroused; every eye fiercely glistened at the recital of their wrongs. He spoke *feelingly* of the *poor innocent Indians.* The faithless conduct of the Government and the wickedness of the Administration ... When he arrived at that part of his discourse relating to the removal of the Indians, his voice faltered.... The tears streamed from his eyes and a groce of handkerchiefs were immediately drawn to his assistance. He resumed his seat amidst the half muffled and half suppressed sobs of the concourse of patriots.... But to be serious. The meeting was all a farce. Mr. Clay had issued his edict; his slaves and minions were summoned to attend.

Trotter had condemned Clay in earlier editorials for supporting the Maysville Turnpike bill in order to force Jackson to make a difficult choice between signing an unconstitutional bill or angering the electorate. The paucity of discussion at the rally over the road itself and the hypocrisy of condemnations over Indian removal from men whose parents benefited from Indian wars confirmed to Trotter how staged the entire event had been.[87]

By July 1830, reality had set in among National Republicans that no federal money would underwrite the Maysville Turnpike. Although rumors had circulated days previous to the veto that Jackson would not sign the bill, Clay claimed to be "shocked and mortified." He made a circuit through Ohio in July and August, rallying citizens against Jackson's indifference to western circumstances. At home, his friends tried to circumvent the veto by mustering grassroots support for their project. "Genl Jackson having put his *veto* on the Maysville Turnpike bill, has produced great excitement here," Thomas Bodley explained to his son. "He has lost about half his friends in

Mason Nicholas Bourbon & Fayette from that & other causes. . . . But the road & Internal improvements will go on, with Spirit in this State, the Presidents opposition to them, *to the contrary notwithstanding*." "What will that man deserve who shall procure the funds by subscription or loan to finish the *Veto* road?" contemplated Leslie Combs, one of Clay's fellow Lexington attorneys. In February 1831, he wrote merchant Andrew January of Maysville to coordinate subscription efforts. January's co-commissioner, John Armstrong, replied on his behalf and assured Combs that "this evening we had a Town meeting & there was not a Dissenting Voice against the corporation taking Stock to the amt of $30,000. Thus Sir you see we are in good Earnest. If the corporation of the other Town will do likewise we have thus enough Independent of Andy Jackson." Despite Armstrong's reassurance, not until the state legislature purchased another five hundred shares of stock was the company able to continue its work. Eventually, the state paid $213,200 into the Maysville Turnpike, half of the total cost.[88]

Armstrong's response to Combs also indicated that the length of the improved road covered only "4 miles, 35 poles" by 1831. In the meantime, the railroad project drew both political and financial attentions away from the Maysville Turnpike. By April, promoters subscribed $600,000 towards a railroad that would connect Lexington to Cincinnati. Not surprisingly, Clay was selected as chair of the new railroad corporation. Even the most vocal supporters of the Turnpike were captivated by the railroad. "We had the most splendid procession, at laying the first rail of the Lexington & Ohio rail road, ever Witnessed in this place, it far exceeded LaFayette's parade," wrote Thomas Bodley. "This event is considered the Regeneration or new birth of *Lexington,* its looking up already." "Last Saturday the corner stone of the railroad was laid," wrote Susan Yandell to her father six months later. "It will infuse new life into every thing, and we shall have the advantages of Nashville or Cincinnati without any of the disadvantages." And then, as an afterthought, she noted, "I believe the Maysville, or Veto turnpike as it is sometimes called is progressing."[89]

The turnpike company continued its efforts, making life tumultuous for those who lived and traveled along the beaten path. The many innkeepers whose institutions lined the Maysville Road found they could not depend upon the company to abide by the restrictions originally placed on the corporation. Less wealthy men like William Mooreland scrambled to find a piece of land on the new cut to relocate his inn. He eventually paid $1,300 for a lot where he opened the "Ten Mile House." More influential men like Thomas Metcalfe used their sway to maintain their businesses. Metcalfe persuaded the company to route the road past his Forest Retreat inn, consequently bypassing Carlisle, the new county seat of Nicholas County. Forest Retreat flourished, Millersburg received the benefits of being the major village between Paris and Washington, and Carlisle floundered. County courts, empowered to condemn lands for the purposes of the turnpike, became the nemeses of property owners who knew too well the fragility of landownership in Kentucky. Travelers, too, dealt with the upheaval caused by the massive project. As soon as the company completed a five-mile stretch, a tollhouse went up and collected rates from travelers who then found the rest of the road in horrid condition. When, in 1833, Englishman Edward Abdy traversed the Maysville Turnpike, about "two-thirds of the road were macadamized, and in excellent order; but the remainder was infinitely worse than any I had yet seen."[90]

By Abdy's visit, the incomplete sections of the road were through the shale hills around Blue Licks. The most challenging problem was a bridge to span the Licking River. An unknown observer described the workmen's first efforts: "the frame of splendid bridge, of a single arch, over that river, and which, had it been finished, would have been one of the finest in the United States." But a spring flood dislodged the abutments and swept the structure away.[91]

By 1834, with the exception of unimproved sections and the Licking River crossing, the Maysville Turnpike assumed "a line so direct, that the slight sinuosities of its course are scarcely discernible to the eye" and a course "so smooth, that the traveler who thinks proper to close his eyes, is not aware of the rapidity with which he is whirled along." Yet, it was this writer's depiction of the environment that was most telling. "Not a spot remains in its original state of wilderness; but everywhere the hand of art is seen to have exerted its energies with an unusual vigor and felicity of execution," he explained. "Every foot of ground has been adorned or rendered fruitful." With the final sections of the Maysville Turnpike macadamized, the conquest of the wilderness and the conversion of civilization begun half a century earlier was complete.[92]

The turnpike's tollhouses symbolized the new order along the old buffalo trace. They provided social milieus where the gossip of the road circulated: discussions about the weather, harvests, town visits, lost livestock, and passing strangers. But under the scrutiny of a company employee, the mobility that had once distinguished westerners as independent people became regulated. Along the Maysville Turnpike, no one could move freely, neither the African American slave nor the wealthiest gentleman. At each tollhouse, a pike obstructed road traffic. The toll gatherer rotated the log for travelers who paid their fares, but otherwise he was empowered to stop anyone from passing and to lay a fine upon those who tried to circumnavigate the pike. There were exemptions. Local residents moving from one part of a farm to another, going to worship or a funeral, or acting in public capacity as militiamen, jurors, or going to vote could pass freely. But for all others, the state assembly had a thorough list of toll rates:

> For every twenty head of sheep or hogs, six and one fourth cents,
>
> for every ten head of cattle, six and one fourth cents;
>
> for every horse or mule laden or unladen with rider, or leader, four cents;
>
> for every sulky, chair, chaise with one horse and two wheels, nine cents;
>
> and with two horses, twelve and a half cents;
>
> for every chair, coach, phaeton, chaise, stage, waggon, coachee or light waggon with two horses and four wheels, sixteen cents;
>
> for either of the carriages last mentioned with four horses, twenty five cents;
>
> for every other carriage of pleasure under whatever name it may go, the like sums according to the number of wheels and of horses drawing the same;
>
> for every sleigh or sled, three cents for each horse drawing the same;
>
> for every cart or waggon, or other carriage of burden, the wheels of which do not
>
> in breadth exceed three inches, five cents for every horse drawing the same;
>
> for every cart or wagon the wheels of which shall exceed in breadth three inches, and not exceed six inches, four cents for each horse drawing the same;
>
> and for any cart or waggon, the wheels of which shall exceed the breadth of six inches, two cents for each horse drawing the same;
>
> and when any such carriage as aforesaid shall be drawn by oxen or mules, in the whole or in part, two oxen shall be estimated as equal to one horse, and Every mule as equal to one horse in charging the aforesaid tolls.[93]

The toll schedule reflected estimated costs of damage to the macadamized surface. Drovers, whose herds of domestic animals assisted road crews by compacting the macadamized surface, paid minimal fees; as did wagons with wide wheels. Wagoners, whose vehicles with less broad wheels tore ruts into the road, evoked higher toll rates. And pleasure carriages, with wheels of narrow gauge and light loads, tended to spit stones out of the roadbed, thereby eliciting the highest rates. No matter the vehicle, however, the cost of road travel was no longer free. The turnpike was a luxury that many could not afford. Consequently, gatekeepers often filled their pantries with agricultural productions that individuals bartered for passage. Many sojourners avoided the turnpike altogether, traveling at times along the remnants of the old Maysville Road even as it fell into disrepair. The old buffalo trace, once indistinguishable from the wilderness itself, deferred to the Maysville Turnpike, inseparable from the capitalistic world it was designed to serve.[94]

But the turnpike was too late; the future of self-made men in central Kentucky had already begun to suffer. Tollhouses marked the ascendance of capitalism, but not until 1838 did subscribers begin to see meager returns on their investments. In spite of the ostentatiousness and actual wealth of some central Kentuckians, observers of the mid-1830s interpreted declension, not vitality, along the Maysville Turnpike. "I should prefer some new & growing place tho' to one so perfectly at a stand as is Lexington," expressed one aspiring entrepreneur. The languishing economic situation of Lexington and the booming cities of Louisville and Cincinnati boded poorly for Samuel Williams and other entrepreneurs who sought more.[95]

FLIGHT OF THE DESTROYING ANGEL

When Darnaby and Ellis produced their 1827 survey for the Maysville Turnpike, for the first time on any map of central Kentucky, mapmakers sketched a graveyard—an ominous foreshadowing. It sat at the line separating Nicholas County from Old Bourbon. Like the days when the two surveyors plodded along the beaten path, by mid-June of 1833, a steady rain had drenched the region for nearly five weeks. Southward through puddles and mud moved throngs of people matched in numbers only by those who had traveled to Cane Ridge in 1801, their quickened pace revealing the urgency to their collective migration. As Francis Goddard, a Bourbon County citizen, remembered, "The Dreadfull Disorder called the Collera was Raging . . . and was verry fatal in the upper counties." On May 23, the disease made its appearance in Maysville. Within two weeks, sixty residents died as others barricaded themselves in houses or took flight. Yet, cholera could not be escaped so easily as it moved across the countryside hidden in the bowels of fleeing victims. Washington, Mayslick, Blue Licks, and Millersburg quickly became the next milieus of death. Those who fled jammed into overcrowded accommodations in the villages and towns, carrying the cholera with them. As Goddard elaborated, "Some came to our county, the people were seised with Great fear and consternation, our Meetinghouses were Crowded." Within two weeks of cholera's appearance in Kentucky, the unwelcome visitor arrived in Paris and Lexington.[96]

The cholera epidemic of 1833 literally debilitated families and communities. Although the environmental changes around the turn of the century and the political changes of the 1820s had worn away much of its luster, the regenerative potential of the Edenic ideal remained part of Kentuckians' romantic vision of their chosen homeland. Cholera destroyed the final element of that ideal: an implacable belief in the healthfulness of the region. Nature seemed to rebel against white and black Kentuckians who had transformed the environment incessantly for decades.[97]

The epidemic of 1833 had a very different ambience from previous epidemics, instilling fear and terror into the hearts of nearly every person along the Maysville Turnpike. In the face of calamity, some settlers again appealed to supernatural explanations, filtered through the cosmology of the Great Revival. As one Bourbonite complained, "All our vain hopes have been dissipated like the morning dew before the sun. Even while penning . . . the Destroying Angel was hovering over our place." Others incorporated their historical heritage into explanations: "The infant settlement of Lexington was dismayed by the disastrous battle of Blue Licks; but under that calamity they were buoyed up by a reliance upon their manhood," recalled physician Charles W. Short. "The cholera brought with it more terror than hordes of savages, for it came on the viewless couriers of the air and bade defiance to human strength."[98]

The previous autumn, cholera had made a brief appearance in Lexington via Cincinnati. By the end of October 1832, the episode caused "very little alarm or excitement among our citizens," as a local newspaper reported. Physicians and the town council posted broadsides with preventative suggestions: "comfortable clothes, *warm feet,* keeping dry, avoiding excessive exertion of body or mind, retiring early to bed, and temperance in eating and drinking." According to Transylvania University professor Charles Caldwell, "it gave us *one single* blow, killed *five intemperates,* frightened our citizens into *strict temperance,* drove away some of our fainthearted pupils, who were just assembling, and then took wing itself, and troubled us no further." The five victims were very poor whites, and unrestrained alcoholic consumption was considered partially responsible for their demise. For most Lexingtonians, this was hardly noteworthy. By late May 1833, the previous autumn had faded from collective memory. Consequently, even as the *Kentucky Gazette* followed the trail of cholera across Europe and North America throughout 1832 and early 1833, few Kentuckians expressed any immediate concern about it. Physicians returned to the advice of the previous year, assuring fellow citizens that temperance and cleanliness would repel the *Vibrio cholerae,* if it attacked.[99]

Less than two months before cholera struck Maysville, disease was a minor concern as residents dealt with more tangible threats. About two o'clock on the morning of 4 April, church bells pealed out the fire alarm. Originating in Henry Martin's dry goods store, a blaze spread to two other stores, a warehouse, and a coffeehouse. When the fire engulfed the latter building, observers fell into a "general panix." As the *Maysville Eagle* related, "Large flakes of fire, after floating high in the atmosphere, descended in succession on the buildings to a distance of two or three hundred yards to the south-east of those on fire.... The scene at this moment was truly grand and terrific. The stoutest heart quailed, and the impression became general, that the entire Eastern portion of our young, but flourishing little city, must soon become a heap of smouldering ruins." As the wind shifted, redeeming the rest of the town, "Never were the extremes of terror and consternation, so rapidly converted into the extremes of gladness and joy! As if by an electric impulse, there was a general exclamation that *'the town would yet be saved!'*" When the cholera reached Maysville in late May, then, it was just the latest episode in a spring of adversity. And maybe because of that, residents seemed just as oblivious to the threat of cholera as were Lexingtonians. Warnings that spread southward along the road assured listeners that the disease was "*principally* confined to the very impudent, the reckless, and the dissapated, and to those slaves who inhabited damp and unwholesome places."[100]

At the southern end of the road, Lexingtonians remained confident that in their town, "celebrated for its healthfulness, fever seldom occurred. It was resorted to by persons from the Southern States during the summer season, as a city remarkable for its healthy situation." More refined residents had little difficulty believing that preventative measures and their own manners of genteel living would protect them from the epidemic. For the less certain, one doctor proffered citizens to "Come forward without delay, enrol your name on our subscription list, and pay us in advance, and we will almost venture to insure it, that cholera does not visit your premises the present year." Yet, within the first week of June, cholera ravaged Lexington, and residents' demeanor changed abruptly. Speaking in the most familiar metaphors they could draw, military veterans compared the onslaught to warfare: One commented how "he had been in many hard fought battles; he had heard the sound of cannon and musket balls passing through the air; he had seen the dead and dying strewn around him, and had heard the groans and shrieks of wounded, but never had felt such an awful dread of impending danger." Another declared that "I would incomparably prefer a seven months' campaign in a furious war than to undergo another seven days such as these."[101]

Environment played a central role in the cholera epidemic. Weeks of rain and warm temperatures created conditions in which *Vibrio cholerae* flourished. The spring storms were "copious and protracted to a degree rarely before witnessed by the oldest inhabitants," recalled medical professor Lunsford Yandell of Transylvania University. On 11 June, his wife complained to her father that "for five Sundays in succession we were caught at church in the rain." Since the artisan-controlled board of trustees had opened up Town Creek, the rivulet again overran the limestone strata and pushed the waste of human privies and livery stables onto city streets, contaminating private and public wells and, consequently, delivering contaminated water to all parts of the town. As the waters rose, they flooded the great brick markethouses that straddled the creek. By 3 June, a large puddle covered East Main Street, and Town Creek had become a pond along Water Street. Lexington was ripe for waterborne disease.[102]

To a lesser extent, the same was true along the length of the Maysville Turnpike. In the village West, reliance on community wells threatened the health of all. On farms and in rural neighborhoods, families faced similar problems as slow-moving small streams and stagnant millponds, underground cisterns and poor land drainage primed conditions for cholera. The persisting belief in miasma further contributed to the problem. Nine years earlier, Samuel McAdow explained how the yellow fever epidemic that swept through Maysville "was attributed entirely to exhalations from damp cellars and local pith." The inconsistent water levels of the region's waterways "sometimes give rise to vegetable putrefectives." Thus, as news of cholera heightened in early 1833, public attention turned to eliminating miasmatic conditions. A local newspaper called for "streets [to] be kept clean and dry, and the houses well white-washed. All filth should be removed from them and no offal or vegetable matter should be allowed to rot in or about them." In March, Maysville's board of councilmen anticipated possible problems and organized a Committee of Public Health to "attend to the removal or taking care of all persons having the smallpox, or other infectious disease, who are strangers or infirm."[103]

Despite the warnings, the epidemic caught citizens unprepared "as if to mock the calculations of man, and set at utter defiance all reasonings on the subject of origin, propogation, prevention or cure." Like the despoiling of the Kentucky wilderness in the 1770s and 1780s that upset American Indian life in the region, the cholera not only threatened the people of Kentucky but assaulted the stability of their communities. Society could not provide for its own fundamental needs: "many of our farmers, we regret to learn, have found it impossible, owing to the prevalence of the Cholera, and other disease, in their own families, and those of their neighbors, to procure the necessary number of laborers to secure their crops of small grain, and have consequently had to abandon a portion of their wheat, rye, and oats." In Lexington, "the markets are suspended and the bakers' shops shut, with one exception. Not a pound of beef is to be got—and very little else. Not even a cracker for sale." Physicians' warnings about the potential dangers of locally grown vegetables and fruits further limited diets. Even as the cholera subsided, the *Lexington Observer and Kentucky Reporter* reminded subscribers not to eat cherries remaining on trees or cucumbers growing in local gardens. Necessities suddenly became luxuries, available only to the few with extracommunal access to water and food supplies. One mother in Louisville wrote her son, "we hear the people in Lexington has been almost suffering for want of provision particularly bread and crackers" and proceeded to send two wagonloads of crackers and flour.[104]

Drinking water likewise became scarce. Because residents believed that miasma could arise from the earth, the bottled waters of Blue Licks became suspect. As they avoided drinking water, many citizens took to brandy, not heeding the conventional wisdom that intemperance in food and drink lowered one's ability to guard against exposure to the poisonous gases. Along the beaten path, "drunkenness was often the consequence."

Since genteel values held intemperance as an affliction of the lower classes, even as the middle class drank alcohol, they reveled in their own sobriety and security. It was with grave concern, therefore, that physician Charles Short announced in mid-June how the victims were "by no means from the ranks which more commonly supply its victims, but from the most respectable, sober and useful citizens. That is so emphatically true that I have not heard of the death of one solitary drunkard."[105]

As preventions failed and fear became rampant, "the panic was terrible. The streets were deserted," wrote Robert Davidson. "The marketplace was desolate. The graveyards were choked. Coffins were laid down at the gates by the score, in confused heaps." Henry Clay described how "all the store and shops are closed, the presses stopped and no one moving in the streets except those concerned with the dead or the sick." Commercial activity came to a halt; only graveyards and apothecaries operated. "I have seen the streets literally deserted, except by funerals, and messengers hastening for physicians!" remarked one citizen. "I have seen the stores and shops entirely closed, and nothing bought or sold from them, but medicines for the sick and shrouds for the dead. . . . Funeral has met funeral in the street." Smoke billowed from barrels of tar and black pitch burning on street corners as citizens attempted to purify the air. Public wells stood unused, their wooden frames rotting in the June rain. The public watch regularly fired canons in attempts to purge the atmosphere. And, as members of local churches died, the "melancholy peal of the bells . . . told of the prevalence of the plague."[106]

Even across the rural landscape, "9 out of 10 families in Ky. would not let a person coming where the cholera has been into their houses! The *Panic* is astonishing!!" The greatest testament to the fear that conquered individuals along the beaten path was the abandonment of hospitality. Multitudes who took to the road and poured into the countryside found little comfort from taverns, neighbors, or even family. But the biggest problem in finding sympathy and charity was that villages like Maysville were "literally depopulated—all who could procure Carriages, Waggons, Carts or Horses, having left." Assuming that cool air and pure waters protected them from the scourge, those who could arrange and afford the transportation and lodging took haven at resorts like Olympian Springs. Like coffin makers and apothecaries, stagecoach drivers profited as the cholera raged.[107]

Those who remained in their homes tracked the daily tolls of disease. On 31 May, the Maysville press reported that "it has now been forty-eight hours since the disease made its appearance, and we have interred TEN persons.... We have now lying dead and to be interred this morning NINE persons." Among them was Mayor Charles Wolf. The river port's markethouse became a morgue where bodies awaited burial. A. P. Thompson related in June how in Maysville "the Cholera appears to spar no age nor sex but sweeps whole families off in a day or two." In the first twenty-four hours, nine-tenths of the town's citizenry had fled. Those who remained behind were either ill or comforters like merchant Andrew January who had lost two of his children "and a third is down, and cannot, I think, survive an hour."[108]

Of course, the cholera traveled with refugees along the new macadamized turnpike. Thompson's hometown of Mayslick "has been spared, no resident of the place as yet has been attacked with the disease and only one death from passengers." The disease elicited so much fear throughout that village by late June, however, it forced "the inhabitants principally to leave the city and ... so terrified and alarmed the country as to prostrate all business, even the cultivation of our crops." There were just fewer victims for cholera to find in Mayslick. By 11 July, Francis Goddard's hometown of Paris lost 73 persons: 44 whites and 29 blacks (two-thirds of whom worked in Henry Duncan's rope manufactory). Even

Gravemarker of Mayor Charles Erb Wolf, Maysville. Located on grounds of Mason County Historical Museum. Photograph by Rod G. Turner.

among those who tried to flee, death often overtook them in the countryside: 8 Paris citizens died in flight. When cholera made its appearance in Lexington on 4 June, one-third of the town's 6,000 residents had already abandoned their homes. Coaches and carriages that had evidenced the discrepancy between the genteel and the common in the past again differentiated those who could escape from those who could not. By the end of July, as the cholera seemed on the decline, Lexington's death toll stood at 489. Charles Short pondered "whether this is but a respite for the Destroying Angel who is whetting his scythe for another harvest, or whether he is about to take flight to some other devoted place, God in his omniscience only knows."[109]

As other residents along the beaten path counted the human cost, Lexingtonians seemed obsessed with tracing lines of distinction through the geography and demography of the disease. "It seems to be confined to the lowest & most populous part of the town," Susan Yandell assured her father. "We seem to live above the infected atmosphere." As if he had stepped out of Daniel Drake's classroom, medical professor John E. Cooke explained the severity of the cholera in Lexington as a consequence of local topography and environmental causation. Two areas of the town contained the majority of cases, according to Cooke: the valley floor of Town Creek adjacent to Main and Water Streets, and the flat lands west of Mulberry Street. The professor theorized that contaminated waters in these areas contributed to the miasma exuding from vegetation and waste.[110]

Little did Cooke realize that his attempt to locate the origins of Lexington's cholera problem also delineated the city's social zones. Contrary to local newspaper reports that there was a "relentless destroyer that was stalking among us unseen, giving scarcely any intimation to the persons whom it had selected for its victims," by the end of the epidemic, careful observers could detect a pattern to the chaos. The poorest classes of persons suffered the greatest loss: 321 of the 491 known victims—47 free blacks, 184 enslaved blacks, and 90 poor whites.[111]

The proximity of residents to work spaces contributed to a high mortality rate among slaves. Domestics, like those of merchant-manufacturer Benjamin Gratz who crowded into small slave quarters behind his home on North Mill Street, usually lived in densely packed homes. Nine of Gratz's slaves died from cholera. Other urban blacks, both free and slave, lived in small neighborhoods adjacent to their factories, artisan shops, and mercantile workplaces. Rolla Blue's neighborhood adjacent to Water Street was hit hard by cholera. Blue survived, but livery keeper Hannibal Straws succumbed, as did Simon Stilfield, "the celebrated Whistler." For many blacks, their occupations as grave and well diggers put them at additional risk. Unlike whites, most blacks could not flee the Destroying Angel and depended more on the medical attentions of overburdened doctors or slaveowners for whom health care was the only way to protect their human investments. Some owners "of rope and bagging factories, who had under their control a great many negroes," took the advice of local physicians and administered emetics to their slaves. What was remarkable, according to one observer, was that "where this course was pursued, scarcely a negro was lost; whereas in one factory, that I know of, where a different course was pursued twenty five died."[112]

Cramped living arrangements doomed poor blacks. "The habits and conditions of the black population in every country are much alike, and well known," wrote Lansford Yandell. "Bond and free, they are generally filthy and careless, poor, and ignorant, and in want of many of the comforts of life, and nearly always live in low, damp, or ill ventilated houses." Like other educated men who associated poor sanitation with poverty and ignorance, Yandell expanded his evaluation to address poor whites: "Many of the whites themselves were badly fed, and miserably lodged in low, crowded houses, surrounded by filth, and supplied with few of the comforts which the sick require." The situations of their residences in squalid areas of town placed poor blacks and whites in the path of the cholera.[113]

Yet, an understanding of the demography of the cholera epidemic is not simply black and white. The disease found its way into the lives of people at every level of society. Only because of the economic downturn of the late 1810s did white artisans and factory workers, who had fled Lexington for work elsewhere, escape the epidemic. Their displacement situated slaves and hired-outs in harm's way. The W. W. Ater Company reported the deaths of five African American workers. Four slaves in Joshua Weir's ropewalk succumbed. Another seven expired in Benjamin Gratz's ropewalk. James Brand's hemp bagging factory was the scene of nineteen slave deaths. Early in the epidemic at the Lexington markethouse, the cholera bacteria moved from the hands of slaves and lower-class white grocers to their fruits and vegetables to the hands of consumers. Even centers of gentility became cholera traps. Residents pinpointed the Phoenix Hotel as the epicenter of the local epidemic. The high turnover of guests raised suspicions about who brought the disease into town. That many early deaths were reported in poorer neighborhoods adjacent to the tavern only heightened fears.[114]

Proportionately, cholera had the greatest impact on white widows and their dependents, comprising over 10.5 percent of deaths. As with poor whites and blacks, white widows of all classes had limited ability to flee the city. The absence of a husband curtailed bartering power for transportation, food supplies, and other necessities. Additionally, when the epidemic subsided, widows and orphans posed a difficult problem for the town council. While an orphanage was founded on Third Street, a proposed widow's home apparently never came to fruition.[115]

Although cholera directly assaulted the more marginalized peoples of Lexington, many "of the most enterprising and interesting of both sexes have been swept away." Cholera ignored lines of distinction. "It is but too true and lamentably true," remarked Charles Short, "that it had proved more malignant, fatal and indiscriminate in the selection of its victims in Lexington than in any other town of the Union—perhaps of the world." The town treasurer and the day watchman both died, as did the cashier of the Bank of the United States. Renowned hotel owner John Postlethwaite was among the many who contracted the illness from the refuge of fleeing guests at his establishment. Thomas Bodley, the passionate advocate for the turnpike company, died two weeks before his son Breckinridge, who had been a surveyor for the railroad project. Those who were not personally stricken still faced the human tragedy of the epidemic. London Ferrell, the fiery African American preacher of the First Baptist Church, stayed in town to bury both black and white, and eventually his wife. Merchant George Trotter, who had only recently acquired the *Kentucky Gazette,* sold the newspaper in mid-October because "particularly [in] the last seven months, it has been impossible for me to devote to it the requisite attention." The *Gazette* failed to publish for two weeks during the epidemic; "the greater part of the hands in the office were attacked with Cholera and the rest were required to attend them." By killing many persons central to the structure of society, the 1833 cholera epidemic threatened the very stability of life in Lexington. It decimated the labor force, incapacitated community leaders, and disrupted operations of important institutions.[116]

While the epidemic did indeed strike harshly in 1833, there is also a coincidence of survivors worth noting. Harriet Beecher taught female students at Lane Academy in Cincinnati, but in the spring of 1833, just prior to the cholera outbreak in Mason County, she traveled to

Washington to visit the Key family. Marshall Key was the grandson of Thomas Marshall, whose Federal Hill home still dominated the village landscape, and his daughter attended Lane Academy. One day, student and teacher observed a slave auction on the village square. Twenty years later, the scene was remembered as a newspaper ad in *Uncle Tom's Cabin:* "EXECUTOR'S SALE—NEGROES!—Agreeably to order of court, will be sold, on Tuesday, February 20, before the Court-house door, in the town of Washington, Kentucky, the following negroes: Haga, aged 60; John, aged 30; Ben, aged 21; Saul, aged 25; Albert, aged 14. Sold for the benefit of the creditors and heirs of the estate of Jesse Blutchford." About the same time as Beecher witnessed the horror of slavery for the first time, Lieutenant Jefferson Davis arrived in Lexington to recruit dragoons for his father-in-law, Zachary Taylor. Davis was familiar with the city, having attended Transylvania University between 1821 and 1824. He was in Lexington throughout the epidemic. He possibly met Joseph C. Breckinridge or Robert S. Todd. Breckinridge was a lawyer and politician who lived with his family at his father's original estate, Cabell's Dale where his mother and wife taught his twelve-year-old son, John. Todd, a local grocer, cotton manufacturer, and "Henry Clay Whig" had just recently moved his family into a large two-story brick home on West Main Street. His fourteen-year-old daughter, Mary, attended Charlotte Mentelle's boarding school on the other side of town. Either by fortune or through caution, Harriet Beecher, Jefferson Davis, John C. Breckinridge, and Mary Todd all lived through the flight of the Destroying Angel along the Maysville Road.[117]

Their particular survivals suggest that Charles Short's proclamation of the egalitarianism of death was not entirely accurate. Lexingtonians of means had a better chance of survival. Cyrus Parker Jones, a free black who spent most of the early nineteenth century distributing funeral notices at the Lexington markethouse, did not work between 20 May and 10 September 1833. Of course, there was plenty of business to be had, but those who could afford funerals and funeral notices had either fled or were hiding behind the walls of their urban and rural retreats.[118]

Epidemic disease was always difficult for early Americans to address because it often involved political or social issues. Lexington, on the eve of the epidemic, was a city with an immature civic structure: "No city police —(at least not visible)—no board of health—no medical reports." Two local physicians were in Boston, three others died in the first ten days of the epidemic, and still two others "felt it their duty from peculiar considerations to remove their families to a distance." As with the November 1832 cholera assault, the mayor and city council acted belatedly. Unable to stabilize society under the threat of epidemic cholera, they appealed to Louisville for assistance. The request took the Louisville town meeting by surprise: it "was believed by the meeting that the necessaries + comforts of life could not be wanting in so rich section of our state as Fayette." Louisville along the banks of the Ohio River, not Lexington deep in the heart of the countryside, had always faced immediate threats of disease; cholera ravaged Louisville only the autumn before. Two Louisville physicians responded to the call for "medical men to visit Lexington immediately and tender their services to the medical gentlemen of Lexington ... [who] in the faithful discharge of their duties had become worn down and exhausted." The Louisville meeting also sent some crackers "under the impression they may be wanted."[119]

Not until 11 July did the Lexington council meet in response to the crisis. A resolution passed to negotiate a three-thousand-dollar loan "for the purpose of meeting the extraordinary expenses produced by the Cholera." Some of this money went towards the operation of a warehouse "where the poor could furnish themselves, free of cost, with all the necessities of life." Another resolution made it "the duty of the officers of the city, to see that no obstructions are made or suffered to remain in the stream that runs through Water-st[reet]," a clear response to the atmosphere-putrefying ponds created by the June rains. In late July, authorities decided that privy pits needed to be at least five feet deep and never allowed to fill within two feet of the surface, that stables should have privies, and that four bushels of unslaked lime must be thrown into each privy pit twice a year. Additionally, the depths of graves were lowered to five feet, and hog owners had to keep their animals from roaming the streets. These actions, when considered with a November curfew on blacks in response to the previous outbreak, may be interpreted as more than reactions to cholera. They were steps towards greater municipal government, not only revising the parameters of certain private customs, but also establishing the authority of the mayor and council in dealing with public problems.[120]

Despite civic actions, the panic that accompanied cholera remained long after the Destroying Angel took flight. When clergy preached in the months during and after the epidemic, they did not address crowds eager to hear the word of the gospel. Instead, congregations packed into churches seeking a reprieve from fear. In Paris, "numbers joined the churches," explained Francis Goddard, "much prayer was offered up, [and] it pleased the allmighty to stop the plague for this season." Unable to draw parishioners out of their homes, the rector of the American Episcopal Church in Lexington issued a broadside asking, "If the calamity which we have suffered does not turn all hearts to the duties of a pious life what can?" He preached that "public calamities are unmitigated curses, save only where the Bible and the Lord's Day combine their influence to convert them into blessings." The "duties and responsibilities of true christians at this time are very great," he continued. "We are bound to take the lead in all those things which may render our public calamities, mercies to us, to our children and to the community at large. This is a time for great efforts for the conversion of souls of men. This is a time for much and fervent prayer for the influences of the Spirit of God." And he warned all to "Let no slight, trivial excuse prevent your reading of a portion of Scripture every day with fervent prayer."[121]

Still, cholera, not clergy, induced individuals to examine faith and come to terms with the Almighty. Rhoda Anderson related to her son that "my faith is so weak, I fear the Lord will forsake me, and then I recalled his precious promises that he will be with us in all our troubles." During the epidemic, Susan Yandell admitted to her mother that "never before has religion been to me such a source of consolation; never have I been able to lay hold on the precious promises contained in the Bible. . . . I can now read my Bible and perceive new beauties in every verse, can understand many parts that have heretofore appeared dark and obscure."[122]

Organized religion had not capitalized on the enthusiasm of the early 1800s. Throughout the 1810s and 1820s, camp meetings erupted across central Kentucky, but they never reached the crescendo of 1801, and they did not bolster church memberships. Suddenly, disease and death offered opportunity to bring new energy to religion. Clergy appealed, "Dare we provoke worse punishment, by refusing to humble under what has already been laid upon us? The thought that religion will hereafter be as much neglected amongst us, as it has been heretofore, is too shocking to be endured!" The Episcopal minister, "writing to a Parish of mourners," disclosed that his "most urgent plea with you is fly to your bibles and the house of God."[123]

And they did. Expectations of impending doom resurrected the revivalism that waned in the 1810s. Revivalist preacher A. Bowed held a meeting in Lexington in April 1834 to avert future cholera attacks. Later that spring, most of the town's churches participated in a four-week-long revival. At the Market Street Presbyterian Church, nineteen people accepted baptism, a small proportion of the nearly four hundred Lexingtonians who joined established denominations in the wake of the cholera.[124]

Natural coincidence heightened fears and helped religious leaders draw new congregants. In November 1833, a resident of Nicholas County proclaimed that "the world is coming to an end!" At three o'clock in the morning, "falling stars" appeared in the night sky: a meteor shower so brilliant and prolonged that it seemed unnatural. One central Kentuckian remembered how Aunt Betsy Conover "ran screaming through the town calling to the people to get up. Returning home she found Uncle Billy still asleep. She called to him, 'Oh! Mr. Conover, are you still asleep? Do wake up. The day of judgment has come.' He said, 'Pshaw Betsy, do come to bed. Don't be a fool. The day of judgment can't come at night,' and he slept on and missed the magnificent sight. But there was no sleep for the villagers. They were too badly frightened. They sent for the preacher and all repaired to the church, sang, prayed and exhorted until day-light." Future Kentucky historian William Perrin witnessed the spectacle as a youth and recalled how the initial terror abated and "our people stood spell bound, gazing still in awe" at the "bad revolting stars." Just three months after Asiatic cholera terrified Kentuckians, another Destroying Angel seemed to be writing across the heavens that the "crack of doom" was at hand.[125]

Out of the "Cholera times," a most unlikely hero emerged. William Solomon, born in Virginia in 1775, was a day laborer who migrated to Kentucky in the 1790s. During the epidemic, as the wealthy fled town for sanctuary elsewhere, Solomon bravely buried the dead in the Episcopal Burying Ground, which the church had purchased the preceding year. As the town recovered in the mid-1830s, William Solomon became a household name and the newest addition to the local pantheon of heroes: a colorful but poor laborer who became a celebrity. With crumpled hat in hand, he sat for portrait painter Samuel Price. Stories circulated that mythologized Solomon, even connecting him to another local celebrity, Henry Clay, as boyhood friends. One such tale explained how, when a political candidate paid Solomon to vote for him, the poor man pocketed the money and voted his conscience. Another related how, when the local magistrate mistook Solomon for a vagrant, the day laborer was sold to an elderly black woman for eighteen cents. Out of his generous heart, he worked for her the rest of her life, delivering seventy-five cents at the end of each day. Still another comically detailed how Solomon, finding a job trimming trees on the courthouse yard, erroneously sawed a limb between where he sat and the trunk. As onlookers picked him up from the ground, they nicknamed him "King Solomon" after the wisest of Israel's kings.[126]

In the wake of a cholera epidemic that distinguished little between refined and vulgar, in an era when economic and social upheaval left all Kentuckians confused about their present stations in life and concerned about their futures, in an economy where the extra-moral workings of capitalism seemed more rewarding than traditional patterns of familial and communal obligation, and in a culture where virtue had faded from public life and the manacles of the market economy kept almost everyone in dependence, the lore of King Solomon returned audiences to the simplicity, sincerity, honor, and independence of days gone by, when pioneer big men had risen to prominence through personal bravery, risk taking, and a sense of communal responsibility.

William "King" Solomon. From Samuel W. Price, The Old Masters of the Bluegrass *(Louisville, Ky.: John P. Morton & Co., 1902).*

As construction on the Maysville Turnpike reached completion in 1835, the life of the Maysville Road came to its conclusion. In time, the remnants of the old buffalo trace would fade into the Kentucky countryside. Between the 1770s and 1820s, the unimproved dirt trail had provided a conduit for settlement and citizenship, commerce and agriculture, gentility and social stratification, manhood and femininity, political and religious contests. It was seldom at the center of Kentuckians' lives, but it always gave definition to their lives and identities.

By the late 1820s and early 1830s, the American West found along the beaten path was fading. As a consequence of economic displacement and cholera, thousands of people abandoned their homes along the Maysville Road and sought out new frontiers. The Virginia gentry who had guided the political development of the region had largely died out. Their sons converted former tobacco plantations into cattle ranches and tenant farms. Smaller farmers had been bought out by larger landowners during the Panic of 1819 and the subsequent relief struggle. Displaced farmers migrated in search of new lands. Artisans sought opportunities in a new village West arising along the Ohio and Mississippi Rivers. The small worlds along the Maysville Road had changed dramatically in six decades.

In 1835, Bourbon County exported no tobacco or wheat, the first sign that a new order had arrived. Both products were susceptible to rotting on the long trips to New Orleans and did not proffer the profit they once had. Instead, merchants and manufacturers, tavernkeepers and hoteliers who had triumphed in the village West were employing their economic and political influence and much of their fortunes to propel the rural West into ranching and large-scale staple-crop agriculture. Those who, by virtue of wealth, had fled the cholera returned to decimated communities and claimed even stronger economic, social, and political positions. It was a middle-class world now, one in which a turnpike evidenced the conquest of nature, genteel homes served as permanent symbols of social hierarchy, tollhouses drew profit from human mobility, watch houses kept whites secure by intimidating blacks, markethouses limited the influence of moral economy over the market economy, the jollity once found in farmers' houses could be purchased at hotels, and an affluent society watched as steamboats and railroads took the American nation along a different path.

# Appendix: Tables

TABLE 1

Origins of Revolutionary Veterans in Mason and Fayette Counties

Origin	Mason	Fayette	% of Statewide
New England	0.0	1.1	0.5
New York	2.3	2.2	2.3
New Jersey	4.7	6.8	5.8
Delaware	0.0	1.1	0.5
Pennsylvania	20.2	17.0	18.3
Maryland	29.7	3.4	16.1
Virginia	40.4	62.5	49.2
North Carolina	2.3	4.5	3.4
South Carolina	0.0	1.1	0.5

Source: Todd Harold Barnett, "The Evolution of 'North' and 'South': Settlement and Slavery on America's Sectional Border, 1650–1810" (Ph.D. diss., University of Pennsylvania, 1993), 262: Table VI-II.

TABLE 2

Centralized Settlement by 1810

State	Hamlets	Villages	Towns	Cities	% of Whites
Kentucky	23	20	9	1	3.60
Georgia	12	9	4	2	5.27
Indiana	1	2	1	0	4.70

Source: *Federal Census 1810: Kentucky, Georgia, Indiana.*

TABLE 3

Regional and Urban Population Growth in the Early Republic

	1790–1800	1800–1810	1810–1820
United States	35.3	36.4	84.7
New England	22.1	19.3	45.8
Boston	36.1	55.4	111.4
Mid-Atlantic	43.8	42.5	105.1
Philadelphia	39.6	41.8	97.9
New York City	82.7	66.5	204.1
Baltimore	96.3	75.6	244.7
South	24.6	17.1	45.9
Richmond	52.5	104.1	158.8
Charleston	23.3	19.4	47.3
Trans-Appalachia[a]	253.3	179.0	885.9
Lexington	115.2	141.0	418.7

Source: *Federal Census 1790, 1800, 1810;* "Table V: Population of the Four Leading U.S. Cities and Their Suburbs showing the decennial rate of increase, 1790-1810," in David T. Gilchrist, ed., *The Growth of Seaport Cities, 1790–1825* (Charlottesville: University of Virginia Press, 1967), 39.

[a] Does not include Old Northwest in 1790 figures.

TABLE 4

Land Ownership along the Maysville Road, 1800

County	No. of Landowners	as % of Households	% Owning 1–200 acres
Fayette	743	42.5	56.5
Bourbon	863	46.1	70.8
Nicholas	209	41.5	81.3
Mason	757	39.3	71.3
Statewide	16,003	49.2	69.4

Source: Joan Wells Coward, *Kentucky in the New Republic: The Process of Constitution Making* (Lexington: University Press of Kentucky, 1979), 55.

TABLE 5

Slave Ownership along the Maysville Road, 1800

County	% of Householders Owning Slaves	Average Slaveholding
Fayette	34.0	6.16
Bourbon	27.1	4.02
Nicholas	12.5	3.94
Mason	21.7	3.96
Statewide	25.2	4.39

Source: Joan Wells Coward, *Kentucky in the New Republic: The Process of Constitution Making* (Lexington: University Press of Kentucky, 1979), 63.

TABLE 6

Purchases of Cloth, 1792–1810

Product	1792	1796	1806	1807	1810
Domestic cloth					
Country linen	20	21	22	25	36
Broadcloth	6	19	0	28	23
Imported cloth	10	31	36	41	77
Trimmings/needles	17	25	26	61	72
Indigo	7	7	3	13	23

Source: John Moylan's store ledger (Lexington), January 1792–November 1794, UKSC; John W. Hunt's daybook (Lexington), July 1790–September 1796, UKSC, microfilm; Daniel Halstead's account books (Lexington), 1806, FHS; William Tureman's daybook, January–May 1807, MCHM; Edmund Martin's journal (Maysville), vol. B, July 1800–February 1811, MCHM.

TABLE 7

Modes of Exchange in Stores along the Maysville Road

Store & Location	Year	% Credit	% Cash	% Barter
Leonard, Mason Co.	1791	12.6	50.4	36.8
Moylan, Lexington	1792	22.9	33.0	44.1
Hunt, Lexington	1796	85.0	9.0	6.0
Halstead, Lexington	1797	54.4	14.7	30.8
Tureman, Washington	1807	80.0	14.0	6.0
Martin, Maysville	1810	78.0	9.0	12.0

Source: Benedick Leonard's store ledger (Washington), December 1790–November 1791; John Moylan's store ledger (Lexington), January 1792–November 1794, UKSC; John W. Hunt's daybook (Lexington), July 1790–September 1796, UKSC, microfilm; Daniel Halstead's account books (Lexington), 1806, FHS; William Tureman's daybook, January–May 1807, MCHM; Edmund Martin's journal (Maysville), vol. B, July 1800–February 1811, MCHM.

TABLE 8

Lexington Trustees by Occupation, 1792–1812

Occupations	Number
Merchant	30
Artisan	19
Tavernkeeper	4
Lawyer	3
Farmer	3
Manufacturer	2
Printer	2
Doctor	1
Unknown	4

Source: Lexington Trustees Minute Book, 1792–1812, UKSC; William A. Leavy, "A Memoir of Lexington and Its Vicinity," *RKHS* 41 (1943): 323–31; Robert Peter, *A History of Fayette County, Kentucky*, ed. William Henry Perrin (Chicago: O. L. Haskin & Co., 1882), chaps. 5, 8.

TABLE 9

Lexington Town Lot Ownership, 1802–1804

Occupation	Under $100	$101–$350	$351–$1000	$1001–$2000	Over $2001
Mercantile		9	5	9	5
Tavern		1	2	3	
Law			5		
Ministry		2	1		
Construction		13	3	1	
Transportation		5			
Civil Servant		2			
Artisan	1	17	10	5	
Laborer	1	1			
Other	1	3	3	2	
Unknown	5	16	9	1	

Source: William A. Leavy, "A Memoir of Lexington and Its Vicinity," *RKHS* 41 (1943): 323–24; Tax Lists for 1820 and 1804, KDLA. One tax district in Lexington and Fayette County listed lot values in pounds rather than in dollars. The standard conversion rate for the era was 4s.6d per $1.00. See John J. McCusker, "How Much Is That in Real Money? A Historical Price Index for Use as a Deflator of Money Values in the Economy of the United States," *Proceeding of the American Antiquarian Society* 101 (1992): 313.

TABLE 10

Means of Purchase by Gender

Year	% of Transactions as Gender	Credit	Cash	Barter
1796	Men	85.6	8.7	5.4
	Women	90.0	6.6	3.3
	as individual	91.6	0.0	8.3
	as dependent	88.8	11.1	0.0
1807	Men	85.2	12.1	2.5
	Women	80.4	14.6	4.8
	as individual	76.9	15.3	7.6
	as dependent	82.1	7.1	7.1
1810	Men	78.2	9.3	12.1
	Women	80.5	8.3	11.1
	as individual	53.8	23.0	23.0
	as dependent	95.6	0.0	4.3

Source: John W. Hunt's daybook (Lexington), July 1790–September 1796, UKSC, microfilm; William Tureman's daybook, January–May 1807, MCHM; Edmund Martin's journal (Maysville), vol. B, July 1800–February 1811, MCHM.

TABLE 11

Free Black Population Growth, 1790–1810

County or town	1790	1800	1810
Mason County	0	88	45
Maysville	0	n/a	2
Washington	0	n/a	10
Nicholas County	n/a	6	21
Bourbon County	0	62	190
Paris	0	n/a	12
Fayette County	30	88	208
Lexington	2	n/a	85

Source: *Federal Census, Kentucky: 1790, 1800, 1810,* Mason, Nicholas, Bourbon, and Fayette Counties.

TABLE 12

Demographics of Cholera Epidemic in Lexington, 1833

Group	No. of Deaths	% of Total Deaths
Men	287	58.4
Women	204	41.5
Whites	262	53.3
Blacks	229	46.6
Slaves	183	37.2
Free	46	9.3
Total	491	

Source: *City of Lexington, Tax Assessor's List,* 18 April 1833, KDLA; *Lexington Observer and Kentucky Reporter,* 22 July 1833, 3–4; Sara Kathryn Dawson, "The 1833 Cholera Epidemic in Lexington: A Study of Devastation and Recovery" (M.A. thesis, University of Kentucky, 1994), Appendix 1.

# NOTES

ABBREVIATIONS

CHS	Cincinnati Historical Society Research Library, The Museum Center, Cincinnati.
Draper	John Lyman Draper Manuscript Collection, State Historical Society of Wisconsin, Madison, microfilm.
Durrett	Reuben T. Durrett Collection, Special Collections, Joseph Regenstein Library, University of Chicago.
FHS	Manuscript Collections, Filson Historical Society (formerly the Filson Club Historical Society), Louisville, Ky.
KDLA	Public Records Division, Kentucky Department for Libraries and Archives, Frankfort.
KHS	Special Collections, Kentucky Historical Society, Frankfort.
LOC	Manuscript Division, Library of Congress, Washington, D.C.
MCHM	Research Library, Mason County Historical Museum, Maysville, Ky.
TU	Special Collections, Transylvania University, Lexington, Ky.
UCSC	Department of Special Collections, Joseph Regenstein Library, University of Chicago.
UKSC	Department of Special Collections and Archives, Margaret I. King Library, University of Kentucky, Lexington.
AHR	*American Historical Review*
FHQ/FCHQ	*Filson Historical Quarterly (Filson Club Historical Quarterly)*
JAH	*Journal of American History*
JER	*Journal of the Early Republic*
JSH	*Journal of Southern History*
RKHS	*Register of the Kentucky Historical Society*
WMQ	*William and Mary Quarterly*

INTRODUCTION

1. Cecil Harp and J. Winston Coleman Jr., "The Old Lexington and Maysville Turnpike," *Kentucky Engineer* 1 (1939): 13–16; R. S. Cotterill, "The Old Maysville Road: Pioneer Trade and Settlement along a Celebrated Kentucky Highway," *Kentucky Magazine* 1 (1916): 616–20; Samuel M. Wilson, "The Old Maysville Road," in *Proceedings of Second Annual Meeting of the Ohio Valley Historical Association* (Columbus: Ohio State Archaeological and Historical Society, 1909), 16; Hubert G. H. Wilhelm, "The Road as a Corridor for Ideas," in *The National Road,* ed. Karl Raitz (Baltimore: Johns Hopkins University Press, 1996), 256–84.
   The Transportation Revolution has generated a large bibliography, including many that have influenced my thinking: George Rogers Taylor, *The Transportation Revolution, 1815–1860* (New York: Rinehart, 1951); Carter Goodrich, *Government Promotion of American Canals and Railroads, 1800–1890* (New York: Columbia University Press, 1960); John Lauritz Larson, *Internal Improvements: National Public Works and the Promise of Popular Government in the Early United States* (Chapel Hill: University of North Carolina Press, 2001); Carol Sheriff, *The Artificial River: The Erie Canal and the Paradox of Progress, 1817–1862* (New York: Hill & Wang, 1996); Henry deLeon Jr. and Jerry Elijah Brown, *The Federal Road through Georgia, the Creek Nation, and Alabama, 1806–1836* (Tuscaloosa: University of Alabama Press, 1989).
2. Joyce Appleby, *Inheriting the Revolution: The First Generation of Americans* (Cambridge, Mass.: Harvard University Press, 2000), 8; Hal S. Baron, *Those Who Stayed Behind: Rural Society in Nineteenth-Century New England* (New York: Cambridge University Press, 1981).
3. The literature on these themes is vast, but the following works most significantly influenced my interpretation of the Early American Republic and of early Kentucky. On the evolution of a republican government, see Drew R. McCoy, *The Elusive Republic: Political Economy in Jeffersonian America* (Chapel Hill: University of North Carolina Press, 1980); Joan Wells Coward, *Kentucky in the New Republic: The Process of Constitution Making* (Lexington: University Press of Kentucky, 1979). On the rumblings of democracy, see Gordon Wood, *Radicalism of the American Revolution* (New York: Alfred A. Knopf, 1992); Neils Sonne, *Liberal Kentucky 1780–1828* (Lexington: University of Kentucky Press, 1968); Patricia Watlington, *The Partisan Spirit: Kentucky Politics, 1779–1792* (New York: Atheneum, 1972). On the appearance of a market revolution, see Charles Sellers, *The Market Revolution: Jacksonian America, 1815–1860* (Ithaca, N.Y.: Cornell University Press, 1990); Stephen A. Aron, *How the West Was Lost: The Transformation of Kentucky from Daniel Boone to Henry Clay* (Baltimore: Johns Hopkins University Press, 1996). On the development of urban life, see Thomas Bender, *Toward an Urban Vision: Ideas and Institutions in Nineteenth-Century America* (Lexington: University Press of Kentucky, 1975); Richard C. Wade, *The*

Urban Frontier: Pioneer Life in Early Pittsburgh, Cincinnati, Lexington, Louisville, and St. Louis (Chicago: University of Chicago Press, 1959). On the refinement of an awakening commercial order, see Richard Lyman Bushman, *The Refinement of America: Persons, Houses, Cities* (New York: Alfred A. Knopf, 1992); C. Dallett Hemphill, *Bowing to Necessities: A History of Manners in America, 1620–1860* (New York: Oxford University Press, 1999); John F. Kasson, *Rudeness & Civility: Manners in Nineteenth-Century Urban America* (New York: Hill and Wang, 1990). On the empowerment of evangelical Christianity, see Nathan O. Hatch, *The Democratization of American Christianity* (New Haven, Conn.: Yale University Press, 1989); Christine Heyrman, *Southern Cross: The Beginnings of the Bible Belt* (New York: Alfred A. Knopf, 1997); Ellen Eslinger, *Citizens of Zion: The Social Origins of Camp Meeting Revivalism* (Knoxville: University of Tennessee Press, 1999). On the institutionalization of racial slavery, see Ira Berlin, *Many Thousands Gone: The First Two Centuries of Slavery in North America* (Cambridge, Mass.: Belknap Press, 1998); Philip D. Morgan, *Slave Counterpoint: Black Culture in the Eighteenth-Century Chesapeake & Lowcountry* (Chapel Hill: University of North Carolina Press, 1998); Marion B. Lucas, *A History of Blacks in Kentucky,* Volume I: *From Slavery to Segregation, 1760–1891* (Frankfort: Kentucky Historical Society, 1992). On the shaping of patriotism, see Benedict Anderson, *Imagined Communities: Reflections on the Origin and Spread of Nationalism* (New York: Verso Press, 1983); David Waldstreicher, *In the Midst of Perpetual Fetes: The Making of American Nationalism, 1776–1820* (Chapel Hill: University of North Carolina Press, 1997). On romanticizing the natural environment, see Carolyn Merchant, *Ecological Revolutions: Nature, Gender, and Science in New England* (Chapel Hill: University of North Carolina Press, 1989); Lawrence J. Friedman, *Inventors of the Promised Land* (New York: Alfred A. Knopf, 1975).
4. Karl Raitz, "Introduction: The National Road and Its Landscapes," in *A Guide to the National Road,* ed. Karl Raitz (Baltimore: Johns Hopkins University Press, 1996), 7; Sheriff, *The Artificial River,* 1–9.
5. Walt Whitman, "Song of the Open Road," in *Leaves of Grass and Other Writings,* ed. Michael Moon (New York: W. W. Norton, 2002), 126–35. In 1856 when he wrote it, Whitman originally titled the piece "Poem of the Road." It was retitled in 1867.

1. ORIGINS OF COMMUNITY

1. Gilbert Imlay letter, 14CC214, Draper; Nancy O'Malley, "Frontier Defenses and Pioneer Strategies in the Historic Settlement Era," in *The Buzzel about Kentuck: Settling the Promised Land,* ed. Craig Thompson Friend (Lexington: University Press of Kentucky, 1999), 64–67.
2. Otis K. Rice, *Frontier Kentucky* (Lexington: University Press of Kentucky, 1993), 107–9; Samuel W. Wilson, *Battle of the Blue Licks* (Lexington, Ky.: n.p., 1927); Bennet K. Young, *History of the Battle of Blue Licks* (Louisville, Ky.: J. P. Morton

& Co., 1897); Michael C. C. Adams, "An Appraisal of the Blue Licks Battle," *FHQ* 75 (2001): 181–204.
3. Arthur C. McFarlan, *Geology of Kentucky* (Lexington: University of Kentucky Press, 1943), 174; Neal O. Hammon, "Pioneer Routes in Central Kentucky," *FCHQ* 74 (2000): 127–31; William E. Myer, *Indian Trails of the Southeast* (Nashville: Blue & Gray Press, 1971), 46, 54–57.
4. R. Barry Lewis, "Mississippian Farmers," in *Kentucky Archaeology*, ed. R. Barry Lewis (Lexington: University Press of Kentucky, 1996), 156–57; William E. Sharp, "Fort Ancient Farmers," in Lewis, ed., *Kentucky Archaeology*, 178–79; Julian J. N. Campbell, "Present and Presettlement Forest Conditions in the Inner Bluegrass of Kentucky" (Ph.D. diss., University of Kentucky, 1980), 160, 180; Henry Williams, *Americans and Their Forests: A Historical Geography* (New York: Cambridge University Press, 1989), 43. Although popularly referred to buffalo, *Bison americanus* are a different species of wild oxen with massive fore parts, rust-colored manes, black upturned horns, and humped backs. The old buffalo trace, then, was a misnomer.
5. Helen Hornbeck Turner, "The Land and Water Communication Systems of the Southeastern Indians," in *Powhatan's Mantle: Indians in the Colonial Southeast*, ed. Peter H. Wood et al. (Lincoln: University of Nebraska Press, 1989), 9.
6. A. Gwynn Henderson, "Middle Fort Ancient Villages and Organization Complexity in Central Kentucky" (Ph.D. diss., University of Kentucky, 1998), 41–45; Lewis, "Mississippian Farmers," 156–57; Sharp, "Fort Ancient Farmers," 178–79; Campbell, "Present and Presettlement Forest Conditions," 164. Of course, the "natural" beauty of many North American regions had been contrived by Native American agricultural and hunting techniques; William Cronon, *Changes in the Land: Indians, Colonists, and the Ecology of New England* (New York: Hill and Wang, 1983), 47–51; Timothy Silver, *A New Face on the Countryside: Indians, Colonists, and Slaves in South Atlantic Forests, 1500–1800* (New York: Cambridge University Press, 1990), 59–64; Stephen A. Aron, "The Significance of the Kentucky Frontier," *RKHS* 91 (1993): 314–22.
7. Campbell, "Present and Presettlement Forest Conditions," 163; J. Winston Coleman Jr., *Stage-Coach Days in the Bluegrass* (1935; reprint, Lexington: University Press of Kentucky, 1995), 17; deposition of Col. Robert Patterson, 1810, 6BB32, Draper.
8. Francis Jennings, *The Founders of America: From the Earliest Migrations to the Present* (New York: W. W. Norton, 1993), 16; Lucien Beckner, "Eskippakithiki: The Last Indian Town in Kentucky," *RKHS* 6 (1932): 355–82; A. Gwynn Henderson, "Dispelling the Myth: Seventeenth- and Eighteenth-Century Indian Life in Kentucky," *RKHS* 90 (1992): 1–25; idem, "The Lower Shawnee Town: Maintaining an Indian Republic on the Ohio," in Friend, ed., *The Buzzel about Kentuck,* 25–55; Colin G. Calloway, "The Peace That Brought Not Peace," in *The American Revolution in Indian Country: Crisis and Diversity in Native American Communities* (New York: Cambridge University Press, 1995), 272–91.

9. Peter H. Wood, "The Changing Population of the Colonial South: An Overview by Race and Region, 1685–1790," in Wood, ed., *Powhatan's Mantle*, 84–88; Samuel Cole Williams, ed., *Lieut. Henry Timberlake's Memoirs, 1756–1765* (1927; reprint, Marietta, Ga.: Continental Book Co., 1948), 113–16.
10. Aron, "Significance of the Kentucky Frontier," 316.
11. Idem, *How the West Was Lost: The Transformation of Kentucky from Daniel Boone to Henry Clay* (Baltimore: Johns Hopkins University Press, 1996), chaps. 1, 2.
12. Henderson, "Lower Shawnee Town."
13. Aron, *How the West Was Lost*, 47–49; Colin G. Calloway, "Maquachake: The Perils of Neutrality in the Ohio Country," in *The American Revolution in Indian Country*, 158–81.
14. Otis K. Rice, *Frontier Kentucky* (Lexington: University Press of Kentucky, 1983), 100–105; Colin G. Calloway, "Corn Wars and Civil Wars: The American Revolution Comes to Indian Country," in *The American Revolution in Indian Country*, 54, 63.
15. Wood, "The Changing Population of the Colonial South," 86.
16. Arthur C. McFarlan, *Geology of Kentucky* (Lexington: University of Kentucky Press, 1943), 174; Hammon, "Pioneer Routes in Central Kentucky," 127–31; Daniel Drake, *Pioneer Life in Kentucky: A Series of Reminisential Letters*, ed. Charles Drake (Cincinnati: Ropbert Clarke & Co., 1870), 12, 25; Mann Butler, "Manners of Western Pioneers," Mann Butler Papers, Durrett; Joseph Underwood to Joseph Lawrence, 10 May 1792, Wood Family Papers, UKSC.
17. Patricia Watlington, *The Partisan Spirit: Kentucky Politics, 1779–1792* (New York: Atheneum, 1972), 11–14.
18. Nancy O'Malley, *Stockading Up: A Study of Pioneer Stations in the Inner Bluegrass Region of Kentucky* (Lexington, Ky.: Program for Cultural Resource Assessment, 1987); Robert M. Rennick, *Kentucky Place Names* (Lexington: University Press of Kentucky, 1984), 92, 193, 197.
19. Allan W. Eckert, *That Dark and Bloody River* (New York: Bantam Books, 1995), xvii; Cronon, *Changes in the Land*, 65–66; Paul C. Nagel, *This Sacred Trust: American Nationality, 1798–1898* (New York: Oxford University Press, 1971), 19. Scholars continue to debate the actual Indian precedents for "Kentucky"; see Robert M. Rennick, "Place Names," in *The Kentucky Encyclopedia*, ed. John E. Kleber (Lexington: University Press of Kentucky, 1992), 725.
20. John Filson, *The Discovery, Settlement, and Present State of Kentucke* (Wilmington, Del.: James Adams, 1784); Richard Slotkin, *The Fatal Environment: The Myth of the Frontier in the Age of Industrialization, 1800–1890* (Middletown, Conn.: Wesleyan University Press, 1985), 67; Gilbert Imlay, *A Topographical Description of the Western Territory of North America* (1797; reprint, New York: Johnson Reprint Co., 1968), 1; Jack P. Greene, "America and the Creation of the Revolutionary Intellectual World of the Enlightenment," in his *Imperatives, Behaviors, and Identities: Essays in Early American Cultural History* (Charlottesville: University Press of Virginia, 1992), 353.

21. Keith Thomas, *Man and the Natural World: Changing Attitudes in England, 1500–1800* (New York: Oxford University Press, 1983), 150; Matt Cartmill, *A View to Death in the Morning: Hunting and Nature through History* (Cambridge, Mass.: Harvard University Press, 1993), 87–89, 107–9; Roderick Nash, *Wilderness and the American Mind* (New Haven, Conn.: Yale University Press, 1967), chap. 2; Imlay, *A Topographical Description*, 28; Arthur K. Moore, *The Frontier Mind: A Cultural Analysis of the Kentucky Frontiersman* (Lexington: University of Kentucky Press, 1957), 39; Chris Castanier, "'Roadworks': The Open Frontier in American Literature of Travel" (Ph.D. diss., Wayne State University, 1992), 9; Aron, *How the West Was Lost*, 112–13.
22. Slotkin, *The Fatal Environment*, 67; George P. Garrison, ed., "Memorandum of M. Austin's Journey from the Lead Mines in the County of Wythe in the State of Virginia to the Lead Mines in the Province of Louisiana West of the Mississippi," *AHR* 5 (1900): 526; Marion B. Stowell, *Early American Almanacs: The Colonial Weekday Bible* (New York: Burt Franklin, 1977), 273–74; Colin G. Calloway, *New Worlds for All: Indians, Europeans, and the Remaking of Early America* (Baltimore: Johns Hopkins University Press, 1997), 189.
23. Stephen A. Aron, "Pioneers and Profiteers: Land Speculation and the Homestead Ethic in Frontier Kentucky," *Western Historical Quarterly* 23 (1992): 182.
24. Richard Tarnas, *The Passion of the Western Mind: Understanding the Ideas That Have Shaped Our World View* (New York: Harmony Books, 1991), 91–170; Huston Smith, *Forgotten Truths: The Primordial Tradition* (New York: Harper & Row, 1976), 1–2.
25. William Cronon, "Introduction: In Search of Nature," in *Uncommon Ground: Toward Reinventing Nature* (New York: W. W. Norton, 1995), 24–25; idem, "The Trouble with Wilderness; or, Getting Back to the Wrong Nature," in *Uncommon Ground*, 73; Jonathan M. Smith, "Ramifications of Region and Sense of Place," in *Concepts of Human Geography*, ed. Carville Earle, Kent Mathewson, and Martin S. Kenzer (Lanham, Md.: Rowman and Littlefield, 1996), 193–95; Richard White, *The Organic Machine* (New York: Hill and Wang, 1995), ix–x; Carolyn Merchant, *Ecological Revolutions: Nature, Gender, and Science in New England* (Chapel Hill: University of North Carolina Press, 1989), chaps. 2–4; Karol Ann Peard Lawson, "An Inexhaustible Audience: The National Landscape Depicted in American Magazines, 1780–1820," *JER* 12 (1992): 305; Joseph Campbell, *The Power of Myth* (New York: Anchor Books, 1988), 115–16.
26. Drake, *Pioneer Life in Kentucky*, 125.
27. Georges-Henri-Victor Collot, *A Journey in North America*, 2 vols. (1826; reprint, Florence, Italy: O. Lange, 1924), 33–34; Michel Sobel, *The World They Made Together: Black and White Values in Eighteenth-Century Virginia* (Princeton, N.J.: Princeton University Press, 1987), chap. 7; Patrick Scott interview, 11CC8, Draper.
28. Unknown, codex #94, 13–14, Durrett; Bernard Bailyn, *The Peopling of British North America* (New York: Alfred A. Knopf, 1986), 115–19; Carolyn Merchant,

"Reinventing Eden: Western Culture as a Recovery Narrative," in Cronon, ed., *Uncommon Ground,* 132–59; "Some Particulars Relative to Kentucky," in *Travels in the Old South,* ed. Eugene Schwaab, 2 vols. (Lexington: University of Kentucky Press, 1973), 1: 53–60; Marion B. Lucas, *A History of Blacks in Kentucky,* Volume I: *From Slavery to Segregation, 1760–1891* (Frankfort: Kentucky Historical Society, 1992), xv; Gail S. Terry, "Sustaining the Bonds of Kinship in a Trans-Appalachian Migration, 1790–1811: The Cabell-Breckinridge Slaves Move West," *Virginia Magazine of History and Biography* 102 (1994): 455–58, 464–67; G. Hubert Smith, ed., "A Letter from Kentucky," *Mississippi Valley Historical Review* 19 (1932): 91; *Travels and Explorations of the Jesuit Missionaries in New France, 1610–1791,* in *The Jesuit Relations and Allied Documents,* ed. Reuben Gold Thwaites, 73 vols. (Cleveland: Burrows Brothers Co., 1899), 47: 145.

29. Thomas, *Man and the Natural World,* 61–64; Jeanette Edwards, "The Need for a 'bit of history': Place and Past in English Identity," in *Locality and Belonging,* ed. Nadia Lovell (New York: Rutledge Press, 1998), 161–63; Harold M. Proshansky, Abee K. Fabian, and Robert Kaminoff, "Place-Identity: Physical World Socialization of Self," in *Giving Places Meaning,* ed. Linda Groat (New York: Academic Press, 1995), 87–114; Drake, *Pioneer Life in Kentucky,* 14.

30. Thomas, *Man and the Natural World,* 20–21; Nash, *Wilderness and the American Mind,* chap. 2; Bernard Sheehan, *Seeds of Extinction: Jeffersonian Philanthropy and the American Indian* (Chapel Hill: University of North Carolina Press, 1973), 90; Leo Marx, *The Machine in the Garden: Technology and the Pastoral Ideal in America* (New York: Oxford University Press, 1964); and Nash, *Wilderness and the American Mind.*

31. James A. Henretta, "Families and Farms: *Mentalité* in Pre-Industrial America," *WMQ* 35 (1978): 3–32; Keith Wrightson, *English Society: 1580–1680* (New Brunswick, N.J.: Rutgers University Press, 1982), 53–55; Michael Merrill, "Cash Is Good to Eat: Self-Sufficiency and Exchange in the Rural Economy of the United States," *Radical History Review* 3 (1977): 42–71; Christopher Clark, "Household Economy, Market Exchange and the Rise of Capitalism in the Connecticut Valley, 1800–1860," *Journal of Social History* 13 (1979–80): 169–89; idem, *The Roots of Rural Capitalism: Western Massachusetts, 1780–1860* (Ithaca, N.Y.: Cornell University Press, 1990), 195–99; Elizabeth A. Perkins, *Border Life: Experience and Memory in the Revolutionary Ohio Valley* (Chapel Hill: University of North Carolina Press, 1998), 150.

For discussions of moral economy, see Gregory Nobles, "Capitalism in the Countryside: The Transformation of Rural Society in the United States," *Radical History Review* 41 (1988): 163–76; E. P. Thompson, "The Moral Economy of the Crowd in Eighteenth-Century England," *Past and Present* 50 (1971): 76–136; Darrett B. Rutman, "Community: A Sunny Little Dream," in *Small Worlds, Large Questions: Explorations in Early American Social History, 1600–1850* (Charlottesville: University Press of Virginia, 1994), 293–94; Alfred F. Young, "Afterward: How Radical Was the American Revolution?" in *Beyond* the *American Revolution: Explorations in the History of American Radicalism,* ed. Alfred F. Young (DeKalb: Northern Illinois University Press, 1993), 317–64.

32. David Hackett Fischer, *Albion's Seed: Four British Folkways in America* (New York: Oxford University Press, 1989), 583; John Mack Faragher, *Sugar Creek: Life on the Illinois Prairie* (New Haven, Conn.: Yale University Press, 1986), 135–36.

33. Fischer, *Albion's Seed,* 765–71; Gail S. Terry, "Family Empires: A Frontier Elite in Virginia and Kentucky, 1740–1815" (Ph.D. diss., College of William and Mary, 1992), 23; H. T. Duncan Interview, 16CC250, Draper; Collot, *A Journey in North America,* 1: 110–11.

34. Smith, "Ramifications of Region and Sense of Place," 195; Perkins, *Border Life,* 132–41, 150; Aron, *How the West Was Lost,* 24, 32–33; William Leavy, "A Memoir of Lexington and Its Vicinity," *RKHS* 40 (1942): 117; John Mack Faragher, *Daniel Boone: The Life and Legend of an American Pioneer* (New York: Henry Holt, 1992), 102, 105; Military Officers for Bourbon, Fayette, and Mason Counties, 1792, Miscellaneous Manuscripts, Durrett.

35. Ellen Eslinger, "The Problems of Governing Virginia's Kentucky Frontier" (paper presented at the Southern Historical Association meeting, 1994), 9.

36. Daniel Blake Smith, "'This Idea in Heaven': Image and Reality on the Kentucky Frontier," in Friend, ed., *The Buzzel about Kentuck,* 82–95; Joan E. Cashin, *A Family Venture: Men and Women on the Southern Frontier* (Baltimore: Johns Hopkins University Press, 1991), 44–49; Jack P. Greene, "The Colonial Southern Frontier," in *An Uncivil War: The Southern Backcountry during the American Revolution,* ed. Ronald Hoffman et al. (Charlottesville: University of Virginia Press, 1985), 648; Perkins, *Border Life,* 142–46; Margaret Ripley Wolfe, "Fallen Leaves and Missing Pages: Women in Kentucky History," *RKHS* 90 (1992): 67–68; Martin Wymore interview, 11CC130, Draper; Castanier, "Roadworks," 12–13.

37. Orders of Lieutenant Colonel George Trotter, 24 August 1804, Charles Scott Papers, UKSC.

38. Cecil Brown, *Old Roads in Kentucky* (Harrodsburg, Ky.: Daniel M. Hutton, n.d.), 2; William Littell, ed., *The Statute Law of Kentucky,* 3 vols. (Frankfort, Ky.: William Hunter, 1811), 1: 576–77, 185–86; Mason County Court records, 26 August 1789, in Nancy O'Malley, *A New Village Called Washington* (Lexington: Kentucky Humanities Council, 1987), 21. For similar patterns in other states, see Isabel Stewart Mitchell, *Roads and Road-Making in Colonial Connecticut* (Hartford, Conn.: Yale University Press, 1933); Donald C. Jackson, "Roads Most Traveled: Turnpikes in Southeastern Pennsylvania in the Early Republic," in *Early American Technology: Making & Doing Things from the Colonial Era to 1850,* ed. Judith A. McGaw (Chapel Hill: University of North Carolina Press, 1994), 197–239. Road repair became the responsibility of citizens under the watch of courts as early as 1555, when the British Parliament passed legislation instructing parishes to appoint surveyors

and orderers; see Wheaton J. Lanes, "The Early Highway in America, to the Coming of the Railroad," in *Highways in Our National Life,* ed. Jean Labatut and Wheaton J. Lane (Princeton, N.J.: Princeton University Press, 1950), 68–69.
39. Nathaniel S. Shaler, *Kentucky: Pioneer Commonwealth* (Boston: Houghton, Mifflin & Co., 1885), 22–23; Aron, *How the West Was Lost,* 28, 56, 57.
40. Robert Johnson to Patrick Henry, 5 December 1786, in Perkins, *Border Life,* 84, also 30: table 1; Thomas L. Purvis, "The Ethnic Descent of Kentucky's Early Population," 263: table 3; Fischer, *Albion's Seed,* 236–46, 633–39; Lee Shai Weissbach, "The Peopling of Lexington, Kentucky: Growth and Mobility in a Frontier Town," *RKHS* 81 (1983): 119, 126, 131; Albert H. Tillson Jr., "The Southern Backcountry: A Survey of Current Research," *Virginia Magazine of History and Biography* 98 (1990): 387–422; Drake, *Pioneer Life in Kentucky,* 179; John Melish, *Travels through the United States of America in the years 1806 and 1807, and 1809, 1810, and 1811,* 2 vols. (Philadelphia: Thomas and George Palmer, 1812), 1: 401, 413; Rowland Berthoff, "A Country Open for Neighborhood," in *Republic of the Dispossessed: The Exceptional Old-European Consensus in America* (Columbia: University of Missouri Press, 1997), 67–68; Gordon S. Wood, *The Radicalism of the American Revolution* (New York: Alfred A. Knopf, 1992), 245; Clarence Mondale, "Place-on-the-Move: Space and Place for the Migrant," in *Mapping American Culture,* ed. Wayne Franklin and Michael Steiner (Iowa City: University of Iowa Press, 1992), 55; R. C. Ballard Thruston, ed., "Letter by Edward Harris, 1797," *FCHQ* 2 (1928): 167; Adam Rankin, *A Review of the Noted Revival in Kentucky, Commenced in the Year of Our Lord, 1801* (Lexington, Ky.: John Bradford, 1801), 7; Thomas Bender, *Community and Social Change in America* (Baltimore: Johns Hopkins University Press, 1978), 71.
41. Thomas Jefferson to Marquis de Chastelleux, 2 September 1785, in *The Papers of Thomas Jefferson,* ed. Julian P. Boyd, 25 vols. (Princeton, N.J.: Princeton University Press, 1953), 8: 468.
42. See Appendix: Table 1. Origins of Revolutionary Veterans in Mason and Fayette Counties. Todd Harold Barnett, "The Evolution of 'North' and 'South': Settlement and Slavery on America's Sectional Border, 1650–1810" (Ph.D. diss., University of Pennsylvania, 1993), 262: table VI–II; Perkins, *Border Life,* 30: table 1.
43. Otto Rothert, ed., "John D. Shane's Interview with Pioneer John Hodge, Bourbon County," *FCHQ* 14 (1940): 177–78; Drake, *Pioneer Life in Kentucky,* 180, 202; François André Michaux, *Travels to the West of the Alleghany Mountains,* in *Early Western Travels, 1748–1846,* ed. Reuben Gold Thwaites, 32 vols. (Cleveland: Arthur H. Clarke, Co., 1904–7), 3: 247; John Hill to Peyton Skipworth, 4 April 1796, Peyton Skipworth Papers, FHS; Bayly Ellen Marks, "The Rage for Kentucky: Emigration from St. Mary's County, 1790–1810," in *Geographic Perspectives on Maryland's Past,* ed. Robert D. Mitchell and Edward K. Muller, Department of Geography Occasional Papers No. 4 (College Park: University of Maryland, 1979), 127.

44. Drake, *Pioneer Life in Kentucky*, 142, 150; Frances L. S. Dugan and Jacqueline P. Bull, eds., *Bluegrass Craftsman: Being the Reminiscences of Ebenezer Hiram Stedman, Papermaker 1808–1885* (Lexington: University of Kentucky Press, 1959), 11; George W. R. Corlis to John Corlis, 22 September 1816, Corlis Family Papers 1757–1852, FHS; Perkins, *Border Life*, 84–85; Ellen Eslinger, *Citizens of Zion: The Social Origins of Camp Meeting Revivalism* (Knoxville: University of Tennessee Press, 1999), chap. 4.
45. Drake, *Pioneer Life in Kentucky*, 160; Jackson Turner Main, *The Social Structure of Revolutionary America* (Princeton, N.J.: Princeton University Press, 1965), 257–58.
46. Tillson, "The Southern Backcountry," 396; Rhys Isaac, *The Transformation of Virginia 1740–1790* (New York: W. W. Norton, 1982), 302–5; Stanley Elkins and Eric McKitrick, *The Age of Federalism: The Early American Republic, 1788–1800* (New York: Oxford University Press, 1993), 472; Rachel N. Klein, *Unification of a Slave State: The Rise of the Planter Class in the South Carolina Backcountry, 1760–1808* (Chapel Hill: University of North Carolina Press, 1990).
47. Joseph Ficklin interview, 16CC259, Draper.
48. Drake, *Pioneer Life in Kentucky*, 31–32, 177.
49. Ibid., 13–14; Fischer, *Albion's Seed*, 577–78.
50. Drake, *Pioneer Life in Kentucky*, 180–82; Fischer, *Albion's Seed*, 760–63.
51. Drake, *Pioneer Life in Kentucky*, 182–84.
52. Frank Lawrence Owsley, *Plain Folk of the Old South* (Baton Rouge: Louisiana State University Press, 1949), chap. 3; Drake, *Pioneer Life in Kentucky*, 14, 26, 55, 179, 181–82, 188–89.
53. Drake, *Pioneer Life in Kentucky*, 202; Fischer, *Albion's Seed*, 538.
54. Drake, *Pioneer Life in Kentucky*, 52, 56, 96–97, 107–8; Grady McWhiney, *Cracker Culture: Celtic Ways in the Old South* (Tuscaloosa: University of Alabama Press, 1988), 81–88; Fischer, *Albion's Seed*, 541–44, 728–30.
55. Drake, *Pioneer Life in Kentucky*, 95; Fischer, *Albion's Seed*, 370, 703–8, 745; James Flint, *Letters from America, 1818–20*, in Thwaites, ed., *Early Western Travels, 1748–1846*, 9: 146–47; Susan Yandell to David Wendel, 16 January 1826, Yandell Family Papers 1823–1887, FHS; Penne Rested, *Christmas in America: A History* (New York: Oxford University Press, 1995), 25.
56. O'Malley, *A New Village Called Washington*, 45; Robert Peter, *History of Bourbon, Scott, Harrison, and Nicholas Counties, Kentucky*, ed. William Henry Perrin (1882; reprint, Cincinnati: Art Guild Reprints, 1968), 92; Joseph Ficklin interview; Leavy, "A Memoir of Lexington and its Vicinity," *RKHS* 40 (1942): 107, 364; *RKHS* 41 (1943): 333, 335, 338; Clay Lancaster, *Vestiges of the Venerable City: A Chronicle of Lexington, Kentucky, Its Architectural Development and Survey of Its Early Streets and Antiquities* (Lexington, Ky.: Lexington–Fayette County Historic Commission, 1978), 16; Joan Wessinger Conley, ed., *History of Nicholas County* (Carlisle, Ky.: Nicholas Co. Historical Society, 1976), 261; Peter Smeltzer, in Eslinger, *Citizens of Zion*, 82. The term "Dutch" commonly referred to

Germans; A. G. Roeber, "'The Origin of Whatever Is Not English among Us': The Dutch-speaking and the German-speaking Peoples of Colonial British America," in *Strangers in the Realm: Cultural Margins of the British Empire,* ed. Bernard Bailyn and Philip D. Morgan (Chapel Hill: University of North Carolina Press, 1991), 220–83.

57. Leavy, "A Memoir of Lexington," *RKHS* 41 (1943): 59, 108, 329, 333, 363, 365, 366; *Kentucky Gazette,* 21 November 1798; Peter, *History of Fayette County, Kentucky,* 389; Lindsey Apple, "The French in Central Kentucky: The Myth and Reality of Assimilation," *Border States* 4 (1983): 7–8; "The Life and Times of Robert B. McAfee and His Family Connections," *RKHS* 25 (1927): 135.

58. Thomas D. Clark, ed., *A Description of Kentucky in North America* (Lexington: University of Kentucky Press, 1945), 27–29; Howard Mumford Jones, *O Strange New World: American Culture, the Formative Years* (New York: Viking Press, 1964), 292–303.

59. Lucas, *History of Blacks in Kentucky,* xi–xv; Terry, "Sustaining the Bonds of Kinship in a Trans-Appalachian Migration," 455–76.

60. Unknown to Polly Breckinridge, 12 October 1804, Breckinridge Family Papers, LOC.

61. Fortesque Cuming, *Sketches of a Tour to the Western Country through the States of Ohio and Kentucky,* in Thwaites, ed., *Early Western Travels, 1748–1846,* 4: 198; Samuel Ayres to William Ayres, 22 February 1790, in Todd H. Barnett, "Virginians Moving West: The Early Evolution of Slavery in the Bluegrass," *FCHQ* 73 (1999): 241.

62. Aron, *How the West Was Lost,* 128–29; Eugene S. Braderman, "Early Kentucky: Its Virginia Heritage," *South Atlantic Quarterly* 38 (1939): 452; Michaux, *Travels to the West,* 3: 200–201; Lucas, *History of Blacks in Kentucky,* 105–106; Thomas M. Doerflinger, *A Vigorous Spirit of Enterprise: Merchants and Economic Development in Revolutionary Philadelphia* (Chapel Hill: University of North Carolina Press, 1986), 349.

63. David Rice, *An Outline of the History of the Church in the State of Kentucky* (Lexington, Ky.: Thomas T. Skillman, 1824), 230–33; Clement Eaton, "Slave-Hiring in the Upper South: A Step toward Freedom," *Mississippi Valley Historical Review* 46 (1960): 663–78; Sylvia R. Frey and Betty Wood, *Come Shouting to Zion: African American Protestantism in the American South and British Caribbean to 1830* (Chapel Hill: University of North Carolina Press, 1998), 128.

64. Lucas, *History of Blacks in Kentucky,* xv, 130–31; *Federal Census 1790: Kentucky;* Philip D. Morgan, *Slave Counterpoint: Black Culture in the Eighteenth Century Chesapeake & Lowcountry* (Chapel Hill: University of North Carolina Press, 1998), 661; Allan Kulikoff, "Uprooted Peoples: Black Migrants in the Age of the American Revolution, 1790–1820," in *Slavery and Freedom in the Age of the American Revolution,* ed. Ira Berlin and Ronald Hoffman (Charlottesville: University Press of Virginia, 1983), 149: table 1; Sobel, *The World They Made Together,* 3.

65. Drake, *Pioneer Life in Kentucky*, 192, 196; John F. Schermerhorn and Samuel J. Mills, *A Correct View of that part of the United States which lies west of the Allegany Mountains; with regard to religion and morals* (Hartford, Conn.: Peter B. Gleason and Co., 1814), 20.
66. *Kentucky Gazette*, 2 October 1798; Eslinger, *Citizens of Zion*, 162–69; Robert H. Bishop, ed., *The Memoirs of David Rice* (Lexington, Ky.: Thos. T. Skillman, 1824), 68; Clifford Geertz, "Religion as Cultural System," in *The Interpretation of Cultures: Selected Essays* (New York: Basic Books Inc., 1973), 89; Paul K. Conkin, *Cane Ridge: America's Pentecost* (Madison: University of Wisconsin Press, 1990), 64; Nathan O. Hatch, *The Democratization of American Christianity* (New Haven, Conn.: Yale University Press, 1989).
67. Marcus J. Borg, *Reading the Bible Again for the First Time* (San Francisco: HarperCollins, 2001), 11–12; Charles B. Talbert, ed., "Looking Backward through One Hundred Years: Personal Recollections of James B. Ireland," *RKHS* 57 (1959): 106.
68. Talbert, ed., "Looking Backward through One Hundred Years," 106; Keith Wrightson, *English Society, 1580–1680* (New Brunswick, N.J.: Rutgers University Press, 1982), 54; Ruth H. Bloch, "Religion and Ideological Change in the American Revolution," in *Religion and American Politics: From the Colonial Period to the 1980s*, ed. Mark A. Noll (New York: Oxford University Press, 1990), 46.

    Historian David Hall has argued that Calvinism was the predominant faith of most Americans prior to the Revolutionary War, and I would extend that generalization to the Early Republic, at least along the Maysville Road; Hall, "Religion and Society," in *Colonial British America*, ed. Jack P. Greene and J. R. Pole (Baltimore: Johns Hopkins University Press, 1984), 323. While I agree with Ruth Bloch and others who suggested that Calvinism was "not a rigid system of doctrines so much as an open-ended and ambiguous effort to resolve a series of fundamental tensions," I also think that the quest for answers did emanate from a rigid social construct organized around the polarities of pure (scriptural) and impure morals; Bloch, "Religion and Ideological Change in the American Revolution," 46. One's situation along that continuum was indicated by behavior, but could manifest as social status, ethnicity, race, gender, and wealth as well; see Marcus J. Borg, *Meeting Jesus Again for the First Time: The Historical Jesus and the Heart of Contemporary Faith* (San Francisco: HarperCollins, 1994), 46–53.
69. Talbert, ed., "Looking Backward through One Hundred Years," 105–6. Folklore often portrayed witches transforming people into horses and riding all night, but because only the form and not the rational attributes of the victim had been transformed, he or she would remember jumping ditches or, as in this case, carrying loads; Josiah Henry Combs, "Sympathetic Magic in the Kentucky Mountains: Some Curious Folk-Survivals," *Journal of American Folk-Lore* 27 (1914): 328–30.

70. William Lynwood Montell, *Ghosts along the Cumberland: Deathlore in the Kentucky Foothills* (Knoxville: University of Tennessee Press, 1975), 6–7; Drake, *Pioneer Life in Kentucky*, 199–200; Talbert, ed.," Looking Backward through One Hundred Years," 106.
71. Lucas, *A History of Blacks in Kentucky*, 130–31; Jonathan Butler, *Awash in a Sea of Faith: Christianizing the American People* (Cambridge, Mass.: Harvard University Press, 1990), 96–97; Eugene D. Genovese, *Roll, Jordon, Roll: The World the Slaves Made* (New York: Vintage Books, 1976), 217–27; Henry F. May, *The Enlightenment in America* (New York: Oxford University Press, 1976), xiv. Alan Taylor argued that beliefs in superstition, especially as they related to the economic and social position of believers, prevailed where economic growth lagged behind aspirations, where religious beliefs were most heterogeneous, and where poorer folk dominated; see Taylor, "The Early Republic's Supernatural Economy: Treasure Seeking in the American Northeast, 1780–1830," *American Quarterly* 38 (1986): 6–33.

    For Euramerican and African American folk beliefs about conjuring and witchcraft in Kentucky, see Sadie F. Price, "Kentucky Folklore," *Journal of American Folk-Lore* 14 (1901): 30–38; Hubert Gibson Shearin, "Some Superstitions in the Cumberland Mountains," *Journal of American Folk-Lore* 24 (1911): 319–22.
72. Talbert, ed.,"Looking Backward through One Hundred Years," 105.
73. Patrick Scott interview, 11CC8, Draper; John L. Brooke, *The Refiner's Fire: The Making of Mormon Cosmology, 1644–1844* (New York: Cambridge University Press, 1994), 27; Thomas E. Barton, *Virginia Folk Legends* (Charlottesville: University of Virginia Press, 1991), 80–85.
74. Bill Cecil-Fronsman, *Common Whites: Class and Culture in Antebellum North Carolina* (Lexington: University Press of Kentucky, 1992), 122; Butler, *Awash in a Sea of Faith*, 223–24; Christine Leigh Heyrman, *Southern Cross: The Beginnings of the Bible Belt* (New York: Alfred A. Knopf, 1997), 63–64, 73–74; Andrew Delbanco, *The Death of Satan: How Americans Have Lost the Sense of Evil* (New York: Farrar, Straus and Giroux, 1995), 96.
75. D. S. McCullough interview, 16CC301, Draper.
76. Robert Peter, *History of Fayette County, Kentucky*, ed. William Henry Perrin (1882; reprint, Easley, S.C.: Southern Historical Press, 1979), 316–17; Bishop, *Memoirs of David Rice*, 141–42; Robert Davidson, *History of the Presbyterian Church in Kentucky* (1847; reprint, Greenwood, S.C.: Attic Press, 1974), chap. 3; Drake, *Pioneer Life in Kentucky*, 110.
77. *Kentucky Gazette*, 13 April 1793; Borg, *Reading the Bible Again for the First Time*, 11–12; Gal. 6:7, *KJV*; Borg, *Meeting Jesus Again for the First Time*, 75–80; Drake, *Pioneer Life in Kentucky*, 192; John Lyle diary, 8 August 1801, KHS; John F. Wilson, "Religion and Revolution in American History," *Journal of Interdisciplinary History* 23 (1993): 597–98.
78. Drake, *Pioneer Life in Kentucky*, 192.

79. The term "village West" is inspired by Darrett Rutman with Anita Rutman, "The Village South," in *Small Worlds, Large Questions: Explorations in Early American Social History, 1600–1850* (Charlottesville: University Press of Virginia, 1994), 231–72.
80. Otis K. Rice, *Frontier Kentucky* (Lexington: University Press of Kentucky, 1975), 37–40; *Kentucky Gazette,* 8 October 1796; Elmer T. Clark et al., eds., *The Journal and Letters of Francis Asbury,* 3 vols. (Nashville: Abingdon Press, 1958), 2: 410–11.
81. Imlay, *A Topographical Description,* 24, 98, 154; Howard C. Douds, "Merchants and Merchandising in Pittsburgh, 1759–1800," *Western Pennsylvania Historical Magazine* 20 (1937): 127; Dwight L. Smith, ed., *The Western Journals of John May: Ohio Company Agent and Business Adventurer* (Cincinnati: Historical and Philosophical Society of Ohio, 1961), 158.
82. Harold B. Gill Jr. and George M. Curtis III, "A Virginian's First Views of Kentucky: David Meade to Joseph Prentiss, August 14, 1796," *RKHS* 90 (1992): 124; Garrison, ed., "Memorandum of M. Austin's Journey," 524–25; Collot, *A Journey in North America,* 1: 33.
83. Richard Lingeman, *Small Town America* (New York: G. P. Putnam's Sons, 1980), 87; Samuel M. Wilson, "The Old Maysville Road," in *Proceedings of the Second Annual Meeting of the Ohio Valley Historical Association* (Columbus: Ohio State Archaeological and Historical Society, 1909), 99; William Faux, *Memorable Days in America,* part 1, in Thwaites, ed., *Early Western Travels, 1748–1846,* 11: 190; W. R. Jillson, ed., "Samuel D. McCullough's Reminiscences of Lexington," *RKHS* 27 (1929): 426; A. S. Morrison, ed., "Isaac Weld 1796," in *Travels to Virginia in Revolutionary Times* (Lynchburg, Va.: J. P. Bell, 1922), 108–9; Isaac Weld, *Travels through the States of North America,* 2 vols. (London: John Stockdale, 1807), 1: 234.
84. Drake, *Pioneer Life in Kentucky,* 9; Lewis Condict, "Journal of a Trip to Kentucky in 1795," *Proceedings of the New Jersey Historical Society* 4 (1919): 124; Eckert, *That Dark and Bloody River,* 513; George W. R. Corlis to John Corlis, 8 June 1816, Corlis-Respess Family Papers, 1698–1984, FHS.
85. Smith, "This Idea in Heaven," 77–98; "Copy of the Original Letter of John Breckinridge to His Friend, Col. Joseph Cabell, in Buckingham, Virginia," *RKHS* 3 (1905): 20; Terry, "Sustaining the Bonds of Kinship in a Trans-Appalachian Migration," 455–76.
86. G. Glenn Clift, *A History of Maysville and Mason County,* 2 vols. (Lexington, Ky.: Transylvania Printing Co., 1936), 1: 45–46; Collot, *A Journey in North America,* 1: 97; Imlay, *A Topographical Description,* 29; James Rood Robertson, *Petitions of the Early Inhabitants of Kentucky* (Louisville, Ky.:
J. P. Morton & Co., 1914), 155–56; Rennick, *Kentucky Place Names,* 193; idem, "The Post Offices of Mason County," *FCHQ* 72 (1998): 290.
87. Charles Brunk Heinemann, *"First Census" of Kentucky 1790* (Washington, D.C.: M. Brumbaugh, 1940), 3; Collot, *A Journey in North America,* 1: 97; Cuming, *Sketches of a Tour,* 4: 169; Condict, "Journal of a Trip to Kentucky," 117;

Michaux, *Travels to the West*, 3: 195; Thomas Ashe, *Travels in America performed in 1806* (London: E. M. Blunt, 1808), 184; Dwight L. Smith and Ray Swick, eds., *A Journey through the West: Thomas Rodney's 1803 Journal from Delaware to the Mississippi Territory* (Athens: Ohio University Press, 1997), 93; Bayard Still, "To the West on Business," *Pennsylvania Magazine of History and Biography* 64 (1940): 18.

88. Collot, *A Journey in North America*, 97; Ashe, *Travels in America*, 185–86; David Meade to Ann Randolph, 1 September 1796, in Bayard Still, ed., "The Westward Migration of a Planter Pioneer in 1796," *WMQ* 21 (1941): 332; Michaux, *Travels to the West*, 3: 195; Needham Parry interview, 14CC2, Draper; John Cleve Symmes, in Charles R. Staples, *The History of Pioneer Lexington, 1779–1800* (Lexington: University of Kentucky Press, 1939), 61; "Narrative of John Heckwelder's Journey to the Wabash in 1792," *Pennsylvania Magazine of History and Biography* 12 (1888): 38; "Some Particulars Relative to Kentucky," 1: 55; Smith and Swick, eds., *A Journey through the West*, 94.
89. Asa Farrar interview, 13CC1, Draper; Collot, *A Journey in North America*, 98; Ashe, *Travels in America*, 185.
90. Thruston, ed., "Letter by Edward Harris, 1797," 165; Meade to Randolph, in Still, ed., "Westward Migration of a Planter Pioneer in 1796," 332; Thomas Chapman, "A Journey through the United States," in Schwaab, ed., *Travels in the Old South*, 1: 24; Collot, *A Journey in North America*, 98; "Some Particulars Relative to Kentucky," 55; Needham Parry interview.
91. O'Malley, *A New Village Called Washington*, 6–7; Drake, *Pioneer Life in Kentucky*, 10; idem, "Some Accounts of the Epidemic Diseases which prevail at Mays-Lick, in Kentucky," *Philadelphia Medical and Physical Journal* 3 (1808): 85–90.
92. Robertson, *Petitions*, 91–92; Collot, *A Journey in North America*, 98; Chapman, "A Journey through the United States," 24; Still, ed., "Westward Migration of a Planter Pioneer," 332; Joseph Scott map, accompanying *United States Gazetteer* (Philadelphia: n.p., 1795), FHS; Michaux, *Travels in the West*, 3: 172; Melish, *Travels through the United States*, 1: 411; Cuming, *Sketches of a Tour*, 4: 196; Still, "To the West on Business," 14; G. Glenn Clift, *"Second Census" of Kentucky* (Frankfort: Kentucky State Historical Society, 1954), vi.
93. Collot, *A Journey in North America*, 1: 105, 107.
94. Michaux, *Travels to the West*, 3: 197; Chapman, "A Journey through the United States," 30; Ned Darnaby interview, 11CC164, Draper.
95. Collot, *A Journey in North America*, 99–100; Valentine Peers to Eleanor Peers, 3 May 1797, Valentine Peers Correspondence, UKSC; Cuming, *Sketches of a Tour*, 4: 173; Flint, *Letters from America*, 9: 128–29; Drake, *Pioneer Life in Kentucky*, 59; Meade to Randolph, in Still, ed., "Westward Migration of a Planter Pioneer in 1796," 340.
96. Drake, *Pioneer Life in Kentucky*, 35, 75; Cuming, *Sketches of a Tour*, 4: 173; Rennick, *Kentucky Place Names*, 193.
97. Adlard Welby, *A Visit to North America*, in Thwaites, ed., *Early Western Travels, 1748–1846*, 12: 218; Collot, *A Journey in North America*, 99; Darrell Haug

Davis, *The Geography of the Bluegrass Region of Kentucky* (Frankfort: Kentucky Geological Survey, 1927), 24–26; John Filson, "Map of Kentucke" (Philadelphia: T. Rook, 1784); "Some Particulars Relative to Kentucky," 55; Michaux, *Travels to the West*, 3: 231; Imlay, *Topographical Description*, 29–30.

98. Perkins, *Border Life*, 77; Collot, *A Journey in North America*, 100–101; Needham Parry interview; James Morrow, ed., "Tours into Kentucky and the Northwest Territory: Three Journals by the Rev. James Smith of Powhatan County, Va., 1783–1797," *Ohio Archeological and Historical Society Publications* 16 (1907): 371; Rennick, *Kentucky Place Names*, 27–28; Daniel Drake, "Notices of the Principal Mineral Springs of Kentucky and Ohio," *The Western Journal of the Medical and Physical Sciences* 2 (1828): 142–67; Robert Johnson map, n.d., in Robertson, *Petitions*, 61–62; Michaux, *Travels to the West*, 3: 197; Still, "Westward Migration of a Planter Pioneer in 1796," 333.

99. Campbell, "Present and Presettlement Forest Conditions," 163; Coleman, *Stage-Coach Days in the Bluegrass*, 17; deposition of Col. Robert Patterson, 1810; McFarlan, *Geology of Kentucky*, 174; Hammon, "Pioneer Routes in Central Kentucky," 127–31.

100. Robertson, *Petitions*, 89–90; Collot, *A Journey in North America*, 100.

101. Cuming, *Sketches of a Tour*, 4: 177; Collot, *A Journey in North America*, 101.

102. Lucien Beckner, "John D. Shane's Copy of Needham Parry Diary of Trip Westward in 1794," *FCHQ* 22 (1948): 227; Cuming, *Sketches of a Tour*, 4: 178; Michaux, *Travels to the West*, 3: 176; Rennick, *Kentucky Place Names*, 197.

103. Imlay, *A Topographical Description*, 32; Campbell, "Present and Presettlement Forest Conditions," 156; Michaux, *Travels to the West*, 3: 229; S. Matthews interview, 11CC158, Draper; C. W. Short, "Prodromus Lexingtoniensis, Secundum Florendioeatum Digeste," *Transylvania Journal of Medicine* 1 (1828): 252; Meade to Randolph, in Still, ed., "Westward Migration of a Planter Pioneer in 1796," 335; Hugh Schiell to John Moylan, 19 June 1784, Wickliffe-Preston Papers, UKSC; Levi Todd interview, 15CC157, Draper.

104. Ashe, *Travels in America*, 189; Peter, *History of Bourbon, Scott, Harrison, and Nicholas Counties, Kentucky*, 89–90; Collot, *A Journey in North America*, 102; Michaux, *Travels to the West*, 3: 198–99; Robertson, *Petitions*, 147–48; Rennick, *Kentucky Place Names*, 226; Chapman, "A Journey through the United States," 25; Cuming, *Sketches of a Tour*, 4: 198; Richard Lee Mason, *Narrative of Richard Lee Mason in the Pioneer West, 1819* (New York: Charles Frederick Heartman, 1819), 29; Valentine Peers to Eleanor Peers, 19 April 1797, Valentine Peers Correspondence, UKSC.

105. Collot, *A Journey in North America*, 102; Cuming, *Sketches of a Tour*, 4: 181–82.

106. Lancaster, *Vestiges of the Venerable City*, 9–11; John W. Reps, *Town Planning in Frontier America* (Princeton, N.J.: Princeton University Press, 1969), 269; idem, *The Making of Urban America: A History of City Planning in the United States* (Princeton, N.J.: Princeton University Press, 1965), 95–103; Conrad M. Arensberg, "American Communities," *American Anthropologist* 57 (1955): 1143–62.

107. Robertson, *Petitions*, 60–62; Clay Lancaster, *Antebellum Architecture of Kentucky* (Lexington: University Press of Kentucky, 1991), 60–61, 127; Rennick, *Kentucky Place Names,* 171; Imlay, *Topographical Description,* 180; Needham Parry interview; Collot, *A Journey in North America,* 102–3; Chapman, "A Journey through the United States," 31; Michaux, *Travels to the West,* 3: 199.
108. Collot, *A Journey in North America,* 104; *Kentucky Gazette,* 19 January 1793; Lexington Trustees Minute Book, 19 May 1794, UKSC, microfilm.
109. A note on categorization of urban locales is in order. Beginning in the 1790s, federal census officials considered a place as a city only when its population exceeded 2,500. By this measure, no place along the old buffalo trace was statistically a city until 1805 when Lexington registered 2,820 free and enslaved residents, approximately 1 percent of the state's total population; Cuming, *Sketches of a Tour,* 4: 185. I accept Stanley Elkins and Eric McKitrick's categorizations of towns as having populations of at least 500 residents, and villages as having 150 to 500 residents; "A Meaning for Turner's Frontier," *Political Science Quarterly* 69 (1954): 341–42. I have accepted Glenn T. Trewartha's definition of a hamlet—see "The Unincorporated Hamlet: One Element of the American Settlement Fabric," *Annals of the Association of American Geographers* 33 (1943): 37—as a place of concentrated population with less than 150 people, although such places also exhibited nondemographic characteristics that differentiated them from farming neighborhoods.

For example, I think that these labels imply certain degrees of economic activity beyond farming, of governmental organization separate from the county court, and of established religion. By 1805, for example, hamlets like Mayslick and Ellisville had concentrated residential patterns, a tavern and a store, and maybe a church, but they lacked a markethouse, a militia, and a courthouse (although Ellisville did house the county court briefly in a log cabin). Villages like Millersburg and Maysville were more likely to support at least two churches, a few mercantiles and artisans' shops, and have some governmental organization, quite possibly in the form of a militia unit. Towns like Washington and Paris had several churches, stores, and artisans' shops, a militia, a courthouse, and a markethouse. Maysville's development into a town around the turn of the century was evidenced in the construction of a markethouse as well. A city like Lexington had similar institutions in greater numbers, but also housed institutions of education and the fine arts. For discussions of varying definitions for such polities, see Carole Shammas, "The Space Problem in Early United States Cities," *WMQ* 57 (2000): 505–43; Rutman, "The Village South," 231–72.
110. Elihu Barker map, n.d., UKSC.
111. R. S. Cotterill, "The Old Limestone Road: Pioneer Trade and Settlements along a Celebrated Kentucky Highway," *Kentucky Magazine* 1 (1916): 616–20; Raymond F. Betts, "'Sweet Meditation through This Pleasant Country': Foreign Appraisals of the Landscape of Kentucky in the Early Years of the Commonwealth," *RKHS* 90 (1992): 29.

112. See Appendix: Table 2. Centralized Settlement by 1810. *Federal Census 1810: Kentucky, Georgia, Indiana*; Carville Earle and Ronald Hoffman, "The Urban South: The First Two Centuries," in *The City in Southern History: The Growth of Urban Civilization in the South,* ed. Blaine A. Brownell and David R. Goldfield (Port Washington, N.Y.: Kennikat Press, 1977), 48–51; David R. Goldfield, *Cotton Fields and Skyscrapers: Southern City and Region, 1607–1980* (Baton Rouge: Louisiana State University Press, 1982), 24–26; Rutman, "The Village South," 231–72.
113. Collot, *A Journey in North America,* 97.
114. See Appendix: Table 3. Regional and Urban Population Growth in the Early Republic. "Table V: Population of the Four Leading U.S. Cities and Their Suburbs showing the decennial rate of increase, 1790–1810," in *The Growth of Seaport Cities, 1790–1825,* ed. David T. Gilchrist (Charlottesville: University of Virginia Press, 1967), 39; Richard C. Wade, *The Urban Frontier: Pioneer Life in Early Pittsburgh, Cincinnati, Lexington, Louisville, and St. Louis* (Chicago: University of Chicago Press, 1959), 7, 17–18, 40, 54; Condict, "Journal of a Trip to Kentucky," 120; David R. Goldfield and Blaine E. Brownell, *Urban America: A History* (Boston: Houghton Mifflin Co., 1990), 84–86.
115. Michaux, *Travels to the West,* 3: 199; Cuming, *Sketches of a Tour,* 4: 183; John R. Stilgoe, *Common Landscape of America, 1580 to 1845* (New Haven, Conn.: Yale University Press, 1982), 88–99.

2. GREAT SETTLERS

1. Raymond F. Betts, "'Sweet Meditation through This Pleasant Country': Foreign Appraisals of the Landscape of Kentucky in the Early Years of the Commonwealth," *RKHS* 90 (1992): 429–39.
2. Thomas D. Clark, *Historic Maps of Kentucky* (Lexington: University Press of Kentucky, 1979), 76; Denis Wood, *The Power of Maps* (New York: Guilford Press, 1992), chap. 1; Michael J. Puglisi, "Muddied Waters: A Discussion of Current Interdisciplinary Backcountry Studies," in David Colin Crass et al., eds., *The Southern Colonial Backcountry: Interdisciplinary Perspectives on Frontier Communities* (Knoxville: University of Tennessee Press, 1998), 43–44.
3. Georges-Henri-Victor Collot, *A Journey in North America,* 2 vols. (1826; reprint, Florence, Italy: O. Lange, 1924), 1: 110–11; Drew R. McCoy, *The Elusive Republic: Political Economy in Jeffersonian America* (New York: W. W. Norton, 1980), 19–21.
4. Collot, *A Journey in North America,* 1: 109–12.
5. Ibid., 1: 105, 107.
6. J. Hector St. John Crèvecoeur, *Letters from an American Farmer* (1782; reprint, Gloucester, Mass.: Peter Smith, 1968), 53.
7. Marion Tinling and Godfrey Davies, eds., *The Western Country in 1793: Reports on Kentucky and Virginia* (San Marino, Calif.: The Castle Press, 1948), vi–xv.

8. Thomas D. Clark, ed., *A Description of Kentucky in North America* (Lexington: University of Kentucky Press, 1945), 92.
9. Collot, *A Journey in North America*, 1: 113–14; Wood, *The Power of Maps*, chap. 1; Puglisi, "Muddied Waters," 43–44; Henry May, *The Enlightenment in America* (New York: Oxford University Press, 1976), xviii.
10. Keith Thomas, *Man and the Natural World: Changing Attitudes in England, 1500–1800* (New York: Oxford University Press, 1983), 182–97.
11. May, *The Enlightenment in America*, 153–251; idem, "The Constitution and the Enlightened Consensus," in *The Divided Heart: Essays on Protestantism and the Enlightenment in America* (New York: Oxford University Press, 1991), 147–60; idem, "The Jeffersonian Moment," in *The Divided Heart*, 161–78; Charles A. Miller, *Jefferson and Nature: An Interpretation* (Baltimore: Johns Hopkins University Press, 1988), 1–13; Thomas, *Man and the Natural World*, 41–50, 65–67; Peter Burke, "Fables of the Bees: A Case-Study in Views of Nature and Society," in *Nature and Society in Historical Context*, ed. Mikulas Teich, Roy Porter, and Bo Gustafsson (New York: Cambridge University Press, 1997), 112–14. The best example of this enlightened approach to nature is Thomas Jefferson's *Notes on the State of Virginia*, in which Jefferson elevated the grandeur of nature while distinguishing it clearly from human society; *Notes on the State of Virginia*, ed. William Peden (Chapel Hill: University of North Carolina Press, 1954).
12. James Madison to Thomas Jefferson, 30 August 1784, in *The Papers of James Madison*, ed. Robert A. Rutland et al., 29 vols. (Chicago: University of Chicago Press, 1962–91), 8: 108; Harry Toulmin, "Comments on America and Kentucky, 1793–1802," *RKHS* 47 (1949): 19. Also see Madison's concerns about the West in "Federalist #38," in Clinton Rossiter, ed., *The Federalist Papers* (New York: Penguin Books, 1961), 231–40.
13. Robert W. Tucker and David C. Hendrickson, *Empire of Liberty: The Statecraft of Thomas Jefferson* (New York: Oxford University Press, 1990), 159–65; John Lauritz Larson, "Jefferson's Union and the Problem of Internal Improvements," in *Jeffersonian Legacies*, ed. Peter S. Onuf (Charlottesville: University Press of Virginia, 1993), 340–69; Joyce Appleby, "Jefferson and His Complex Legacy," in *Jeffersonian Legacies*, 7–8; Bernard W. Sheehan, *Seeds of Extinction: Jeffersonian Philanthropy and the American Indian* (Chapel Hill: University of North Carolina Press, 1973), 95; Thomas Jefferson to James Madison, 20 December 1787, in Rutland et al., eds., *Papers of James Madison*, 10: 338.
14. Benjamin Rush, in Meredith Howard Pyne, "The American Philosophical Society and the Early History of Forestry in America," *Proceedings of the American Philosophical Society* 89 (1945): 454, 463; Andrew R. L. Cayton, *The Frontier Republic: Ideology and Politics in the Ohio Country, 1780–1825* (Kent, Ohio: Kent State University Press, 1986), 15; Jack P. Greene, "Independence, Improvement, and Authority: Toward a Framework for Understanding the Histories of the Southern Backcountry during the Era of the American

Revolution," in *An Uncivil War: The Southern Backcountry during the American Revolution,* ed. Ronald Hoffman, Thad W. Tate, and Peter J. Albert (Charlottesville: University Press of Virginia, 1985), 12; David Freeman Hawke, *Benjamin Rush: Revolutionary Gadfly* (Indianapolis: Bobbs-Merrill Co., 1971), 304–6.

15. Philip Freneau and Hugh Henry Brackenridge, "The Rising Glory of America," in *The Poems of Philip Freneau,* ed. Fred Lewis Pattee, 3 vols. (Princeton: University Library, 1902), 1: 74; Lawrence J. Friedman, *Inventors of the Promised Land* (New York: Alfred A. Knopf, 1975), 14; David Meade to Ann Randolph, in Bayard Still, ed., "The Westward Migration of a Planter Pioneer in 1796," *WMQ* 21 (1941): 327; Thomas Ashe, *Travels in America Performed in 1806* (London: E. M. Blunt, 1808), 68; John Filson, *The Discovery, Settlement, and Present State of Kentucke* (Wilmington, Del.: James Adams, 1784), 50.

16. Thomas Jefferson to James Madison, 20 December 1787, in Rutland et al., eds., *Papers of James Madison,* 10: 338; Gilbert Imlay, *A Topographical Description of the Western Territory of North America* (1797; reprint, New York: Johnson Reprint Co., 1968), 1; Jack P. Greene, "America and the Creation of the Revolutionary Intellectual World of the Enlightenment," in *Imperatives, Behaviors & Identities: Essays in Early American Cultural History* (Charlottesville: University Press of Virginia, 1992), 352, 365; McCoy, *The Elusive Republic,* chaps. 2, 3; Fredrika Johanna Teute, "Land, Liberty, and Labor in the Post-Revolutionary Era: Kentucky as the Promised Land" (Ph.D. diss., Johns Hopkins University, 1988), 40; Lowell Harrison, *Kentucky's Road to Statehood* (Lexington: University Press of Kentucky, 1992), chap. 1; Andrew R. L. Cayton, "'Separate Interests' and the Nation-State: The Washington Administration and the Origins of Regionalism in the Trans-Appalachian West," *JAH* 79 (1992): 42–47; "The Farmer," 1775, in *The Papers of Benjamin Franklin,* ed. Leonard W. Labaree et al., 30 vols. (New Haven, Conn.: Yale University Press, 1961), 5: 470–71; Crèvecoeur, *Letters from an American Farmer,* 36.

17. Thomas, *Man and the Natural World,* 244, 253; Thomas Bender, *Toward an Urban Vision: Ideas and Institutions in Nineteenth-Century America* (Lexington: University Press of Kentucky, 1975), 7; James Madison, "Federalist #10," in Rossiter, ed., *The Federalist Papers,* 102–3; Greene, "America and the Creation of the Revolutionary Intellectual World of the Enlightenment," 348–49; Edward S. Casey, *The Fate of Place: A Philosophical History* (Berkeley: University of California Press, 1997), 162–79; Charles A. Miller, *Jefferson and Nature: An Interpretation* (Baltimore: Johns Hopkins University Press, 1988), 206–7; Jefferson, *Notes on the State of Virginia,* ed. Peden, 164–65; Gordon S. Wood, *The Creation of the American Republic, 1776–1787* (Chapel Hill: University of North Carolina Press, 1969), 282–91.

18. Daniel Drake, *Pioneer Life in Kentucky: A Series of Reminisential Letters,* ed. Charles Drake (Cincinnati: Robert Clarke & Co., 1870), 34–35.

19. Jonathan M. Smith, "Ramifications of Region and Senses of Place," in *Concepts of Human Geography,* ed. Carville Earle, Kent Mathewson, and Martin S. Kenzer

(Lanham, Md.: Rowman & Littlefield, 1996), 198–99; Adam Smith, *An Inquiry into the Nature and Causes of the Wealth of Nations,* ed. R. H. Campbell et al., 2 vols. (London: Clarenden Press, 1976), 1: 282–83, 376–80; Appleby, "Jefferson and His Complex Legacy," 7; McCoy, *The Elusive Republic,* chap. 3; Allan Kulikoff, "The Transition to Capitalism in Rural America," *WMQ* 46 (1989): 125; idem, *The Agrarian Origins of American Capitalism* (Charlottesville: University Press of Virginia, 1992), 126.

20. Robert Johnson to Patrick Henry, 5 December 1786, in Elizabeth A. Perkins, *Border Life: Experience and Memory in the Revolutionary Ohio Valley* (Chapel Hill: University of North Carolina Press, 1998), 84; David Hackett Fischer and James C. Kelly, *Bound Away: Virginia and the Westward Movement* (Charlottesville: University Press of Virginia, 2000), 214–15.

21. Lowell H. Harrison, "A Virginian Moves to Kentucky, 1793," *WMQ* 15 (1958): 201; John Randolph, in Hugh A. Garland, *The Life of John Randolph of Roanoke,* 2 vols. (New York: D. Appleton and Co., 1850), 2: 15.

22. Elizabeth Preston Meredith, in Fredrika Johanna Teute, "Land, Liberty, and Labor in the Post-Revolutionary Era: Kentucky as the Promised Land" (Ph.D. diss., Johns Hopkins University, 1988), 144; David Meade, in Harold B. Gill Jr. and George M. Curtis III, eds., "A Virginian's First Views of Kentucky: David Meade to Joseph Prentiss, August 14, 1796, *RKHS* 90 (1992): 122.

23. *Laws of Kentucky,* 2 vols. (Lexington, Ky.: John Bradford, 1799), 1: 371–77, 2: 131–32; Robert M. Ireland, "Aristocrats All: The Politics of County Government in Ante-bellum Kentucky," *The Review of Politics* 32 (1970): 382–83; idem, *The County in Kentucky History* (Lexington: University Press of Kentucky, 1976), 3–4; Samuel M. Wilson, "The Old Maysville Road," in *Proceedings of the Second Annual Meeting of the Ohio Valley Historical Association* (Columbus: Ohio State Archaeological and Historical Society, 1909), 100–101.

24. David Meade to Ann Randolph, 1 September 1796, in Bayard Still, ed., "The Westward Migration of a Planter Pioneer in 1796," *WMQ* 21 (1941): 333–34; David Barrow diary, 30 July 1795, FHS. Log cabins, particularly lower-class "Virginia houses" with their earthen floors and wooden chimneys, were widely considered to be forms of temporary housing in the eighteenth century; C. Carson et al., "Impermanent Architecture in the Southern American Colonies," *Winterthur Portfolio* 16 (1981): 135–96.

25. James C. Klotter, *The Breckinridges of Kentucky, 1760–1981* (Lexington: University Press of Kentucky, 1986), 6, 25–27; Lowell H. Harrison, *John Breckinridge: Jeffersonian Republican* (Louisville, Ky.: Filson Club, 1969), 48–49, 109–10; Gail S. Terry, "Family Empires: A Frontier Elite in Virginia and Kentucky, 1740–1815" (Ph.D. diss., College of William and Mary, 1992), 93; Teute, "Land, Liberty, and Labor in the Post-Revolutionary Era," chap. 3; Albert H. Tillson Jr., *Gentry and Commonfolk: Political Culture on a Virginia Frontier, 1740–1789* (Lexington: University Press of Kentucky, 1991), 134–35; Marion Nelson Winship, "Kentucky *in* the New Republic: A Study of Distance and

Connection," in *The Buzzel about Kentuck: Settling the Promised Land,* ed. Craig Thompson Friend (Lexington: University Press of Kentucky, 1999), 101–23; Harry S. Laver, "'Chimney Corner Constitutions': Democratization and Its Limits in Frontier Kentucky," *RKHS* 95 (1997): 339, 366–67; John Randolph to James Monroe, 18 February 1804, James Monroe Papers, LOC; Lowell Harrison, "John Breckinridge and the Vice Presidency, 1804; A Political Episode," *FCHQ* 26 (1952): 155–65; Bernard Mayo, *Henry Clay: Spokesman of the New West* (Boston: Houghton Mifflin Co., 1937), 61, 185; Ellen Eslinger, "The Problems of Governing Virginia's Frontier" (paper presented at the Southern Historical Association meeting, 1994), 9; Richard E. Ellis, *The Jeffersonian Crisis: Courts and Politics in the Young Republic* (New York: W. W. Norton, 1971), 141; Rhys Isaac, *The Transformation of Virginia, 1740–1790* (New York: W. W. Norton, 1982), 38–39; Joan Wells Coward, *Kentucky in the New Republic: The Process of Constitution Making* (Lexington: University Press of Kentucky, 1979), 124–59; Matthew G. Schoenbachler, "The Origins of Jacksonian Politics: Central Kentucky, 1790–1840" (Ph.D. diss., University of Kentucky, 1996), 66–68.

26. Thomas, *Man and the Natural World,* 207; William Lynwood Montell and Michael Lynn Morse, *Kentucky Folk Architecture* (Lexington: University Press of Kentucky, 1976), 41; Rexford Newcomb, *Architecture in Old Kentucky* (Urbana: University of Illinois Press, 1953), 84.

27. Clay Lancaster, *Antebellum Architecture of Kentucky* (Lexington: University Press of Kentucky, 1991), 23, 141–42; Kim A. McBride and W. Stephen McBride, "From Colonization to the Twentieth Century," in *Kentucky Archaeology,* ed. R. Barry Lewis (Lexington: University Press of Kentucky, 1996), 191.

28. Lancaster, *Antebellum Architecture of Kentucky,* 70–71; David Hackett Fischer, *Albion's Seed: Four British Folkways in America* (New York: Oxford University Press, 1989), 264–74.

29. Lancaster, *Antebellum Architecture of Kentucky,* 70–71; Richard L. Bushman, *The Refinement of America: Persons, Houses, Cities* (New York: Alfred A. Knopf, 1992), 273–75.

30. Fischer, *Albion's Seed,* 655–62; Lancaster, *Antebellum Architecture of Kentucky,* chap. 1; Robert H. Weibe, *The Opening of American Society: From the Adoption of the Constitution to the Eve of Disunion* (New York: Alfred A. Knopf, 1984), 13; Terry G. Jordon and Matt Kaups, *The American Backwoods Frontier: An Ethnic and Ecological Interpretation* (Baltimore: John Hopkins University Press, 1989).

31. Wood, *Creation of the American Republic,* 57, 71; Tinling and Davies, eds., *The Western Country in 1793,* vi–xv; Clark, ed., *A Description of Kentucky in North America,* 92; Otto Rothert, ed., "John D. Shane's Interview with Pioneer John Hodge, Bourbon County," *FCHQ* 14 (1940): 177–78; Willard Rouse Jillson, *Pioneer Kentucky* (Frankfort: State Journal Co., 1934), 71–107; Frances L. S. Dugan and Jacqueline P. Bull, eds., *Bluegrass Craftsman: Being the Reminiscences of Ebenezer Hiram Stedman, Papermaker, 1808–1885* (Lexington: University of Kentucky Press, 1959), 15 n 3; Fischer, *Albion's Seed,* 812–16; Bernard Bailyn,

Voyagers to the West: A Passage in the Peopling of America on the Eve of the Revolution (New York: Vintage Books, 1986), 27.

I very much agree with E. P. Thompson that "class happens when some men, as a result of common experiences (inherited or shared), feel and articulate the identity of their interests as between themselves, and as against other men whose interests are different from (and usually opposed to) theirs"; Thompson, *The Making of the English Working Class* (New York: Vintage Books, 1963), 9. Class consciousness among Kentucky's gentry was not defined by wealth as much as it was based upon differences with pioneer culture which they perceived as detrimental to their own plans for western development.

32. Joseph Ficklin interview, 16CC257, Draper; Perkins, *Border Life,* 83.
33. Stanley Elkins and Eric McKitrick, "A Meaning for Turner's Frontier," *Political Science Quarterly* 69 (September 1954): 325–26; Robert R. Dykstra and William Silag, "Doing Local History: Monographic Approaches to the Smaller Community," *American Quarterly* 37 (1985): 420–21.
34. My argument about the weakness of democracy among pioneers only applies to their culture along the Maysville Road. Elsewhere—for example, in the Transylvania Company's colony in the southern Bluegrass—pioneers did push for democracy precisely because they shared similar socioeconomic conditions *and* because they found a common enemy in Richard Henderson and the company's proprietors; see Stephen A. Aron, *How the West Was Lost: The Transformation of Kentucky from Daniel Boone to Henry Clay* (Baltimore: Johns Hopkins University Press, 1996), 64–68.
35. *Kentucky Gazette,* 5 July 1788; Harry Innes to Thomas Jefferson, 27 August 1791, in *The Papers of Thomas Jefferson,* ed. Julian Bond, 25 vols. (Princeton, N.J.: Princeton University Press, 1950), 22: 86–87.
36. Laver, "Chimney Corner Constitutions," 339, 366–67; George Nicholas to James Madison, 5 September 1792, Madison MSS; Michael Zuckerman, "Tocqueville, Turner, and Turds: Four Stories of Manners in Early America" *JAH* 85 (1998): 13–42.
37. Kentucky Constitution (1792), art. 12, sec. 1; Thomas Bender, *Community and Social Change in America* (Newark, N.J.: Rutgers University Press, 1978), 5–11; Darrett B. Rutman and Anita H. Rutman, *A Place in Time: Middlesex County, Virginia, 1650–1750* (New York: W. W. Norton, 1984), chap. 1; Edward S. Casey, *The Fate of Place: A Philosophical History* (Berkeley: University of California Press, 1997), 162–79; Wood, *Creation of the American Republic,* 282–91; Thomas, *Man and the Natural World,* 41–50; John Lauritz Larson, "Jefferson's Union and the Problem of Internal Improvements," in *Jeffersonian Legacies,* ed. Peter S. Onuf (Charlottesville: University Press of Virginia, 1993), 343.
38. Teute, "Land, Liberty, and Labor in the Post-Revolutionary Era," chap. 3; William Littell, ed., *The Statute Law of Kentucky,* 4 vols. (Frankfort, Ky.: William Hunter, 1811–19), 1: 573–78, 646–48; 2: 52, 406–8; 3: 104–8; 4: 243–45; Tax List for 1804, Fayette County, KDLA.

39. Laver, "Chimney Corner Constitutions," 351–52; Coward, *Kentucky in the New Republic*, 30–31; Teute, "Land, Liberty, and Labor in the Post-Revolutionary Era," 128–31; McCoy, *The Elusive Republic*, 51, 114–19.
40. Christopher R. Abel, "Rich Man, Poor Man: Perceptions and Realities for Poor Whites in Antebellum Kentucky, 1787–1825" (unpublished manuscript, 1997), 11–12.
41. E. Anthony Rotundo, *American Manhood: Transformations in Masculinity from the Revolution to the Modern Era* (New York: Basic Books, 1993), 15–16; Wood, *Creation of the American Republic*, 68–69; Jack P. Greene, "The Concept of Virtue in Late Colonial British America," in *Imperatives, Behaviors, and Identities*, 208–35; Lowell H. Harrison, "Isaac Shelby," in John E. Kleber, ed., *The Kentucky Encyclopedia* (Lexington: University Press of Kentucky, 1992), 815–16; Lance Banning, *The Jeffersonian Persuasion: Evolution of a Party Ideology* (Ithaca, N.Y.: Cornell University Press, 1978), 75–77, 88–90.
42. Harry S. Laver, "Muskets and Plowshares: Kentucky's Militia, the Creation of Community, and the Construction of Masculinity, 1790–1850" (Ph.D. diss., University of Kentucky, 1998), 157–74; David Waldstreicher, *In the Midst of Perpetual Fetes: The Making of American Nationalism, 1776–1820* (Chapel Hill: University of North Carolina Press, 1997); Don Higginbotham, "The American Militia: A Traditional Institution with Revolutionary Responsibilities," in *Reconsiderations on the Revolutionary War: Selected Essays* (Westport, Conn.: Greenwood Press, 1978), 83–103.
43. Robert Peter, *History of Fayette County, Kentucky*, ed. William Henry Perrin (1882; reprint, Easley, S.C.: Southern Historical Press, 1979), 404; Simon P. Newman, *Parades and the Politics of the Street: Festive Culture in the Early American Republic* (Philadelphia: University of Pennsylvania Press, 1997), 40, 116–17.
44. Waldstreicher, *In the Midst of Perpetual Fetes*, 233; Cynthia A. Kierner, "Genteel Balls and Republican Parades: Gender and Early Southern Civic Rituals, 1677–1826," *Virginia Magazine of History and Biography* 104 (1996): 185–210; Newman, *Parades and the Politics of the Street*, 102.
45. Mary Beth Norton, *Liberty's Daughters: The Revolutionary Experience of American Women, 1750–1800* (Boston: Little, Brown & Co., 1980), 155–255; idem, "The Evolution of White Women's Experience in Early America," *AHR* 89 (June 1984): 593–619; Ruth H. Block, "The Gendered Meanings of Virtue in Revolutionary America," *Signs* 13 (1987): 37–58; Linda J. Kerber, *Women of the Republic: Intellect & Ideology in Revolutionary America* (Chapel Hill: University of North Carolina Press, 1980), 69–114, 283–85; Rotundo, *American Manhood*, 18; *Kentucky Gazette*, 19 July 1808; Glenn Wallach, *Obedient Sons: The Discourse of Youth and Generations in American Culture, 1630–1860* (Amherst: University of Massachusetts Press, 1997), 51–54.
46. Mary Ann Corlis to Susan Corlis, 22 February 1816, Corlis-Respess Family Papers, 1648–1984, FHS.

47. William T. Barry to John Barry, 2 January 1807, William Taylor Barry letters, 1798–1835, FHS; J. Winston Coleman Jr., "John Bradford and The Kentucky Gazette," *FCHQ* 34 (1960): 24–34; Rosiland Remer, *Printers and Men of Capital: Philadelphia Book Publishers in the New Republic* (Philadelphia: University of Pennsylvania Press, 1996), 25.
48. John Adams, "Novanglus," 30 January 1775, in *The Papers of John Adams*, ed. Robert J. Taylor et al., 10 vols. (Cambridge, Mass.: Harvard University Press, 1977–1996), 2: 242; Joyce Appleby, *Capitalism and a New Social Order: The Republican Vision of the 1790s* (New York: New York University Press, 1982), 102.
49. Donald R. Wright, *African Americans in the Early Republic, 1789–1831* (Arlington Heights, Ill.: Harlan Davidson, 1993), 79; Coward, *Kentucky in the New Republic*, 136–38; David Rice, *An Outline of the History of the Church in the State of Kentucky* (Lexington, Ky.: Thomas T. Skillman, 1824), 233; R. H. Stanton and E. C. Phister, arr., *The Charter of the City of Maysville and the Amendments together with the By-laws and Ordinances adopted by the City Council* (Maysville, Ky.: Samuel Pike, 1849), 66–67; *Acts Passed at the Second Session of the First General Assembly of the Commonwealth of Kentucky* (Lexington, Ky.: John Bradford, 1792), 20; *Kentucky Gazette*, 27 September 1798; Littell, *The Statue Law of Kentucky*, 2: 113–23; Kentucky Constitution, 1799, sec. 2, art. 8; Lexington Trustees Minutes Book, 27 May 1807, UKSC; Marion B. Lucas, *A History of Blacks in Kentucky*, Volume 1: *From Slavery to Segregation 1760–1891* (Frankfort: Kentucky Historical Society, 1992), 105–6.
50. Richard Lee Mason, *Narrative of Richard Lee Mason in the Pioneer West 1819* (New York: Charles Frederick Heartman, 1819), 28; Edward Strutt Abdy, *Journal of a Residence and Tour in the United States*, 3 vols. (London: John Murray, 1835), 2: 341; J. Winston Coleman Jr., *Slavery Times in Kentucky* (Chapel Hill: University of North Carolina Press, 1940), 96–97; George Trotter to William Worsley, 3 December 1810, 5CC9, Draper; Tinling and Davis, eds., *The Western Country in 1793*, 79; James Flint, *Letters from America*, in *Early Western Travels 1748–1846*, ed. Reuben Gold Thwaites, 32 vols. (Cleveland: Arthur H. Clarke Co., 1904–7), 9: 141; Michel Sobel, *The World They Made Together: Black and White Values in Eighteenth-Century Virginia* (Princeton, N.J.: Princeton University Press, 1987), 157–58; Todd H. Barnett, "Virginians Moving West: The Early Evolution of Slavery in the Bluegrass," *FCHQ* 73 (1999): 224.
51. Donald G. Mathews, *Religion in the Old South* (Chicago: University of Chicago Press, 1977), 3–10; idem, "The Second Great Awakening as an Organizing Process, 1780–1830: An Hypothesis," *American Quarterly* 21 (1969): 23–43; Richard Rankin, *Ambivalent Churchmen and Evangelical Churchwomen: The Religion of the Episcopal Elite in North Carolina, 1800–1860* (Columbia: University of South Carolina Press, 1993), 4; Herbert Leventhal, *In the Shadow of the Enlightenment: Occultism and Renaissance Science in Eighteenth-Century America* (New York: New York University Press, 1976), 234–35; A. Gregory

Schneider, "From Democratization to Domestication: The Transitional Orality of the American Methodist Rider," in *Communication & Change in American Religious History,* ed. Leonard I. Sweet (Grand Rapids, Mich.: William B. Eerdmans Publishing Co., 1993), 145–46; Christine Heyrman, *Southern Cross: The Beginnings of the Bible Belt* (New York: Alfred A. Knopf, 1997), 8.

The Episcopal *Book of Common Prayer,* revised in 1789, prescribed Episcopalians to "worship, and discipline, in such manner as they might judge most convenient for their future prosperity; consistently with the constitution and laws of the country." It was a Christian ethos that purposefully dovetailed with the republican ideology of the era, permitting followers to practice self-restraint and experiment theologically, while enjoying the pleasures of life; *A Book of Common Prayer, and Administration of the Sacraments, and other Rites and Ceremonies of the Church* (1789; reprint, New York: New York Bible and Common Prayer Book Society, 1865), preface.

52. Heyrman, *Southern Cross,* 11; Elizabeth King Smith and Mary LeGande Didlake, *Christ Church 1796–1946* (Richmond, Va.: Whitlett & Shipperson, 1946), 5; Frances Keller Swinford and Rebecca Smith Lee, *The Great Elm Tree: Heritage of the Episcopal Diocese of Lexington* (Lexington, Ky.: Faith House Press, 1969), 19–20; Clay Lancaster, *Vestiges of the Venerable City: A Chronicle of Lexington, Kentucky—Its Architectural Development and Survey of Its Early Streets and Antiquities* (Lexington, Ky.: Lexington–Fayette County Historic Commission, 1978), 26; Drake, *Pioneer Life in Kentucky,* 193.

53. American Episcopal Church Papers, 1802–1805, UKSC; Lancaster, *Vestiges of the Venerable City,* 27; Susan Yandell to Mrs. D. Wendell, n.d., Yandell Family Papers 1823–1887, FHS; Penne Restad, *Christmas in America: A History* (New York: Oxford University Press, 1995), 31; William Price to Edward Payne, 20 December 1787, William Price File 1755–1808, FHS.

54. William T. Barry to John Barry, 1801, William Taylor Barry Papers 1798–1835, FHS; *The Kentucky Almanac* (Lexington, Ky.: John Bradford, 1804), 12.

55. Heryman, *Southern Cross,* 7, 9; Garry Wills, *Under God: Religion and American Politics* (New York: Touchstone, 1990), 21.

56. Harry Toulmin, "Comments on America and Kentucky, 1793–1802," *RKHS* 47 (1949): 17; Robert Davidson, *History of the Presbyterian Church in Kentucky* (1847; reprint, Greenwood, S.C.: Attic Press, 1974), 124; John Lyle diary, 5 October 1801, KHS.

57. Michael Harris, "Lexington, Ky., 1807" (unpublished manuscript, n.d.), 8–9; "Extract of a Letter from a Gentleman to his Friend at the City of Washington, dated Lexington, Kentucky, March 9, 1801," in *The Baptists, 1783–1830: A Collection of Source Material,* ed. William W. Sweet (New York: H. Holt & Co., 1931), 610; *Kentucky Gazette,* 27 June, 4 July, 25 July 1795; Thomas Barr to Samuel Trotter, 18 September 1796, in Neils Sonne, *Liberal Kentucky 1780–1828* (Lexington: University of Kentucky Press, 1968), 27; Robert McAfee, "The Life and Times of Robert A. McAfee and His Family and

Connections," *RKHS* 25 (1927): 222–23; David Barrow diary, 1795, FHS; Eric Foner, *Tom Paine and Revolutionary America* (New York: Oxford University Press, 1976), 245–49.

58. David Barrow diary, 30 July 1795; Jonathan Butler, *Awash in a Sea of Faith: Christianizing the American People* (Cambridge, Mass.: Harvard University Press, 1990), 218–20; May, *The Enlightenment in America,* 263–64; Stanley Elkins and Eric McKitrick, *The Age of Federalism: The Early American Republic, 1787–1800* (New York: Oxford University Press, 1993), 329; Robert Stuart, "Reminiscences, Respecting the Establishment and Progress of the Presbyterian Church in Kentucky," *Journal of the Presbyterian Historical Society* 23 (1945): 166.

59. Joshua Wilson's valediction, 1802; Joshua Wilson to Sally Wilson, 14 March 1808, both in Joshua Lacy Wilson Papers, Durrett.

60. "The Methodist," in Charles A. Johnson, *The Frontier Camp Meeting: Religion's Harvest Time* (Dallas: Southern Methodist University Press, 1955), 204; Robert H. Bishop, *An Outline of the History of the Church in the State of Kentucky, During a Period of Forty Years: Containing the Memoirs of Rev. David Rice* (Lexington, Ky.: Thomas T. Skillman, 1824), 330–31; Rankin, *Ambivalent Churchmen and Evangelical Churchwomen,* 3–7, 18–21.

61. David Barrow diary, 30 July 1795; Schermerhorn and Mills, *A Correct View of that part of the United States which lies west of the Allegany Mountains,* 20; Josiah Morrow, "Tours into Kentucky and the Northwest Territory: Three Journals by the Rev. James Smith of Powhatan County, Va., 1783–1796–1797," *Ohio Archeological and Historical Society Publications* 16 (1907): 374; Donald F. Carmony, ed., "Spencer Records' Memoirs of the Ohio Valley Frontier, 1766–1795," *Indiana Magazine of History* 55 (1959): 377.

62. Davidson, *History of the Presbyterian Church in Kentucky,* 124; John Lyle diary, 5 October 1801.

63. Sonne, *Liberal Kentucky,* chap. 2; William Suddith interview, 12CC62, Draper; Willard Rouse Jillson, *A Transylvania Trilogy* (Frankfort: Kentucky State Historical Society, 1932).

64. *Kentucky Gazette,* 2 January, 5 May 1800; Sonne, *Liberal Kentucky,* chap. 2; John D. Wright Jr., *Transylvania: Tutor to the West* (Lexington, Ky.: Transylvania University, 1975), 37–39; Samuel Brown to Dr. James Speed, 5 January 1800, Joseph Hamiton Daveiss Papers, 1780–1856, FHS.

65. Testimony of Rice Jones, Transylvania University, in Sonne, *Liberal Kentucky,* 70–71, 68, 72; Thomas D. Clark, ed., *The Voice of the Frontier: John Bradford's Notes on Kentucky* (Lexington: University Press of Kentucky, 1993), 238; Wright, *Transylvania,* 42–43.

66. Joseph Ficklin interview, 16CC257, Draper; Margaretta Brown to John Brown, 6 August 1800, 8 March 1802, Brown Family Papers, 1799–1846, FHS; Lacy K. Ford, "Frontier Democracy: The Turner Thesis Revisited," *JER* 13 (1993): 155.

67. William Lytle to John Breckinridge, 10 January 1797, Papers of the Breckinridge Family, LOC; John Melish, *Travels through the United States of*

America in the Years 1806, 1807, and 1809, 1810, and 1811, 2 vols. (Philadelphia: Thomas and George Palmer, 1812), 1: 413; Thomas L. Purvis, "The Ethnic Descent of Kentucky's Early Population: A Statistical Investigation of European and American Sources of Emigration, 1790–1820," *RKHS* 80 (1982): 263: table 3; John B. Sanderlin, "Ethnic Origins of Early Kentucky Land Grantees," *RKHS* 85 (1987): 103–10; Patricia Watlington, *The Partisan Spirit: Kentucky Politics, 1779–1792* (New York: Atheneum, 1972), 223–34; Fischer, *Albion's Seed,* 236–46, 633–39; Lee Shai Weissbach, "The Peopling of Lexington, Kentucky: Growth and Mobility in a Frontier Town," *RKHS* 81 (1983): 119, 126, 131; and McCoy, *The Elusive Republic,* 18–19.

68. Robert Pettus Hay, "A Jubilee for Freemen: The Fourth of July in Frontier Kentucky, 1788–1816," *RKHS* 64 (1996): 171–73; *Kentucky Gazette,* 25 July 1799.
69. Hay, "A Jubilee for Freemen," 175–79; *Kentucky Gazette,* 14 July 1812, 17 July 1813; *Lexington Reporter,* 14 July 1810.
70. Darrett B. Rutman, "Community: A Sunny Little Dream," in *Small Worlds, Large Questions: Explorations in Early American Social History, 1600–1850* (Charlottesville: University Press of Virginia, 1994), 293–94; Watlington, *A Partisan Spirit,* 144–47; Mary K. Bonsteel Tachau, "The Whiskey Rebellion in Kentucky: A Forgotten Episode of Civil Disobedience," *JER* 2 (1982): 239–59; *Kentucky Gazette,* 26 September 1795; Matthew Schoenbachler, "Republicanism in the Age of Democratic Revolution: The Democratic-Republican Societies of the 1790s," *JER* 18 (1998): 258; Stephen A. Aron, *How the West Was Lost: The Transformation of Kentucky from Daniel Boone to Henry Clay* (Baltimore: Johns Hopkins University Press, 1996), 83–84; Patricia Watlington, "Discontent in Frontier Kentucky," *RKHS* 65 (1967): 77–93. For a differing opinion on the degree of expressed social discord, see Honor Sachs, "Rethinking Resistance: Popular Dissidence in the Kentucky Backcountry, 1774–1784" (paper presented at the Society for Historians of the Early American Republic, 1999).
71. Charles R. Staples, *The History of Pioneer Lexington, 1779–1800* (Lexington: University of Kentucky Press, 1939), 44, 49, 65, 75.
72. Robert D. Mitchell, "Revisionism and Regionalism," in *Appalachian Frontiers: Settlement, Society, & Development in the Preindustrial Era* (Lexington: University Press of Kentucky, 1991), 16.
73. Gregory H. Nobles, *American Frontiers: Cultural Encounters and Continental Conquest* (New York: Hill and Wang, 1997), 87–91.
74. Otis K. Rice, *Frontier Kentucky* (Lexington: University Press of Kentucky, 1975), chap. 5; Robert A. Gross, *The Minutemen and Their World* (New York: Hill and Wang, 1976), chap. 5. Malcolm J. Rohrbough suggested that no connection existed between expansion into the west and the outbreak of the Revolution; see *The Trans-Appalachian Frontier* (New York: Oxford University Press, 1978),

10. Considering its attempts at greater commercial regulation with the Stamp Act and Townshend Duties, however, the British government most certainly would have enforced the Proclamation of 1763 more forcefully had westward migration accelerated and conflict with the Indians increased.
75. Nancy Shoemaker, "Land Marks: How American Indians and Europeans Wrote Their Presence on the Landscape of Seventeenth- and Eighteenth-Century North America East of the Mississippi" (paper presented to the Omohundro Institute of Early American History and Culture, 1998). For a discussion of the concept of "social memory," see Scot A. French, "What Is Social Memory?" *Southern Cultures* 2 (1995): 9–18.
76. John H. Long, ed., *Kentucky: Atlas of Historical County Boundaries* (New York: Charles Scribner's Sons, 1995), 8–9.
77. R. S. Cotterill, "John Fleming: Pioneer of Fleming County," *RKHS* 49 (1951): 193–202; Huntley Dupree, "The Political Ideas of George Nicholas," *RKHS* 39 (1941): 201–23.
78. Carolyn Murray-Wooley, *The Founding of Lexington, 1775–1776* (Lexington, Ky.: Lexington–Fayette County Historical Commission, 1975), Appendix A: "Regarding the Site of the Naming of Lexington & Some Historians' Accounts"; Ellen Eslinger, "The Problems of Governing Virginia's Kentucky Frontier" (paper presented at the Southern Historical Association, 1994); Arthur K. Moore, *The Frontier Mind: A Cultural Analysis of the Kentucky Frontiersman* (Lexington: University of Kentucky Press, 1957), 39; Bernard Bailyn, *The Ideological Origins of the American Revolution* (Cambridge, Mass.: Harvard University Press, 1967), chap. 5; Wood, *The Creation of the American Republic*, chaps. 1, 2; Frederick Jackson Turner, *The Frontier in American History* (New York: Henry Holt and Co., 1920), 122.
79. Robert M. Rennick, *Kentucky Place Names* (Lexington: University Press of Kentucky, 1984), 226, 310.
80. J. Winston Coleman, "John Bradford and *The Kentucky Gazette*," *FCHQ* 34 (1960): 24–34; John Bradford, in George W. Ranck, *History of Lexington, Kentucky* (Cincinnati: Robert Clarke, 1872), 125.
81. Ronald R. Formisano, *The Transformation of Political Culture: Massachusetts Parties 1790s–1840s* (New York: Oxford University Press, 1983), 58; *Kentucky Almanac* (Lexington, Ky.: John Bradford, 1794), UKSC; Appleby, *Capitalism and a New Social Order*, 55; Teute, "Land, Liberty, and Labor in the Post-Revolutionary Era," 62, 72; *Kentucky Gazette*, 5 July 1788; Hay, "A Jubilee for Freemen," 177; Jeffrey L. Pasley, *"The Tyranny of Printers": Newspaper Politics in the Early American Republic* (Charlottesville: University Press of Virginia, 2001), chap. 3; Bernard Bailyn and Philip D. Morgan, eds., *Strangers within the Realm: Cultural Margins of the First British Empire* (Chapel Hill: University of North Carolina Press, 1991), 1; Bender, *Community and Social Change in America*, 79–82; Benedict Anderson, *Imagined Communities: Reflections on the Origin and Spread of Nationalism* (New

York: Verso Press, 1983), 7; Jack P. Greene, *Peripheries and Center: Constitutional Development in the Extended Polities of the British Empire and the United States, 1607–1788* (Athens: University of Georgia Press, 1986), xii.

82. Drake, *Pioneer Life in Kentucky,* 177; Perkins, *Border Life,* 168; Wayne Franklin and Michael Steiner, "Taking Place: Toward the Regrounding of American Studies," in *Mapping American Culture,* ed. Wayne Franklin and Michael Steiner (Iowa City: University of Iowa Press, 1992), 5; Fortesque Cuming, *Sketches of a Tour to the Western Country through the States of Ohio and Kentucky,* in Thwaites, ed., *Early Western Travels, 1748–1846,* 4: 177.

83. William Price to Isaac Shelby, 5 July 1794, Miscellaneous Manuscripts, UCSC; *The* (Lexington) *Reporter,* 20 July 1811; Fischer and Kelly, *Bound Away,* 140; *Kentucky Gazette,* 23 July 1811, 10 July 1815; Hay, "A Jubilee for Freemen," 177; Sarah J. Purcell, "The Commemoration of Heroes during the Revolutionary War" (h-net essay, 24 November 1997, http://Revolution.h-net.msu.edu); Simon P. Newman, *Parades and the Politics of the Streets: Festive Culture in the Early American Republic* (Philadelphia: University of Pennsylvania Press, 1997), 89; Laver, "Muskets and Plowshares," 158; John R. Stilgoe, *Common Landscape of America, 1580 to 1845* (New Haven, Conn.: Yale University Press, 1982), 107; Joyce Appleby, ed., *Recollections of the Early Republic: Selected Autobiographies* (Boston: Northeastern University Press, 1997), ix–xxi.

84. Henry Nash Smith, *Virgin Land: The American West as Symbol and Myth* (Cambridge, Mass.: Harvard University Press, 1950), chap. 6; J. J. Polk, *Autobiography of Dr. J. J. Polk* (Louisville, Ky.: John P. Morton & Co., 1867), 14–15; Madison, "Federalist #38," in Rossiter, ed., *The Federalists Papers,* 239; "Madison's Observations on Jefferson's Draft of a Constitution for Virginia (1788)," in *The Papers of Thomas Jefferson,* 6: 308–9; *Kentucky Gazette,* 25 January 1803; Winship, "Kentucky *in* the New Republic," 101–23; Kevin R. Gutman, "From Interposition to Nullification: Peripheries and Center in the Thought of James Madison," *Essays in History* 36 (1994): 90–113.

85. Gary J. Kornblith and John M. Murrin, "The Making and Unmaking of an American Ruling Class," in *Beyond the American Revolution: Explorations in the History of American Radicalism,* ed. Alfred F. Young (DeKalb: Northern Illinois University Press, 1993), 55; Basil Hall, in James Sterling Young, *The Washington Community 1800–1828* (New York: Columbia University Press, 1966), 91, also chap. 5.

86. *The Reporter,* 15 July 1809; (Frankfort) *Palladium,* 13 July 1811; Jeanette Edwards, "The Need for a 'bit of history': Place and Past in English Identity," in *Locality and Belonging,* ed. Nadia Lovell (New York: Rutledge, 1998), 161.

3. UPSETTING THE BALANCE

1. John Corlis to George William Respess Corlis, 15 January 1816; Joseph Corlis to John Corlis, 25 February 1816; George William Respess Corlis to John Corlis, 8 June 1816; George William Respess Corlis to Susan Corlis, 27 May 1816, all in Corlis-Respess Family Papers 1698–1984, FHS.
2. George William Respess Corlis to Susan Corlis, 27 May 1816.
3. Timothy Flint, in Thomas R. Cox et al., *This Well-Wooded Land: Americans and Their Forests from Colonial Times to the Present* (Lincoln: University of Nebraska Press, 1985), 54; *Kentucky Gazette,* 2 October 1798; J. M. Powell, *Mirrors of the New World: Image and Image-Makers in the Settlement Process* (Hamden, Conn.: Archon Books, 1977), 64–65.
4. Willard Rouse Jillson, "John Filson's Book and Map: Kentucke, 1784," *RKHS* 28 (1930): 283; Joyce Appleby, "Commercial Farming and the 'Agrarian Myth' in the Early Republic," *JAH* 68 (1982): 833–49.
5. Stephen A. Aron, *How the West Was Lost: The Transformation of Kentucky from Daniel Boone to Henry Clay* (Baltimore: Johns Hopkins University Press, 1996), chaps. 2, 3; Richard Lyman Bushman, "Markets and Composite Farms in Early America," *WMQ* 55 (1998): 351–74.
6. Harry Toulmin, "Comments on America and Kentucky," *RKHS* 47 (1949): 114.
7. Journal of Charles Julian, 1800–1818, UKSC.
8. Fredrika Johanna Teute, "Land, Liberty, and Labor in the Post-Revolutionary Era: Kentucky as the Promised Land" (Ph.D. diss., Johns Hopkins University, 1988), 632; Lacy K. Ford, "Frontier Democracy: The Turner Thesis Revisited," *JER* 13 (1993): 155; Gilbert Imlay, *A Topographical Description of the Western Territory of North America* (1797; reprint: New York: Johnson Reprint Co., 1968), 173.

    Despite his rhetoric, even Toulmin was drawn to Julian's vision. He did not even make a permanent investment in his Kentucky home, preferring to rent the thirty-acre farmstead for £18 sterling a year, a luxury few yeoman farmers could afford.
9. See Appendix: Table 4. Land Ownership along the Maysville Road, 1800. Joan Wells Coward, *Kentucky in the New Republic: The Process of Constitution Making* (Lexington: University Press of Kentucky); Teute, "Land, Liberty, and Labor," 632; James Henretta, "Families and Farms: *Mentalité* in Pre-Industrial America," *WMQ* 35 (1978): 6–9; Stephen A. Aron, "Pioneers and Profiteers: Land Speculation and the Homestead Ethic in Frontier Kentucky," *Western Historical Quarterly* 23 (1992): 179–98; Neal O. Hammon, "Settlers, Land Jobbers, and Outlyers: A Quantitative Analysis of Land Acquisition on the Kentucky Frontier," *RKHS* 84 (1986): 241–46; Samuel M. Wilson, *Kentucky Land Warrants, for the French, Indian, & Revolutionary Wars* (1917; reprint, n.p.: Southern Historical Press, 1994), 1; Neal O. Hammon, *Early Kentucky Land Records, 1773–1780* (Louisville, Ky.: Filson Club, 1992), 282, 288–89; Thomas D. Clark,

Agrarian Kentucky (Lexington: University Press of Kentucky, 1977), 6–10; Patricia Watlington, *The Partisan Spirit: Kentucky Politics, 1779–1792* (New York: Atheneum, 1972), 11–34; Lee Soltow, "Kentucky Wealth at the End of the Eighteenth Century," *Journal of Economic History* 43 (1983): 632–33, and appendix; idem, *Distribution of Wealth and Income in the United States in 1798* (Pittsburgh, Pa.: University of Pittsburgh Press, 1989), 77. For a discussion of the effects of landlessness on a large part of Kentucky's population, see Teute, "Land, Liberty, and Labor in the Post-Revolutionary Era," 620–24.

10. Aron, "Pioneers and Profiteers," 179–98; Bayard Still, ed., "The Westward Migration of a Planter Pioneer in 1796," *WMQ* 21 (1941): 341; Malcolm J. Rohrbough, *The Land Office Business: The Settlement and Administration of American Public Lands, 1789–1837* (New York: Oxford University Press, 1968), 23, 48; Kim Gruenwald, *River of Enterprise: The Commercial Origins of Regional Identity in the Ohio Valley, 1790–1850* (Bloomington: Indiana University Press, 2002), chap. 4; Alan Taylor, *William Cooper's Town: Power and Persuasion on the Frontier of the Early Republic* (New York: Alfred A. Knopf, 1995), 328; Georges-Henri-Victor Collot, *A Journey in North America*, 2 vols. (1826; reprint, Florence, Italy: O. Lange, 1924), 1: 101; Imlay, *A Topographical Description*, 175; Toulmin, "Comments on America and Kentucky," 100; John Melish, *Travels through the United States of America in the Years 1806 and 1807, and 1809, 1810, and 1811*, 2 vols. (Philadelphia: Thomas and George Palmer, 1812), 1: 404.

11. Lee Shai Weissbach, "The Peopling of Lexington, Kentucky: Growth and Mobility in a Frontier Town," *RKHS* 81 (1983): 115–33; Ellen Eslinger, "Migration and Kinship on the Trans-Appalachian Frontier: Strode's Station, Kentucky," *FCHQ* 62 (1988): 52–66; Gail S. Terry, "Family Empires: A Frontier Elite in Virginia and Kentucky, 1740–1815" (Ph.D. diss., College of William and Mary, 1992), 282; Robert E. Bieder, "Kinship as a Factor in Migration," *Journal of Marriage and the Family* 35 (1973): 429–39; Lowell H. Harrison, "A Virginian Moves to Kentucky, 1793," *WMQ* 15 (1958): 201; Toulmin, "Comments on America and Kentucky," 115; Michael Merrill and Sean Wilentz, eds., *The Key of Liberty: The Life and Democratic Writings of William Manning, "A Laborer," 1747–1814* (Cambridge, Mass.: Harvard University Press, 1993), 12; Joan E. Cashin, *A Family Venture: Men and Women on the Southern Frontier* (Baltimore: Johns Hopkins University Press, 1991), chap. 4; "Some Particulars Relative to Kentucky," in *Travels in the Old South,* ed. Eugene Schwaab, 2 vols. (Lexington: University Press of Kentucky, 1973), 1: 59–60; Paul G. E. Clemens and Lucy Simler, "Rural Labor and the Farm Household in Chester County, Pennsylvania, 1750–1820," in *Work and Labor in Early America,* ed. Stephen Innes (Chapel Hill: University of North Carolina Press, 1988), 110; Zhaohui Hong, "Changing Interpretations of Land Speculation in Western Development: Historians and the Shaping of the American West" (Ph.D. diss., University of Maryland, 1992), 8; Teute, "Land, Liberty, and Labor in the Post-

Revolutionary Era," 632; Stephen A. Aron, "Significance of the Frontier in the Transition to Capitalism," *The History Teacher* 27 (1994): 271–72.
12. John C. Wallace Journal, 1786–1802, FHS; Donald F. Carmony, ed., "Spencer Record's Memoirs of the Ohio Valley Frontier, 1766–1795," *Indiana Magazine of History* 55 (1959): 370.
13. Robert Breckinridge to Samuel Beall, 18 April 1792, Beall-Booth Papers, FHS; John Breckinridge to S. Meredith, n.d., Papers of the Breckinridge Family, LOC; Teute, "Land, Liberty, and Labor in the Post-Revolutionary Era," 294, 405; Allan Kulikoff, *From British Peasants to Colonial American Farmers* (Chapel Hill: University of North Carolina Press, 2000), 288. Historians have not yet determined the statewide extent of tenancy, which may have run as high as 25 percent by 1802; Teute, "Land, Liberty, and Labor in the Post-Revolutionary Era," 294.
14. *Kentucky Gazette,* 7 November, 21 November 1795; Lewis C. Gray, *History of Agriculture in the Southern United States to 1860,* 2 vols. (1932; reprint: Gloucester, Mass.: Peter Smith, 1958), 1: 646–57.
15. R. C. Ballard Thruston, ed., "Letter by Edward Harris, 1797," *FCHQ* 2 (1928): 167; *Kentucky Gazette,* 21 November 1795; Charles C. Bolton, *Poor Whites of the Antebellum South: Tenants and Laborers in Central North Carolina and Northeast Mississippi* (Durham, N.C.: Duke University Press, 1994).
16. Alexander B. Groshart, ed., *The Poems and Literary Prose of Alexander Wilson, the American Orinthologist,* 2 vols. (Paisley, Scotland: Alexander Gardner, 1876), 1: 184–85; Marion Tinling and Godfrey Davis, eds., *The Western Country in 1793: Reports on Kentucky and Virginia* (San Marino, Calif.: Castle Press, 1948), 68; John Hill to Peyton Skipworth, 4 April 1796, Peyton Skipworth Papers, FHS; William Faux, *Memorable Days in America,* part I, in *Early Western Travels, 1748–1846,* ed. Reuben Gold Thwaites, 32 vols. (Cleveland: Arthur H. Clarke, Co., 1904–7), 11: 187; Daniel Drake, *Pioneer Life in Kentucky: A Series of Reminisential Letters,* ed. Charles Drake (Cincinnati: Robert Clarke & Co., 1870), 206–7.
17. John T. Lyle interview, 14CC17, Draper; *Kentucky Gazette,* 7 February 1803.
18. "A Friend to the Distressed," *Kentucky Gazette,* 19 December 1795; *Acts Passed at the First Session of the Second General Assembly for the Commonwealth of Kentucky, 1793* (Lexington, Ky.: John Bradford, 1793), 45–46. Apparently, binding poor children into apprenticeship could be done originally without the knowledge of parents; in 1808, the Kentucky Supreme Court determined that county courts must first inform the next of kin before pursuing such a course; *Curry v. Jenkins,* 3 Hardin 501, *Kentucky Reports* (1808).
19. Gordon S. Wood, *Radicalism of the American Revolution* (New York: Alfred A. Knopf, 1992), 170–71; Jack P. Greene, "Independence, Improvement, and Authority: Toward a Framework for Understanding the Histories of the Southern Backcountry during the Era of the American Revolution," in *An*

Uncivil War: The Southern Backcountry during the American Revolution, ed. Ronald Hoffman, Thad W. Tate, and Peter J. Albert (Charlottesville: University Press of Virginia, 1985), 14; Asa Farrar interview, 13CC3, Draper; *Kentucky Gazette*, 25 April 1789.

20. Joan Wells Coward, *Kentucky in the New Republic: The Process of Constitution Making* (Lexington: University Press of Kentucky, 1979), 55: table 1; Ellen Eslinger, *Citizens of Zion: The Social Origins of Camp Meeting Revivalism* (Knoxville: University of Tennessee Press, 1999), 67–74.

21. Groshart, ed., *The Poems and Literary Prose of Alexander Wilson*, 185; Clarence Mondale, "Place-on-the-Move: Space and Place for the Migrant," in *Mapping American Culture*, eds. Wayne Franklin and Michael Steiner (Iowa City: University of Iowa Press, 1992), 54; Drake, *Pioneer Life in Kentucky*, 176; Mann Butler, in Thomas D. Clark, *A History of Kentucky* (Lexington, Ky.: John Bradford Press, 1960), 279.

22. Nathaniel Hart to James McDowell, 18 June 1809, James McDowell Papers, State Historical Society of Wisconsin, Madison; *Martin v. McKinney*, 2 Speed 321, *Kentucky Reports* (1804).

23. Drake, *Pioneer Life in Kentucky*, 57–58; James Rood Robertson, *Petitions of the Early Inhabitants of Kentucky to the General Assembly of Virginia, 1769–1792* (Louisville, Ky.: J. P. Morton & Co., 1914), 148–49; *Bourbon County Court Order Book A*, Kentucky Department for Libraries and Archives, Frankfort, 245, 288–89; Eslinger, *Citizens of Zion*, 47–48; Aron, *How the West Was Lost*, 117–21.

24. Clemens and Simler, "Rural Labor and the Farm Household," 111; *Acts Passed at the Second Session*, 36; *Laws of Kentucky*, 2 vols. (Lexington, Ky.: John Bradford, 1799), 1: 411–13; William Littell, *The Statute Law of Kentucky*, 3 vols. (Frankfort, Ky.: William Hunter, 1809), 3:78–80; G. Glenn Clift, *History of Maysville and Mason County*, 2 vols. (Lexington, Ky.: Transylvania Printing Co., 1936), 1: 137; U. B. Phillips, *Life and Labor in the Old South* (Boston: Little, Brown & Co., 1929), 64; *Kentucky Gazette*, 28 February 1804; Harry Watson, "'The Common Rights of Mankind': Subsistence, Shad, and Commerce in the Early Republican South," *JAH* 83 (1996): 13–43; Gary Kulik, "Dams, Fish, and Farmers: Defense of Public Rights in Eighteenth-Century Rhode Island," in Steven Hahn and Jonathan Prude, eds., *The Countryside in the Age of Capitalist Transformation: Essays in the Social History of Rural America* (Chapel Hill: University of North Carolina Press, 1985), 25–50.

25. André Michaux, *Journal of Travels into Kentucky*, in Thwaites, ed., *Early Western Travels 1748–1846*, 3: 237; Fortesque Cuming, *Sketches of a Tour to the Western Country*, in Thwaites, ed., *Early Western Travels 1748–1846*, 4: 170; Lucien Beckner, ed., "Reverend John D. Shane's Notes on Interviews, in 1844, with Mrs. Hinds and Patrick Scott of Bourbon County," *FCHQ* 10 (1936): 169; *Kentucky Gazette*, 16 September, 11 November 1797; James T. Lemon, *The Best Poor Man's Country: A Geographical Study of Southeastern Pennsylvania* (Baltimore:

Johns Hopkins University Press, 1972), 152; François André Michaux, *Travels to the West of the Alleghany Mountains,* in Thwaites, ed., *Early Western Travels, 1748–1846,* 3: 196; Faux, *Memorable Days in America,* 187.

26. Michaux, *Travels to the West,* 3: 205, 239; Faux, *Memorable Days in America,* 187; *Laws of Kentucky,* 2 vols. (Lexington, Ky.: John Bradford, 1799), 2: 411–13; Daniel Halstead to Alex Sanders, 1805, Daniel Halstead Account Books, 1797–1830, FHS; *Federal Census, 1810: Kentucky,* Mason County; Gordon S. Wood, *The Radicalism of the American Revolution* (New York: Alfred A. Knopf, 1992), 306.

27. Thruston, "Letter by Edward Harris," 166; James Flint, *Letters from America,* in Thwaites, ed., *Early Western Travels 1748–1846,* 9: 147; Joseph Hornsby diary 1798–1804, FHS; Journal of Charles Julian, UKSC; John McClelend File, KHS; Christopher Clark, *The Roots of Rural Capitalism: Western Massachusetts, 1780–1860* (Ithaca, N.Y.: Cornell University Press, 1990), chaps. 2, 3; Jeanne Boydston, *Home & Work: Housework, Wages, and the Ideology of the Early Republic* (New York: Oxford University Press, 1990), chaps. 1, 2; Mutch, "Yeoman and Merchant in Pre-Industrial America," 279–302. Historians note persistent needs for the services of courthouse, militia, and market even in the absence of towns; Darrett B. and Anita H. Rutman, *A Place in Time: Middlesex County, Virginia, 1650–1750* (New York: W.W. Norton, 1984), chap. 7; Carville Earle, *The Evolution of a Tidewater Settlement System: All Hallow's Parish, Maryland, 1650–1783* (Chicago: Department of Geography, University of Chicago, 1975); Joseph A. Ernst and H. Roy Merrens, "'Camden's turrets pierce the skies!': The Urban Process in the Southern Colonies during the Eighteenth Century," *WMQ* 30 (1973): 549–73; H. Roy Merrens, *Colonial North Carolina in the Eighteenth Century: A Study in Historical Geography* (Chapel Hill: University of North Carolina Press, 1964); Charles J. Farmer, *In the Absence of Towns: Settlement and Country Trade in Southside Virginia, 1730–1800* (Lanham, Md.: Rowman & Littlefield, 1993).

28. Hornsby Diary, FHS; Tinling and Davis, eds., *The Western Country in 1793,* 81; Journal of Charles Julian; *The Kentucky Farmers' Almanac* (Lexington, Ky.: William W. Worsley, 1810–11), UKSC.

29. Drake, *Pioneer Life in Kentucky,* 54; Hornsby Diary, 1 January 1803.

30. See Appendix: Table 5. Slave Ownership along the Maysville Road, 1800. Coward, *Kentucky in the New Republic,* 37, 63; "A Statement shewing the number of Slave Holders and Slaves in the several Counties, 1797," Bullitt Family Papers, FHS; Gray, *History of Agriculture in the Southern United States,* 1: 482.

31. Coward, *Kentucky in the New Republic,* 37, 63; Michaux, *Travels to the West,* 200–201; Bushman, "Markets and Composite Farms in Early America," 357.

32. Michaux, *Travels to the West,* 245–47; Drake, *Pioneer Life in Kentucky,* 45, 75; "Some Particulars Relative to Kentucky," 56; Still, ed., "Westward Migration of a Planter Pioneer," 335; Henry Williams, *Americans and Their Forests: A Historical*

Geography (New York: Cambridge University Press, 1989), 71; Carolyn Murray-Wooley and Karl Raitz, *Rock Fences of the Bluegrass* (Lexington: University Press of Kentucky, 1992), 77–82.

33. Valentine Peers to Eleanor Peers, 26 July 1801, Valentine Peers Correspondence, UKSC; Edmund Martin journal (Maysville), vol. B, July 1808–February 1811, MCHM.
34. Tinling and Davis, ed., *The Western Country in 1793,* 96–97; Mann Butler, "Manners and Habits of the Western Pioneers," UCSC, 25–26; Toulmin, "Comments on America and Kentucky," 106; Groshart, ed., *The Poems and Literary Prose of Alexander Wilson,* 195; Adrienne D. Hood, "The Material World of Cloth: Production and Use in Eighteenth-Century Rural Pennsylvania," *WMQ* 53 (1996): 43–66.
35. See Appendix: Table 6. Purchases of Cloth, 1792–1810. Flint, *Letters from America,* 9: 129.
36. "Some Particulars Relative to Kentucky," 56; Elizabeth A. Perkins, "The Consumer Frontier: Household Consumption in Early Kentucky," *JAH* 78 (1991): 506–7.
37. Laurel Thatcher Ulrich, *The Age of Homespun: Objects and Stories in the Creation of a Myth* (New York: Alfred A. Knopf, 2001), 289–90; Clark, *Roots of Rural Capitalism,* 95–117; Carole Shammas, "How Self-Sufficient Was Early America?" *Journal of Interdisciplinary History* 13 (1982–83): 268. During the 1790s, at least four bluedyers transformed plain, white homespun into calico for the citizens of Fayette County alone; J. Winston Coleman Jr., ed., *Lexington's First City Directory 1806* (Lexington, Ky.: Winburn Press, 1953); William A. Leavy, "A Memoir of Lexington and Its Vicinity," *RKHS* 41 (1943): 310–46.
38. E. P. Thompson, "The Moral Economy of the Crowd in Eighteenth-Century England," *Past and Present* 50 (1971): 76–136; Henretta, "Families and Farms," 3–22; Gregory Nobles, "Capitalism in the Countryside: The Transformation of Rural Society in the United States," *Radical History Review* 41 (1988): 163–76; Alfred F. Young, "Afterward: How Radical Was the American Revolution?" in Alfred F. Young, ed., *Beyond the American Revolution: Explorations in the History of American Radicalism* (DeKalb: Northern Illinois University Press, 1993), 317–64, esp. 319–22; Aron, *How the West Was Lost;* Matthew G. Schoenbachler, "The Origins of Jacksonian Politics: Central Kentucky, 1790–1840" (Ph.D. diss., University of Kentucky, 1996).
39. *Kentucky Gazette,* 13 September 1803; Mutch, "Yeoman and Merchant in Pre-Industrial America," 279–302; Richard B. Sheridan, "The Domestic Economy," in Jack P. Greene and J. R. Pole, eds., *Colonial British America: Essays in the New History of the Early Modern Era* (Baltimore: Johns Hopkins University Press, 1984), 43–85; James T. Lemon, "Spatial Order: Households in Local Communities and Regions," in Greene and Pole, eds., *Colonial British America,* 86–122; Stephen Innes, *Creating the Commonwealth: The Economic Culture of Puritan New England* (New York: W. W. Norton, 1995).

40. *Kentucky Gazette,* 11 October, 25 October 1803; Peter S. Onuf, "Liberty, Development, and Union: Visions of the West in the 1780s," *WMQ* 43 (1986): 193–203; J. E. Crowley, *This Sheba, Self: The Conceptualization of Economic Life in Eighteenth-Century America* (Baltimore: Johns Hopkins University Press, 1974), 2–6; Michael Merrill and Sean Wilentz, eds., *The Key of Liberty: The Life and Democratic Writings of William Manning, "A Laborer," 1747–1814* (Cambridge, Mass.: Harvard University Press, 1993), 12–13; T. H. Breen, "Narrative of Commercial Life: Consumption, Ideology, and Community on the Eve of the American Revolution," *WMQ* 50 (1993) 482; Perkins, "The Consumer Frontier," 486–510.
41. *Kentucky Gazette,* 14 March 1804.
42. Journal of Charles Julian; Toulmin, "Comments on America and Kentucky," 113.
43. Inside cover of Mason County, Kentucky, Account Book 1797–1799, FHS; Breen, "Narrative of Commercial Life"; Richard L. Bushman, "Opening the American Countryside," in *The Transformation of Early American History: Society, Authority, and Ideology,* eds. James A. Henretta, Michael Kammen, and Stanley N. Katz (New York: Alfred A. Knopf, 1992), 255; "Some Particulars Relative to Kentucky," 59; James L. Huston, "The American Revolutionaries, the Political Economy of Aristocracy, and the American Concept of the Distribution of Wealth," *AHR* 98 (1993): 1081; Wood, *The Radicalism of the American Revolution,* 36–38.
44. Gordon Wilson, ed., "An Ornithologist Visits Kentucky," *RKHS* 56 (1958): 238; Toulmin, "Comments on America and Kentucky," 19.
45. Otis K. Rice, *Frontier Kentucky* (Lexington: University of Kentucky Press, 1975), chaps. 5–7; Ellen Eslinger, "The Problem of Governing Virginia's Kentucky Frontier" (paper presented to the Southern Historical Association, 1994); Malcolm J. Rohrbough, *The Trans-Appalachian Frontier: People, Societies, and Institutions 1775–1850* (New York: Oxford University Press, 1978), 9; Richard L. Bushman, *The Refinement of America: Persons, Houses, Cities* (New York: Alfred A. Knopf, 1992), 404; Teute, "Land, Liberty, and Labor in the Post-Revolutionary Era," 34–40; Marion Nelson Winship, "The Portable Planter: Virginia Gentry Travel Across the Appalachians, 1790–1810" (paper presented to the Southern Historical Association, 1992), 1; Gail S. Terry, "Family Empires: A Frontier Elite in Virginia and Kentucky, 1740–1815" (Ph.D. diss., College of William and Mary, 1992); Greene, "Independence, Improvement, and Authority," 3–36; Andrew R. L. Cayton, *The Frontier Republic: Ideology and Politics in the Ohio Country, 1780–1825* (Kent, Ohio: Kent State Univ. Press, 1986); Albert H. Tillson Jr., "The Southern Backcountry: A Survey of Current Research," *Virginia Magazine of History and Biography* 98 (1990): 387–422.

   For the development of a commercial economy in the early West, see Perkins, "The Consumer Frontier," 508; John Mack Faragher, *Women and Men on the Overland Trail* (New Haven, Conn.: Yale University Press, 1979), 13; Rohrbough, *The Trans-Appalachian Frontier,* 41; Aron, "The Significance of the

Kentucky Frontier," *RKHS* 91 (1993): 318; Andrew R. L. Cayton, "Marietta and the Ohio Company," in *Appalachian Frontiers: Settlement, Society, & Development in the Preindustrial Era,* ed. Robert D. Mitchell (Lexington: University Press of Kentucky, 1991), 187–200.

Hints of the transition from a "frontier" economy to a more commercial orientation are found in the contrast between Benedick Leonard's 1791 Washington store ledger and post-1793 ledgers. Of forty barter transactions, eighteen were animal furs and skins. There is not one fur transaction in the other ledgers; see Benedick Leonard Ledger, 27 December 1790–21 November 1791, MCHM.

46. Gary J. Kornblith and John Murrin, "The Making and Unmaking of an American Ruling Class," in *Beyond the Revolution: Explorations in the History of American Radicalism,* ed. Alfred F. Young (DeKalb: Northern Illinois University Press, 1993), 28.
47. *Kentucky Gazette,* 13 September 1803; Bushman, *The Refinement of America,* 208–9; Robert H. Weibe, *The Opening of American Society: From the Adoption of the Constitution to the Eve of Disunion* (New York: Alfred A. Knopf, 1984), 11–12.
48. Thomas Irwin to Wilson Hunt, 3 February 1794, John Wesley Hunt Papers, FHS; Patrick Moore to Thomas Irwin, 26 April 1795; Wilson Hunt to Abijah and John W. Hunt, 15 April 1796, both in Abijah and John W. Hunt Papers, FHS.
49. John A. Seitz to unknown, 22 June 1796, Sullivan-Gates Papers, FHS.
50. Robert E. Mutch, "Yeoman and Merchant in Pre-Industrial America: Eighteenth Century Massachusetts as a Case Study," *Societas* 7 (1977): 292; Peter Dobkin Hall, *The Organization of American Culture, 1700–1900: Private Institutions, Elites, and the Origins of American Nationality* (New York: New York University Press, 1982), 73–74; James Bliss to Thomas Bodley, 11 October 1798, Bodley Family Papers, 1773–1939, FHS; Perkins, "The Consumer Frontier," 486–510; Aron, "The Significance of the Frontier in the Transition to Capitalism," in "The Transition to Capitalism in America: A Panel Discussion," *The History Teacher* 27 (1994): 263–88; idem, "Pioneers and Profiteers: Land Speculation and the Homestead Ethic in Frontier Kentucky," *Western Historical Quarterly* 23 (1992): 179–98.
51. *Kentucky Gazette,* 24 January 1798.
52. Greene, "Independence, Improvement, and Authority," 11; Hall, *The Organization of American Culture,* chap. 4; Bernard Bailyn, *The New England Merchants in the Seventeenth Century* (Cambridge, Mass.: Harvard University Press, 1955); Huston, "The American Revolutionaries, the Political Economy of Aristocracy, and the American Concept of the Distribution of Wealth, 1765–1900," 1087.
53. *Kentucky Gazette,* 28 March 1799, 11 March 1800, 1 May 1804, 24 September 1805; Robert Peter, *History of Fayette County, Kentucky,* ed. William Henry Perrin (1882; reprint, Easley, S.C.: Southern Historical Press, 1979), 373; *Lexington Directory Taken for Charles's Almanac for 1806,* UKSC.

54. Toulmin, "Comments on America and Kentucky, 1793–1802," 110; Frances L. S. Dugan and Jacqueline P. Bull, eds., *Bluegrass Craftsman: Being the Reminiscences of Ebenezer Hiram Stedman, Papermaker 1808–1885* (Lexington: University of Kentucky Press, 1959), 1.
55. Gary B. Nash, "The Social Evolution of Preindustrial American Cities, 1700–1820," in Raymond A. Mohl, ed., *The Making of Urban America* (Wilmington, Del.: Scholarly Resources, 1988), 38; Ronald Schultz, "Alternative Communities: American Artisans and The Evangelical Appeal, 1780–1830," in *American Artisans: Crafting Social Identity*, ed. Howard B. Rock, Paul A. Gilje, and Robert Asher (Baltimore: Johns Hopkins University Press, 1995), 66–67; Huston, "The American Revolutionaries, the Political Economy of Aristocracy, and the American Concept of the Distribution of Wealth, 1765–1900," 1083; Aron, *How the West Was Lost*, 140; Sean Wilentz, *Chants Democratic: New York City & the Rise of the American Working Class, 1788–1850* (New York: Oxford University Press, 1984), 92–93.
56. *Lexington Directory Taken for Charles's Almanac for 1806; Directory for 1818 for the Town of Lexington, Kentucky, from Worsley and Smith's Almanac*, UKSC; Cuming, *Sketches of a Tour*, 172–73; Jackson Turner Main, *The Social Structure of Revolutionary America* (Princeton, N.J.: Princeton University Press, 1965), 134–35.
57. Dugan and Bull, eds., *Bluegrass Craftsman*, 23; Toulmin, "Comments on America and Kentucky," 110.
58. W. J. Rorabaugh, "'I Thought I Should Liberate Myself from the Thraldom of Others': Apprentices, Masters, and the Revolution," in Young, ed., *Beyond the American Revolution*, 187; Charles R. Staples, *The History of Pioneer Lexington 1779–1806* (Lexington: University of Kentucky Press, 1939), 232; Peter Mason to Robert Stubblefield, 7 September 1807, Stubblefield Family Papers, UKSC; *Acts Passed at the First Session of the Second General Assembly for the Commonwealth of Kentucky, 1793* (Lexington, Ky.: John Bradford, 1793), 45–46.
59. Benjamin Wood to Joseph Lawrence, 27 June 1792, Wood Family Papers, UKSC; Cuming, *Sketches of a Tour*, 186; David Sayre interview, 15CC32, Draper; Peter, *History of Fayette County*, 289; Melish, *Travels through the United States*, 2: 188.
60. Tax List, Fayette County, 1804, KDLA; Thomas Hulme, *A Journal Made During a tour in the Western Countries of America: September 30, 1818–August 7, 1819*, in Thwaites, ed., *Early Western Travels 1748–1846*, 10: 66.
61. Rorabaugh, "'I Thought I Should Liberate Myself from the Thraldom of Others,'" 196; *Lexington Observer and Reporter*, 20 July 1811; *Kentucky Gazette*, 19 January 1801; Nash, "Social Evolution of Preindustrial American Cities," 35.
62. Leavy, "A Memoir of Lexington and Its Vicinity," *RKHS* 40 (1942): 117–31; David Hackett Fischer, *Albion's Seed: Four British Folkways in America* (New York: Oxford University Press, 1989), 755–56.
63. Craig T. Friend, "Trotter & Sons: Merchants and the Early West," in *The Human Tradition in the Early Republic*, ed. Michael Morrison (Wilmington,

Del.: Scholarly Resources Inc., 2000), 35–52; Gregory Nobles, "The Rise of Merchants in Rural Massachusetts: A Case Study of Eighteenth-Century Northhampton, Massachusetts," *JSH* 24 (1990): 5–23; W. T. Baxter, *The House of Hancock: Business in Boston, 1724–1775* (Cambridge, Mass.: Harvard University Press, 1945); Bernard Bailyn, *The New England Merchants in the Seventeenth Century* (Cambridge, Mass.: Harvard University Press, 1955); Edward C. Papenfuse, *In Pursuit of Profit: The Annapolis Merchants in the Era of the American Revolution 1763–1805* (Baltimore: Johns Hopkins University Press, 1975); Thomas M. Doerflinger, *A Vigorous Spirit of Enterprise: Merchants and Economic Development in Revolutionary Philadelphia* (Chapel Hill: University of North Carolina Press, 1986); idem, "Commercial Specialization in Philadelphia's Merchant Community, 1750–1791," *Business History Review* 57 (1983): 20–49; Glenn Porter and Harold C. Livesay, *Merchants and Manufactures: Studies in the Changing Structure of Nineteenth-Century Marketing* (Baltimore: Johns Hopkins University Press, 1971), 7; Gary A. O'Dell, "The Trotter Family, Gunpowder, and Early Kentucky Entrepreneurship, 1784–1833," *RKHS* 88 (1990): 394–430; William Vincent Byers, ed., *B. & M. Gratz: Merchants in Philadelphia, 1754–1798* (Jefferson City, Mo.: Hughs Stephens Printing Co., 1916), 258.

64. Willard R. Jillson, ed., "Samuel D. McCullough's Reminiscences of Lexington," *RKHS* 27 (1929): 426; William Lytle to John Breckinridge, 10 January 1797, Papers of the Breckinridge Family, LOC; Abijah Hunt to John W. Hunt, 30 April 1795, 18 April 1797, Abijah and John W. Hunt Papers.

65. *Acts Passed at the First Session of the Second General Assembly* (Lexington, Ky.: John Bradford, 1793), 54; Thomas T. Davis debate over Direct Tax, 5th Cong., 2nd sess., 13 June 1798, *Annals of Congress*, 43 vols. (Washington, D.C.: Gales and Seaton, 1851), 8: 1917; Lowell H. Harrison, ed., "A Letter Concerning Economic Conditions in Kentucky in 1802," *RKHS* 53 (1955): 258; Doerflinger, *A Vigorous Spirit of Enterprise*, 138; Lee Soltow, *Distribution of Wealth and Income in the United States in 1798* (Pittsburgh: University of Pittsburgh Press, 1989), 83.

66. John Moylan advertisement, *Kentucky Gazette*, 2 February 1793; John Clay to John W. Hunt, 3 June 1805; Abijah Hunt to John W. Hunt, 6 February 1800; Bartlett & Cox to John W. Hunt, 2 May 1804, all in Hunt-Morgan Papers, UKSC; James A. Ramage, "The Hunts and Morgans: A Study of a Prominent Kentucky Family" (Ph.D. diss., University of Kentucky, 1972), 112; Michael Allen, *Western Rivermen, 1763–1861: Ohio and Mississippi Boatmen and the Myth of the Alligator Horse* (Baton Rouge: Louisiana State University Press, 1990), 62.

67. Wilson P. Hunt to John W. Hunt, 2 May, 14 September 1804; 16 January 1805, Hunt-Morgan Papers.

68. *Kentucky Gazette*, 4 January 1794; François André Michaux, *Travels to the West of the Alleghany Mountains 1802*, in Thwaites, ed., *Early Western Travels, 1748–1846*, 3: 203; Clark, *Roots of Rural Capitalism*, 28–38, 163–79; T. H. Breen, "Narrative

of Commercial Life: Consumption, Ideology, and Community on the Eve of the American Revolution," *WMQ* 50 (1993): 471–501.
69. *Kentucky Gazette,* 13 September, 30 September, 4 October 1803.
70. Byers, ed., *B.& M. Gratz,* 28; Merrill and Wilentz, eds., *The Key of Liberty,* 12; Michael Merrill, "Cash Is Good to Eat: Self-Sufficiency and Exchange in the Rural Economy of the United States," *Radical History Review* 3 (1977): 54–61. For a contrast of interpretation, see Perkins, "The Consumer Frontier," 507.
71. Zodak Cramer, *The Navigator,* in *Who's Who on the Ohio River,* ed. Ethel C. Leahy (Cincinnati: E. C. Leahy Publishing Co., 1931), 91, 107; Clark, *The Roots of Rural Capitalism;* Allan Kulikoff, *Tobacco and Slaves: The Development of Southern Cultures in the Chesapeake, 1680–1800* (Chapel Hill: University of North Carolina Press, 1986); John T. Schlotterbeck, "The 'Social Economy' of an Upper South Community: Orange and Greene Counties, Virginia, 1815–1860," in *Class, Conflict, and Consensus: Antebellum Southern Community Studies,* eds. Orville Vernon Burton and Robert C. McMath (Westport, Conn.: Greenwood Press, 1992), 3–28; Winifred Barr Rothenberg, "The Market and Massachusetts Farmers," *Journal of Economic History* 41 (1981): 283–314; Robert D. Mitchell, *Commercialism and Frontier: Perspectives on the Early Shenandoah Valley* (Charlottesville: University Press of Virginia, 1977).
72. Thomas Sloo to John W. Hunt, 28 February 1800; Barton Hart & Co. to John W. Hunt, 16 September 1810, both in Hunt-Morgan Papers.
73. Michaux, *Travels to the West,* 203; Howard C. Douds, "Merchants and Merchandising in Pittsburgh, 1759–1800," *Western Pennsylvania Historical Magazine* 20 (1937): 129–30; Ramage, "The Hunts and Morgans," 26; *Directory for Pittsburgh & its Vicinity, 1813* and *Philadelphia Directory for 1808,* in *United States City Directories through 1860* (New Haven, Conn.: Research Publications, 1960–), microfiche; Meeker, Cockran & Co. to James Trotter, 5 December 1796; Bickham, Gellig & Co. to Samuel and George Trotter, 9 September 1803; James Adams to Samuel and George Trotter, 3 July 1810; James Adams to Samuel Trotter, 12 January 1811; James Adams to Samuel and George Trotter, 15 April 1811, all in Samuel and George Trotter Company Business Correspondence, CHS; Collot, *A Journey in North America,* 33; Cuming, *Sketches of a Tour,* 4: 169; Michaux, *Travels to the West,* 202; Byers, ed., *B.& M. Gratz,* 245; Doerflinger, *A Vigorous Spirit of Enterprise,* 86.
74. William Suddith interview, 12CC62, Draper; "Some Particulars Relative to Kentucky," in *Travels in the Old South,* ed. Eugene Schwaab, 2 vols. (Lexington: University of Kentucky Press, 1973), 1: 60. A series of letters between a Mr. Shaw and Governor Isaac Shelby relay the idea that neither gold nor silver was scarce in Kentucky; see N. Shaw to Isaac Shelby, 20 August 1794, 2 September 1794, Shelby Family Papers, LOC; Benedick Leonard's store ledger (Washington), December 1790–November 1791; John Moylan's store ledger (Lexington), January 1792–November 1794, UKSC; John W. Hunt's

daybook (Lexington), July 1790–September 1796, UKSC, microfilm; Daniel Halstead's account books (Lexington), 1806, FHS; William Tureman's daybook, January–May 1807, MCHM; Edmund Martin's journal (Maysville), vol. B, July 1800–February 1811, MCHM.

75. See Appendix: Table 7. Modes of Exchange in Stores along the Maysville Road. Samuel Smith debate over Direct Tax, 13 June 1798; 5th Cong., 2nd sess., *Annals of Congress* (Washington, D.C.: Gales and Seaton, 1851), 8: 1923.

76. See Appendix: Table 8. Lexington Trustees by Occupation, 1792–1812. Likewise, in the 1790s and early 1800s, the general assembly approved local laws that addressed street cleanings, town lots, horses and hogs within town limits, and shooting; see William Littell, ed., *The Statute Law of Kentucky*, 3 vols. (Frankfort, Ky.: William Hunter, 1809–11), vol. 1.

77. Mutch, "Yeoman and Merchant in Pre-Industrial America," 293; Richard C. Wade, *The Urban Frontier: Pioneer Life in Early Pittsburgh, Cincinnati, Lexington, Louisville, and St. Louis* (Chicago: University of Chicago Press, 1959), 78; Lexington Trustees Minute Book, 12 September 1790–19 May 1794, UKSC, microfilm.

78. Toulmin, "Comments on America and Kentucky, 1793–1802," 105; Carl Bridenbaugh, *Cities in Revolt: Urban Life in America, 1743–1776* (New York: Alfred A. Knopf, 1955), 80–82.

79. William J. Novak, *The People's Welfare: Law and Regulation in Nineteenth-Century America* (Chapel Hill: University of North Carolina Press, 1996), 95–105; Thomas F. DeVoe, *The Market Book, Containing a Historical Account of the Public Markets in the Cities of New York, Boston, Philadelphia, and Brooklyn*, 2 vols. (New York: Burt Franklin, 1862), 1: 330–455; Carl Bridenbaugh, *Cities in the Wilderness: The First Century of Urban Life in America, 1625–1742* (New York: Ronald Press Co., 1938), 192–95; Carole Shammas, *The Pre-Industrial Consumer in England and America* (Oxford, Eng.: Cambridge University Press, 1990), 266–67; Allan Kulikoff, *The Agrarian Origins of American Capitalism* (Charlottesville: University Press of Virginia, 1992), 19–21; Rothenberg, *From Market-Places to a Market Economy*, chap. 1; Hutson, "The American Revolutionaries, the Political Economy of Aristocracy, and the American Concept of the Distribution of Wealth, 1765–1900," 1087.

80. Lexington Trustees Minute Book, 23 April, 6 July 1795; 4 January, 14 January 1796; 23 April 1796; 1 May 1797; Cuming, *Sketches of a Tour*, 181; Clay Lancaster, *Vestiges of the Venerable City: A Chronicle of Lexington, Kentucky, Its Architectural Development and Survey of Its Early Streets and Antiquities* (Lexington, Ky.: Lexington–Fayette County Historic Commission, 1978), 15.

81. Lexington Trustees Minute Book, 18 April 1803; Wade, *The Urban Frontier*, 79–87; Cuming, *Sketches of a Tour*, 181–83; John W. Hunt's daybook; Thomas Chapman, "A Journey through the United States," in *Travels in the Old South*, 1: 31. Although the presence of bacon among these commodities may seem to compromise the argument, two factors should be taken into account. First, the John Hunt ledger records the summer months of July, August, and September

when citizens shunned the sale of meat items in the open market; see Michaux, *Travels to the West,* 203. Second, as a highly salted food item, bacon was popular among merchants because they could use it in regional and distant trade, and supplement the diets of boatmen transporting their goods; see Edmund Martin journals, 1808–1810.

82. Nancy O'Malley, *A New Village Named Washington* (Washington, Ky.: Old Washington, Inc., 1992), 15.
83. *Kentucky Gazette,* 24 January 1798; Mutch, "Yeoman and Merchant in Pre-Industrial America," 291; Lexington Trustees Minute Book, 21 October 1800, 26 December 1808, 4 December 1809, 15 December 1810.
84. Toulmin, "Comments on America and Kentucky," 106; Zodak Cramer, *The Navigator,* in *Who's Who on the Ohio River,* ed. Ethel C. Leahy (Cincinnati: E. C. Leahy Publishing Co., 1931), 91.
85. *Kentucky Gazette,* 17 January, 24 January 1798; Ruth Bogin, "Petitioning and the New Moral Economy of Post-Revolutionary America," *WMQ* 45 (1988): 425; Fernand Braudel, *Capitalism and Material Life, 1400–1800,* trans. Miriam Kochan (New York: Harper & Row, 1973), 400.
86. William Tureman's daybook, 1806, MCHM; Shammas, "How Self-Sufficient Was Early America?" 262.
87. Craig T. Friend, "'Fond Illusions' and Environmental Transformations Along the Maysville–Lexington Road," *RKHS* 94 (1996): 1–23.
88. David Meade to Judge Prentis, 14 August 1796, Webb-Prentis Family Papers, Albert and Shirley Small Special Collections Library, University of Virginia, Charlottesville; John Breckinridge to James Breckinridge, 9 May 1797, Papers of the Breckinridge Family, LOC; Harry Toulmin, *A Description of Kentucky in North America: To Which are Prefixed Miscellaneous Observations Respecting the United States,* ed. Thomas D. Clark (Lexington: University of Kentucky Press, 1945), 19.
89. George P. Garrison, ed., "A Memorandum of M. Austin's Journey from the Lead Mines of the County of Wythe in the State of Virginia to the Lead Mines in the Province of Louisiana West of the Mississippi, 1796–1797," *AHR* 5 (1900): 524–25.
90. "Some Particulars Relative to Kentucky," 1: 55; Thruston, ed., "Letter by Edward Harris, 1797," 165; Arthur C. McFarlan, *Geology of Kentucky* (Lexington: University of Kentucky Press, 1943), 169–72.
91. Meade to Randolph, in Still, ed., "The Westward Migration of a Planter Pioneer in 1796," 335; *Travels and Explorations of the Jesuit Missionaries in New France 1610–1791,* in *The Jesuit Relations and Allied Documents,* ed. Reuben Gold Thwaites, 73 vols. (Cleveland: Arthur H. Clarke Co., 1896–1901), 46: 145; "Some Particulars Relative to Kentucky," 55.
92. Garrison, ed., "Memorandum of M. Austin's Journey," 524–25; Michaux, *Travels to the West,* 3: 228–30; Williams, *Americans and Their Forests,* 60; Julian J. N. Campbell, "Present and Presettlement Forest Conditions in the Inner Bluegrass

of Kentucky" (Ph.D. diss., University of Kentucky, 1980), 171; William Clinkenbeard interview, 11CC58, Draper; "Report of the North American Land Company," 27 August 1795, Miscellaneous Manuscripts, UCSC; Cramer, *The Navigator*, 91.

93. "Some Particulars Relative to Kentucky," 55; John H. Long, ed., *Kentucky: Atlas of Historical County Boundaries* (New York: Charles Scribner's Sons, 1995), 63, 148, 316.

94. *Federal Census 1790: Kentucky*, Mason, Bourbon, and Fayette Counties; Long, *Kentucky*, 63–66, 148–50, 316–20, 359–60.

95. Campbell, "Present and Presettlement Forest Conditions," 184; Williams, *Americans and Their Forests*, 56; Michaux, *Travels to the West*, 230; Drake, *Pioneer Life in Kentucky*, 133, 165; Keith Thomas, *Man and the Natural World: Changing Attitudes in England, 1500–1800* (New York: Oxford University Press, 1983), 194. This ideological association of sensible cultivation to the new republican political order permeated American thought during the Early Republic; see Karol Ann Peard Lawson, "An Inexhaustible Audience: The National Landscape Depicted in American Magazines, 1780–1820," *JER* 12 (1992): 303–30.

96. Michaux, *Travels to the West*, 228; *Beckley v. Bryan and Ransdale*, 2 Speed 91, *Kentucky Reports* (1801).

97. Campbell, "Present and Presettlement Forest Conditions," 172–73; "Some Particulars Relative to Kentucky," 55.

98. Imlay, *A Topographical Description*, 30; William Clinkenbeard interview, 11CC60, Draper; John Floyd to William Preston, 19 January 1780, 17CC120, Draper; Campbell, "Present and Presettlement Forest Conditions," 163; H. E. Everman, *The History of Bourbon County, 1785–1865* (Paris, Ky.: Bourbon Press, 1977), 5; Michaux, *Travels to the West*, 321.

99. Williams, *Americans and Their Forests*, 60; Nancy O'Malley, *"Stockading Up": A Study of Pioneer Stations in the Inner Bluegrass Region of Kentucky* (Lexington, Ky.: Program for Cultural Resource Assessment, 1987), 310; Meade to Randolph, in Still, ed., "Westward Migration of a Planter Pioneer in 1796," 336; Alfred Crosby, *Ecological Imperialism: The Biological Expansion of Europe, 900–1900* (New York: Cambridge University Press, 1986).

100. Patricia Dalton Haragan, *Weeds of Kentucky and Adjacent States: A Field Guide* (Lexington: University Press of Kentucky, 1991), 21, 52, 188; Campbell, "Present and Presettlement Forest Conditions," 184; John H. Ellis, *Medicine in Kentucky* (Lexington: University Press of Kentucky, 1977), 24–25; Richard Lee Mason, *Narrative of Richard Lee Mason in the Pioneer West 1819* (New York: Charles Frederick Heartman, 1819), 28.

101. Cuming, *Sketches of a Tour*, 4: 174; Campbell, "Present and Presettlement Forest Conditions," 143; Williams, *Americans and Their Forests*, 45; Lyman Carter and Katherine S. Bort, "The History of Kentucky Bluegrass and White Clover in the United States," *Journal of the American Society of Agronomy* 8 (1916): 256–66; Crosby, *Ecological Imperialism*, 158; Ursula March Davidson, "The Original

Vegetation of Lexington, Ky., and Vicinity" (M.A. thesis, University of Kentucky, 1950).
102. Joshua McQueen interview, 13CC21, Draper; Ted Franklin Belue, *The Long Hunt: Death of the Buffalo East of the Mississippi* (Mechanicsburg, Pa.: Stackpole Books, 1996), chaps. 8, 9.
103. Nancy O'Malley, *A Cultural Evaluation of Archaeological Resources along Paris Pike, Bourbon and Fayette Counties, Kentucky* (Lexington, Ky.: Program for Cultural Resource Assessment, 1987), 69; idem, *Archaeological Test Excavations at Two Sites along Paris Pike, Bourbon County, Kentucky* (Lexington, Ky.: Program for Cultural Resource Assessment, 1993), 153; Ellen Eslinger, *Citizens of Zion: The Social Origins of Camp Meeting Revivalism* (Knoxville: University of Tennessee Press, 1999), 10–11; Aron, *How the West Was Lost*, 25–26.
104. David Barrow diary, FHS; O'Malley, *Archaeological Test Excavations*, 153; Michaux, *Travels to the West*, 235; Collot, *A Journey in North America*, 1: 111.
105. Still, ed., "Westward Migration of a Planter Pioneer," 335; Williams, *Americans and Their Forests*, 70–71; Carolyn Murray-Wooley and Karl Raitz, *Rock Fences of the Bluegrass* (Lexington: University Press of Kentucky, 1992), 77–82; Craig Thompson Friend, "'Work & Be Rich': Economy and Culture on the Bluegrass Farm," in *The Buzzel about Kentuck: Settling the Promised Land*, ed. Craig Thompson Friend (Lexington: University Press of Kentucky, 1999), 125–51.
106. *Kentucky Gazette*, 25 February 1812; William Taylor Barry to Catherine Barry, 17 November 1819, William Taylor Barry Letters, 1798–1835, FHS; William A. Leavy to Samuel Brown, 17 June 1820, Samuel Brown Papers 1817–1825, FHS; Clay Lancaster, *Antebellum Architecture of Kentucky* (Lexington: University Press of Kentucky, 1991), 30, 62–64; Groshart, *Poems and Literary Prose of Alexander Wilson*, 1: 191; Bushman, *The Refinement of America*, 134, 256–59; Fred Knitten and Henry Glassie, "Building in Wood in the Eastern United States: A Time-Place Perspective," in *Material Culture Studies in America*, ed. Thomas J. Schlereth (Nashville: American Association for State & Local History, 1982), 237–50; *Kentucky Gazette*, 26 December 1795. Lee Soltow demonstrated how, in another developing region, the higher the value on a residence, the less likelihood that it would be constructed of log; see "Table 5: Characteristics of Dwellings, Classified by Value in Mifflin County, Pennsylvania, in 1798," in "Housing Characteristics on the Pennsylvania Frontier: Mifflin County Dwelling Values in 1798," *Pennsylvania History* 47 (1980): 68.
107. Meade to Randolph, in Still, ed., "Westward Migration of a Planter Pioneer," 335; Meredith Howard Pyne, "The American Philosophical Society and the Early History of Forestry in America," *Proceedings of the American Philosophical Society* 89 (1945): 456.
108. William Clinkenbeard interview, 11CC58, Draper; William Cronon, *Changes in the Land: Indians, Colonists, and the Ecology of New England* (New York: Hill and Wang, 1983), 122–23; Timothy Silver, *A New Face on the Countryside: Indians, Colonists, and Slaves in South Atlantic Forests, 1500–1800* (New York:

Cambridge University Press, 1990), 134–35; Richard Lee, *Forest Hydrology* (New York: Columbia University Press, 1980), 133–43.

109. Daniel Drake, *A Systematic Treatise, Historical, Etiological, and Practical on the Principal Diseases of the Interior Valley of North America*, 2 vols. (Vol. 1, Cincinnati: Winthrop B. Smith & Co., 1850; Vol. 2, Philadelphia: Lippincott Crombe and Co., 1854), 1: 253–54; Michaux, *Travels to the West*, 198; William Taylor Barry to John Barry, 19 March 1806, William Taylor Barry Letters 1798–1835, FHS.

110. *Acts Passed at the Second Session of the First General Assembly of the Commonwealth of Kentucky* (Lexington, Ky.: John Bradford, 1792), 38.

111. W. R. Jillson, ed., "Samuel McCollough's Reminiscences of Lexington," *RKHS* 27 (1929): 417; Thomas Ashe, *Travels in America Performed in 1806* (London: E. M. Blunt, 1808), 191; William Taylor Barry to Catherine Barry, 3 Oct. 1819, William Taylor Barry Letters, 1798–1835; Joseph Wood to Joseph Lawrence, 11 March 1792, Wood Family Papers, UKSC; Thomas Chapman, "A Journey through the United States," in Schwaab, ed., *Travels in the Old South*, 1: 31; Mason, *Narrative of Richard Lee Mason*, 28.

112. Jillson, ed., "Samuel McCollough's Reminiscences of Lexington," 417, 430; Ashe, *Travels in America*, 191; William Taylor Barry to Catherine Barry, 3 October 1819, William Taylor Barry Letters, 1798–1835, FHS; Joseph Wood to Joseph Lawrence, 11 March 1792; Chapman, "A Journey through the United States," 1: 31; Mason, *Narrative of Richard Lee Mason*, 28; Collot, *A Journey in North America*, 102–3.

113. James A. Padgett, ed., "The Letters of Hubbard Taylor to President James Madison," *RKHS* 36 (1938): 117; *Kentucky Gazette*, 1 February 1794; Thruston, ed., "Letter by Edward Harris," 168; *Journal of the Senate at the Third Session of the General Assembly for the Commonwealth of Kentucky* (Lexington, Ky.: John Bradford, 1793); *Kentucky Gazette*, 14 December 1793; Michaux, *Travels to the West*, 205; John Duffy, *The Sanitarians: A History of American Public Health* (Urbana: University of Illinois Press, 1990), 48, 56–59, 68.

114. Duffy, *The Sanitarians*, 56–59, 68; Silver, *A New Face on the Countryside*, 192; Crosby, *Ecological Imperialism*, 138; Chapman, "A Journey through the United States," 31; Clay Lancaster, *Vestiges of the Venerable City: A Chronicle of Lexington, Kentucky* (Lexington, Ky.: Lexington–Fayette County Historic Commission, 1978), 9; Charles R. Staples, *The History of Pioneer Lexington, 1779–1806* (1939; reprint, Lexington: University Press of Kentucky, 1996), 323.

115. Lewis Condict, "Journal of a Trip to Kentucky in 1795," *Proceedings of the New Jersey Historical Society* 4 (1919): 120; Ellis, *Medicine in Kentucky*, 2; Staples, *History of Pioneer Lexington*, 316.

116. Williams, *Americans and Their Forests*, 59.

117. Daniel Drake, "Some Accounts of the Epidemic Diseases which prevail at Mays-Lick, in Kentucky," *The Philadelphia Medical and Physical Journal* 3

(1808): 85–90; Williams, *Americans and Their Forests,* 59; Joyce Appleby, *Inheriting the Revolution: The First Generation of Americans* (Cambridge, Mass.: Belknap Press, 2000), 16.

118. Edwin H. Ackerknecht, *A Short History of Medicine* (Baltimore: Johns Hopkins University Press, 1982), 223; John D. Wright Jr., *Transylvania: Tutor to the West* (Lexington, Ky.: Transylvania University, 1975), 81; Ronald L. Numbers and John Harley Warner, "The Maturation of American Medical Science," in *Sickness & Health in America,* eds. Judith Warner Leavitt and Ronald L. Numbers (Madison: University of Wisconsin Press, 1985), 121; John Shackleford, "On the Epidemic Fever of 1834" (M.D. thesis, Medical Department of Transylvania University, 1824), TU; Samuel T. McAdow, "On the Epidemic of 1824" (M.D. thesis, Medical Department of Transylvania University, 1825), TU; John A. English, "An Inaugural Dissertation on the Epidemic Cholera, as it appeared in the City of Lexington in June and July 1833" (M.D. thesis, Medical Department of Transylvania University, 1834), TU; Drake, *A Systematic Treatise.*

119. Bill Cecil-Fronsman, *Common Whites: Class and Culture in Antebellum North Carolina* (Lexington: University Press of Kentucky, 1992), 120–22; John Harley Warner, "The Idea of Southern Medical Distinctiveness: Medical Knowledge and Practice in the Old South," in Leavitt and Numbers, eds., *Sickness & Health in America,* 56–58; Charles E. Rosenberg, "The Therapeutic Revolution: Medicine, Meaning, and Social Change in Nineteenth-Century America," *Perspectives in Biology and Medicine* 20 (1977): 487; Alan Taylor, "The Early Republic's Supernatural Economy: Treasure Seeking in the American Northwest, 1780–1830," *American Quarterly* 38 (1986): 17.

120. Ellis, *Medicine in America,* 1; Cuming, *Sketches of a Tour,* 175; John D. Peers to Valentine Peers, 13 August 1808, Valentine Peers Collection; *The Kentucky Almanac* (Lexington, Ky.: John Bradford, 1794, 1800, 1804), UKSC; Daniel Drake, "Notices of the Principal Mineral Springs of Kentucky and Ohio," *The Western Journal of the Medical and Physical Sciences* 2 (1828): 142–67; Jon Butler, *Awash in a Sea of Faith: Christianizing the American People* (Cambridge, Mass.: Harvard University Press, 1990), 231–32; Paul Starr, *The Social Transformation of American Medicine: The Rise of a Sovereign Profession and the Making of a Vast Industry* (New York: Basic Books, 1982), chap. 1.

121. *Kentucky Almanac* (1794); Drake, *Pioneer Life in Kentucky,* 201.

122. Herbert Leventhal, *In the Shadow of the Enlightenment: Occultism and Renaissance Science in Eighteenth-Century America* (New York: New York University Press, 1976), 27; *The Farmer's Almanac* (Lexington, Ky.: W. Essex & Son, 1816), UKSC; *Kentucky Almanac* (1808).

123. *Kentucky Almanac* (1794); *The Farmers' Almanac* (Lexington, Ky.: James W. Palmer, 1822), UKSC; Drake, *Pioneer Life in Kentucky,* 200.

4. NEW AMERICANS?

1. J. Winston Coleman Jr., *Lafayette's Visit to Lexington: An Account of the General's Sojourn in the Bluegrass, May, 1825* (Lexington, Ky.: Winburn Press, 1969), 9–12.
2. David Meade to Norborne Beall, 15 December 1818, Beall-Booth Family Papers 1778–1953, FHS; Lewis Perry, *Boats against the Current: American Culture between Revolution and Modernity, 1820–1860* (New York: Oxford University Press, 1993), 15–17.
3. Benedict Anderson, *Imagined Communities: Reflections on the Origin and Spread of Nationalism* (New York: Verso, 1983), 5–6; Joseph Campbell, *The Power of Myth* (New York: Anchor Books, 1988), 163; "Soldiers to the Field!!!" Broadside, August 1823, in Matthew G. Schoenbachler, "The Origins of Jacksonian Politics: Central Kentucky, 1790–1840" (Ph.D. diss., University of Kentucky, 1996), 193; "To the People of Kentucky" Broadside, undated, Old Court–New Court Broadside collection, KHS.
4. John Lyle diary, 8 August 1801, KHS; Levi Purviance, *The Biography of Elder David Purviance* (Dayton, Ohio: B. F. & G. W. Ellis, 1848), 301. Estimates of the numbers at Cane Ridge may be found in Paul K. Conkin, *Cane Ridge: America's Pentecost* (Madison: University of Wisconsin Press, 1990), 88; Ellen Eslinger, *Citizens of Zion: The Social Origins of Camp Meeting Revivalism* (Knoxville: University of Tennessee Press, 1999), 208–10; Z. F. Smith, "The Great Revival of 1800: The First Camp-Meeting," *RKHS* 7 (1909): 19–35.
5. Conkin, *Cane Ridge*, 76–78, 80–81; George W. Ranck, "'The Traveling Church': An Account of the Baptist Exodus from Virginia to Kentucky in 1781," *RKHS* 79 (1981): 240–65; Ellen Eslinger, "Some Notes on the History of Cane Ridge Prior to the Great Revival," *RKHS* 91 (1993): 1–23.
6. Patrick Scott interview, 11CC8, Draper.
7. Rexford Newcomb, *Architecture in Old Kentucky* (Urbana: University of Illinois Press, 1953), 36; Clay Lancaster, *Antebellum Architecture in Kentucky* (Lexington: University Press of Kentucky, 1991), 27; Conkin, *Cane Ridge*, 77–78
8. Conkin, *Cane Ridge*, 74; Barton W. Stone, *History of the Christian Church in the West*, in *The Cane Ridge Reader*, ed. Hoke S. Dickinson (Cincinnati: Art Guild Reprints, 1972), 4.
9. Andrew Todd to Levi Todd, 6 April 1792, 16CC48, Draper; Robert H. Bishop, *An Outline of the History of the Church in the State of Kentucky, During a Period of Forty Years: Containing the Memoirs of Rev. David Rice* (Lexington, Ky.: Thomas T. Skillman, 1824), 336; Scott interview.
10. Nathan O. Hatch, *Democratization of American Christianity* (New Haven, Conn.: Yale University Press, 1989), chap. 6; Arthur K. Moore, *The Frontier Mind: A Cultural Analysis of the Kentucky Frontiersman* (Lexington: University of Kentucky Press, 1957), 38; Stephen A. Aron, "'The Poor Men to Starve': The Lives and Times of Workingmen in Early Lexington," in *The Buzzel about Kentuck: Settling the Promised Land,* ed. Craig Thompson Friend (Lexington: University Press of Kentucky, 1999), 175–93.

11. Josiah Morrow, ed., "Tours into Kentucky and the Northwest Territory: Three Journals by the Rev. James Smith of Powhatan County, Va., 1793–1797," *Ohio Archeological and Historical Society Publications* 16 (1907): 374; "A Lover of the True Rights of Man," *Kentucky Gazette,* 22 November 1803; John Lyle to Rev. John Lyle, 7 October 1800, Lyle Family Papers, UKSC; Donald G. Mathews, "The Second Great Awakening as an Organizing Process, 1780–1830: An Hypothesis," *American Quarterly* 21 (1969): 27; Herekiah Harriman to John Finklin, 5 June 1800, Hixson Collection, MCHM.
12. John B. Boles, *The Great Revival, 1787–1805* (Lexington: University Press of Kentucky, 1972), 51; "An Account of the Revival of Religion Which Began in the Eastern Part of the State of Kentucky in May, 1801," Robert Patterson Papers, Draper; Joan Wessinger Conley, ed., *History of Nicholas County* (Carlisle, Ky.: Nicholas Co. Historical Society, 1976), 262; John Rogers, *The Biography of Elder Barton W. Stone,* in Dickson, ed., *Cane Ridge Reader,* 37.
13. Adam Rankin, *A Review of the Noted Revival in Kentucky* (Lexington, Ky.: John Bradford, 1802), 47–63; "Extract of a letter from Colonel Robert Patterson of Lexington, Kentucky, to Rev. John King," *The Methodist Magazine* 26 (1803): 84.
14. Rogers, *The Biography of Elder Barton Warren Stone,* 37; "Extract of a letter from Colonel Robert Patterson of Lexington, Kentucky, to Rev. John King," 84.
15. J. B. Jackson, "The Order of a Landscape: Reason and Religion in Newtonian America," in *The Interpretation of Ordinary Landscapes: Geographical Essays,* ed. D. W. Meinig (New York: Oxford University Press, 1979), 156–57; Campbell, *The Power of Myth,* 119–20.
16. Roger Finke and Rodney Stark, "How the Upstart Sects Won America: 1776–1850," *Journal for the Scientific Study of Religion* 28 (1989): 28, 32; Paul G. Nagel, *This Sacred Trust: American Nationality 1798–1898* (New York: Oxford University Press, 1971), 12; Christopher Waldrep, "The Making of a Border State Society: James McGready, the Great Revival, and the Prosecution of Profanity in Kentucky," *AHR* 99 (1994): 772–73; Christine Heyrman, *Southern Cross: The Beginnings of the Bible Belt* (New York: Alfred A. Knopf, 1997), chaps. 2–5; D. Newell Williams, "Barton W. Stone's Revivalist Theology," in *Cane Ridge in Context: Perspectives on Barton W. Stone and the Revival,* ed. Anthony L. Dunnavant (Nashville: Disciples of Christ Historical Society, 1992), 80–81; Allan Kulikoff, "The Transition to Capitalism in Rural America," *WMQ* 46 (1989): 137; Ruth H. Bloch, "Religion and Ideological Change in the American Revolution," in *Religion and American Politics: From the Colonial Period to the 1980s,* ed. Mark A. Noll (New York: Oxford University Press, 1990), 44–61; Mathews, "The Second Great Awakening as an Organizing Process," 27.
17. Heyrman, *Southern Cross,* 268 n 3; Eslinger, *Citizens of Zion,* xvi; "An Account of the Revival of Religion Which Began in the Eastern Part of the State of Kentucky in May, 1801," Robert Patterson Papers, Draper.
18. Marcus J. Borg, *The God We Never Knew: Beyond Dogmatic Religion to a More Authentic Contemporary Faith* (New York: HarperCollins, 1997), 37–41; Clarence Mondale, "Place-on-the-Move: Space and Place for the Migrant," in *Mapping*

American Culture, ed. Wayne Franklin and Michael Steiner (Iowa City: University of Iowa Press, 1992), 184.

19. John Lyle diary, 8 August 1801; Conkin, *Cane Ridge,* 109; George Baxter to Mr. Davis, 4 September 1803, in Eslinger, *Citizens of Zion,* 223; Boles, *The Great Revival,* 67.

20. John Lyle diary, 6 October 1801; W. P. Strickland, ed., *The Autobiography of Rev. James B. Finley, or Pioneer Life in the West* (Cincinnati: Methodist Book Concern, 1854), 364; Conkin, *Cane Ridge,* 99, 108–9; Robert M. Calhoun, "The Evangelical Persuasion," in *Religion in a Revolutionary Age,* ed. Ronald Hoffman and Peter J. Albert (Charlottesville: University Press of Virginia, 1994), 157.

21. Strickland, ed., *The Autobiography of Rev. James B. Finley,* 166–67; unknown, in *Increase in Piety, or the Revival of Religion in the United States of America* (Newburyport, Conn.: Angier March, 1802), 53.

22. Conkin, *Cane Ridge,* 106; Borg, *The God We Never Knew,* chap. 2. The "New Light" was the colloquial name given the *pomoxis annularis Rafinesque,* the crappie species of sunfish first identified in the early 1800s by a Transylvania University natural scientist; David Starr Jordon and Barton Warren Evermann, *American Food and Game Fishes* (New York: Doubleday, Page & Co., 1902), 334.

23. Andrew Delbanco, *The Death of Satan: How Americans Have Lost the Sense of Evil* (New York: Farrar, Straus, and Giroux, 1995), 57.

24. Allan Kulikoff, "The Transition to Capitalism in Rural America," *WMQ* 46 (1989): 137; Waldrep, "The Making of a Border State Society," 773.

25. George Baxter to Archibald Alexander, 1 January 1802, in *Increase in Piety,* 63–64; Terah Templin to Joshua L. Wilson, 2 February 1803, Joshua Lacy Wilson Papers, Durrett.

26. Nagel, *This Sacred Trust,* 12; Bertram Wyatt-Brown, *Southern Honor: Ethics and Behavior in the Old South* (New York: W. W. Norton, 1982), 298; Waldrep, "The Making of a Border State Society," 772, 774; Finley to Witherspoon, 20 September 1801, in Eslinger, *Citizens of Zion,* 221.

27. Tamat Frankiel, "Ritual Sites in the Narrative of American Religion," in *Retelling U.S. Religious History,* ed. Thomas A. Tweed (Berkeley: University of California Press, 1997), 67; Waldrep, "The Making of a Border State Society," 773.

28. Conkin, *Cane Ridge,* 102; Strickland, ed., *The Autobiography of Rev. James B. Finley,* 166, 366; John Lyle diary, 29 September 1801; Adam Rankin, *Ambivalent Churchmen and Evangelical Churchwomen: The Religion of the Episcopal Elite in North Carolina, 1800–1860* (Columbia: University of South Carolina Press, 1993), 7.

29. Conkin, *Cane Ridge,* 85, 106, 113; John Lyle diary, 5 October 1801; William L. Hiemstra, "Early Frontier Revivalism in Kentucky," *RKHS* 59 (1961): 137–38; Waldrep, "The Making of a Border State Society," 773.

30. Elizabeth Corlis to John Corlis, 1 September 1827, Corlis-Respess Family Papers 1698–1984, FHS; Frances Keller Swinford and Rebecca Smith Lee, *The Great Elm Tree: Heritage of the Episcopal Diocese of Lexington* (Lexington, Ky.: Faith

House Press, 1969), 16–19; Arthur Campbell to Rev. Charles Cummings, 25 December 1802, Kings Mountain Papers, Draper.
31. Eslinger, *Citizens of Zion*, 170; Eric Foner, *Tom Paine and Revolutionary America* (New York: Oxford University Press, 1976), 124; Stanley Elkins and Eric McKitrick, "A Meaning for Turner's Frontier," *Political Science Quarterly* 69 (1954): 325–26; Gordon S. Wood, *The Radicalism of the American Revolution* (New York: Alfred A. Knopf, 1992), 294, 331, 333.
32. Bishop, ed., *Outline of the History of the Church*, 43–44; John Lyle diary, 2 August 1802; William Suddith interview, 12CC62, Draper; Elizabeth Corlis to John Corlis, 1 September 1827; Mary Ann Corlis to John Corlis, 4 September 1827; John Corlis to Francis Corlis, 4 September 1827, all in Corlis-Respess Family Papers; Stephen A. Aron, *How the West Was Lost: The Transformation of Kentucky from Daniel Boone to Henry Clay* (Baltimore: Johns Hopkins University Press, 1996), 186–87.
33. Conkin, *Cane Ridge*, 123, 129–31.
34. John B. Boles, *Religion in Antebellum Kentucky* (Lexington: University Press of Kentucky, 1976), 29; Eslinger, *Citizens of Zion*, 231; Waldrep, "The Making of a Border State Society," 771–72; Aron, *How the West Was Lost*, 175; John F. Schermerhorn and Samuel J. Mills, *A Correct View of that part of the United States which lies west of the Allegany Mountains; with regard to religion and morals* (Hartford, Conn.: Peter B. Gleason and Co., 1814), 18–19.
35. Aron, *How the West Was Lost*, 150–56; Christopher Waldrep, "Opportunity on the Frontier South of the Green," in *The Buzzel about Kentuck: Settling the Promised Land*, ed. Craig Thompson Friend (Lexington: University Press of Kentucky, 1999), 158–60.
36. John Lyle diary, 8 August 1801, 15 May 1803; Conkin, *Cane Ridge*, 122–23; Samuel S. Hill, "Cane Ridge Had a Context: Let's See What They Were," in Dunnavant, ed., *Cane Ridge in Context*, 118; Bishop, *Outline of the History of the Church*, 353.
37. Strickland, ed., *The Autobiography of Rev. James B. Finley*, 166–67.
38. Blair A. Pogue, "'I Cannot Believe the Gospel That Is So Much Preached': Gender, Belief, and Discipline in Baptist Religious Culture," in Friend, ed., *The Buzzel about Kentuck*, 226, 228; Rankin, *Ambivalent Churchmen and Evangelical Churchwomen*, 31–34. South of the Green River, revival-inspired grand juries indicted nearly 1 percent of their counties' populations for blasphemy and profanity; Waldrep, "The Making of a Border State Society," 777–78.
39. Pogue, "'I Cannot Believe the Gospel That Is So Much Preached,'" 233–34; Waldrep, "The Making of a Border State Society," 769–70, 777–78, 779: table.
40. Blair Pogue, "The Feminization of the Early American West? Protestant Evangelical Women, 1780–1860" (paper presented at the Southern Historical Association meeting, 1994), 10.
41. William A. Leavy, "Memoir of Lexington and Its Vicinity," *RKHS* 41 (1943): 117–18; James Fishback, *The Philosophy of the Human Mind, in Respect to Religion; or, A Demonstration from the Necessity of Things That Religion Entered the*

World by Revelation (Lexington, Ky.: Thomas T. Skillman, 1813), 290; James McGready, "The Experience and Privileges of the True Believer," in *The Posthumous Works of the Reverend and Pious James McGready, Late Minister of the Gospel, in Henderson County,* ed. James Smith, 2 vols. (Louisville: n.p., 1831), 1: 146; (Lexington) *Western Monitor,* 26 August 1814; Janet Moore Lindman, "Acting the Manly Christian: White Evangelical Masculinity in Revolutionary Virginia," *WMQ* 57 (2000): 393–416.

42. Bishop, *Outline of the History of the Church,* 353; William W. Woodward, in Hiemstra, "Early Frontier Revivalism in Kentucky," 139. For a sense of the equality afforded blacks in some revivalist congregations, see "Record of Communicants & baptized persons in Caneridge Congregation, Dec. 9, 1811," in John Lyle diary.
43. Bryan's Station Church Book, in Ellen Eslinger, "The Beginnings of Afro-American Christianity among Kentucky Baptists," in Friend, ed., *The Buzzel about Kentuck,* 200–201, 212 n 10; Lucas, *A History of Blacks in Kentucky,* 118; Eslinger, *Citizens of Zion,* 9:30; Heyrman, *Southern Cross,* 222–23.
44. Hatch, *Democratization of American Christianity,* 174.
45. Joshua Wilson's valediction, 1802; *Kentucky Gazette,* 10 October 1805.
46. John D. Wright, *Transylvania: Tutor to the West* (Lexington, Ky.: Transylvania University, 1975), 47–48.
47. David Kaser, *Joseph Charless: Printer in the Western Country* (Philadelphia: University of Pennsylvania Press, 1963), 47, 51; Michael Harris, "Lexington, Ky., 1807" (unpublished manuscript, 1991), 8–9.
48. Alexander Wilson to Alexander Larson, 28 April 1810, in Alexander B. Groshart, ed., *The Poems and Literary Prose of Alexander Wilson, the American Orinthologist,* 2 vols. (Paisley, Scotland: Alexander Gardner, 1876), 1: 195.
49. *Kentucky Gazette,* 19 July 1808.
50. G. Glenn Clift, *The "Corn Stalk" Militia of Kentucky* (Frankfort: Kentucky Historical Society, 1957), iv; Richard G. Stone Jr., *A Brittle Sword: The Kentucky Militia, 1776–1912* (Lexington: University Press of Kentucky, 1976), 12–15; Elizabeth A. Perkins, *Border Life: Experience and Memory in the Revolutionary Ohio Valley* (Chapel Hill: University of North Carolina Press, 1998), 132–41.
51. Stone, *A Brittle Sword,* 34, 38; David Crouch interview, 12CC229, Draper; "An Act more effectually to provide for the National Defence, by establishing an Uniform Militia throughout the United States," 8 May 1792, Second Congress, Sess. 1, Sec. 1.
52. *Kentucky Gazette,* 19 June 1823; "An Act more effectually to provide for the National Defence," Sec. I.
53. E. Anthony Rotundo, *American Manhood: Transformations in Masculinity from the Revolution to the Modern Era* (New York: Basic Books, 1993), 18–20; Steven Watts, *The Republic Reborn: War and the Making of Liberal America, 1790–1820* (Baltimore: Johns Hopkins University Press, 1987), 65; C. Dallett Hemphill, *Bowing to Necessities: A History of Manners in America, 1620–1860* (New York:

Oxford University Press, 1999), chap. 4; "An Oration, Pronounced by Ephraim M. Ewing, a Student in the Transylvania University," *Kentucky Gazette,* 13 November 1810; Amos Kendell, in William Stickney, ed., *Autobiography of Amos Kendall* (Boston: Lee & Shepard, 1872), 126.

54. *Kentucky Gazette,* 30 October 1810; Steven Watts, *The Republic Reborn: War and the Making of Liberal America, 1790–1820* (Baltimore: Johns Hopkins University Press, 1987), 77.

55. John F. Kasson, *Rudeness & Civility: Manners in Nineteenth-Century Urban America* (New York: Hill and Wang, 1990), 19–33; Joseph Ficklin interview, 16CC257, Draper; David Hackett Fischer, *Albion's Seed: Four British Folkways in America* (New York: Oxford University Press, 1989), 382–89; Wood, *The Radicalism of the American Revolution,* 196.

56. James Russell to Thomas Bodley, 10 April 1799, Bodley Family Papers, 1773–1939, FHS; Leavy, "A Memoir of Lexington and Its Vicinity," *RKHS* 40 (1942): 254; Peter Dobkin Hall, *The Organization of American Culture, 1700–1900: Private Institutions, Elites, and the Origins of American Nationality* (New York: New York University Press, 1982), 74.

57. Orval W. Baylor, *John Pope: Kentuckian* (Cynthiana, Ky.: Hobson Press, 1943), 3, 9, 27; John Pope to Ninian Edwards, 9 January 1808, in *The Edwards Papers; Being a Portion of the Collection of the Letters, Papers, and Manuscript of Ninian Edwards,* ed. E. B. Washburn, 3 vols. (Chicago: Chicago Historical Society, 1884), 3: 34. For material evidence of Pope's success in attaining a gentleman's status, see W. Stephen and Kim A. McBride, *Preliminary Archaeological Investigations at the Pope House, Lexington, Kentucky* (Lexington, Ky.: Program for Cultural Resources Assessment, 1991), 16, 32.

58. Joseph C. Cabell to John Breckinridge, 20 May 1800, Papers of the Breckinridge Family, LOC; Wright, *Transylvania,* 44, 52; Henry Clay to William Prentiss, 13 February 1807, in *The Papers of Henry Clay,* ed. James F. Hopkins et al., 10 vols. (Lexington: University of Kentucky Press, 1959–1991), 4: 281; William T. Barry to John Barry, 23 February 1804; William T. Barry to John Barry, 19 March 1806, both in William Taylor Barry Letters, 1798–1835; *Kentucky Gazette,* 16 August 1803.

59. Wood, *The Radicalism of the American Revolution,* 196–97; Rotundo, *American Manhood,* 20; Hemphill, *Bowing to Necessities,* 67.

60. Anya Jabour, *Marriage in the Early Republic: Elizabeth and William Wirt and the Companionate Ideal* (Baltimore: Johns Hopkins University Press, 1998), 3; Linda K. Kerber et al., "Beyond Roles, Beyond Spheres: Thinking about Gender in the Early Republic," *WMQ* 46 (1989): 565–85; Joan Huff Wilson, "The Illusion of Change: Women and the American Revolution," in *The American Revolution: Explorations in the History of American Radicalism,* ed. Alfred F. Young (DeKalb: Northern Illinois Press, 1976), 387, 427–28; Gail S. Terry, "Family Empires: A Frontier Elite in Virginia and Kentucky, 1740–1815" (Ph.D. diss., College of William and Mary, 1992); Claudia Goldin, "The Economic Status of Women

in the Early Republic: Quantitative Evidence," *Journal of Interdisciplinary History* 16 (1986): 402–3; Eliza Todd to Nancy, 9 February 1804, 11CC118, Draper; Rotundo, *American Manhood,* 22–25. Nancy F. Cott discovered a comparable female ethos in New England, beginning somewhat earlier, but definitely arising from dramatic new patterns of economic development; see *The Bonds of Womanhood: "Woman's Sphere" in New England, 1780–1835* (New Haven: Yale University Press, 1977), 1–5.

61. Humphrey Marshall, *The History of Kentucky, Exhibiting An Account of the Modern Discovery; Settlement; Progressive Improvement; Civil and Military Transactions; and the Present State of the Country,* 2 vols. (1812; reprint, Frankfort: George S. Robinson, 1824), 1: 33, 2: iii–iv.

62. Richard L. Bushman, *The Refinement of America: Persons, Houses, Cities* (New York: Alfred A. Knopf, 1992), 186–97; Gordon S. Wood, *The Creation of the American Republic, 1776–1787* (Chapel Hill: University of North Carolina Press, 1969), 421–23; James Blythe, in Neils Sonne, *Liberal Kentucky, 1780–1828* (Lexington: University of Kentucky Press, 1968), 118–23; Lucretia Hart Clay, in Robert V. Remini, *Henry Clay: Statesman for the Union* (New York: W. W. Norton, 1992), 30; *Kentucky Gazette,* 13 October 1800.

63. *Kentucky Gazette,* 9 July 1811, 18 February 1824; Watts, *The Republic Reborn,* 104–5.

64. Drake, *Pioneer Life in Kentucky,* 28; *Kentucky Gazette,* 9 July 1811.

65. Leavy, "Memoir of Lexington and Its Vicinity," *RKHS* 41 (1943): 317; *Kentucky Reporter,* 7 December 1811, 13 July 1813; A. C. Quisenberry, *Kentucky in the War of 1812* (Frankfort: Kentucky State Historical Society, 1915), 155–56, 171–73; Norman K. Risjord, "1812: Conservatives, War Hawks, and the Nation's Honor," *WMQ* 18 (1961): 196–210.

66. *Kentucky Gazette,* 13 November 1810, 18 February 1812, 13 March 1815; "Speech on Proposed Repeal of the Non-Intercourse Act," 22 February 1810, in Hopkins et al., eds., *The Papers of Henry Clay,* 1: 450; David Waldstreicher, *In the Midst of Perpetual Fetes: The Making of American Nationalism, 1776–1820* (Chapel Hill: University of North Carolina Press, 1997), 289; George T. Blakey, "Rendezvous with Republicanism: John Pope vs. Henry Clay in 1816," *Indiana Magazine of History* 62 (1966): 245.

67. *Kentucky Reporter,* 7 December 1811; Harry S. Laver, "Muskets and Plowshares: Kentucky's Militia, the Creation of Community, and the Construction of Masculinity, 1790–1850" (Ph.D. diss., University of Kentucky, 1998), 180; Richard Johnson to James Madison, 24 July 1812, in James A. Padgett, ed., "The Letters of Colonel Richard M. Johnson of Kentucky," *RKHS* 38 (1940): 189; Millersburg, Ky., residents to Isaac Shelby, 31 August 1812, UKSC.

68. James Wincester, in Laver, "Muskets and Plowshares," 187; John Allen to Jane Allen, 19 August 1812, Allen-Butler Family Papers, 1799–1864, UKSC; Thomas Bedford to Elizabeth Kennedy, 12 November 1812, Edwin Green Bedford

Papers, 1812–1902, UKSC; John Lyle to Isaac Shelby, 15 October 1812, John Lyle Papers, UKSC.
69. L. L. Todd to Mary Russell, 14 December 1812, Wickliffe-Preston Papers, UKSC; *Kentucky Gazette,* 3 August 1813, 28 February, 11 July 1814; Linda J. Kerber, *Women of the Republic: Intellect & Ideology in Revolutionary America* (Chapel Hill: University of North Carolina Press, 1980), 111.
70. Quisenberry, *Kentucky in the War of 1812,* 146–47; James Wallace Hammack Jr., *Kentucky & the Second American Revolution: The War of 1812* (Lexington: University Press of Kentucky, 1976), 104–5.
71. Lyrics in Quisenberry, *Kentucky in the War of 1812,* 151–54.
72. Perkins, *Border Life,* 132; Michael Allen, *Western Rivermen, 1763–1861: Ohio and Mississippi Boatmen and the Myth of the Alligator Horse* (Baton Rouge: Louisiana State University Press, 1990), 7–8; (New York) *Courier,* in Moore, *The Frontier Mind,* 86, also chap. 5; Christian Schultz, *Travels on an Inland Voyage,* 2 vols. (New York: Isaac Riley, 1810), 2: 145–46; Elliott J. Gorn, "Gouge and Bite, Pull Hair and Scratch': The Social Significance of Fighting in the Southern Backcountry," *AHR* 90 (1985): 28–29; Stephen Asperheim, "Indian Warfare and Cultural Nationalism in Kentucky, 1790–1795" (paper presented at the Society for Historians of the Early American Republic, 1999).
73. *Kentucky Gazette,* 24 August 1813, 24 February 1826; Cynthia A. Kierner, "Genteel Balls and Republican Parades: Gender and Early Southern Civic Rituals, 1677–1826," *Virginia Magazine of History and Biography* 104 (1996): 209; Francis L. S. Dugan and Jacqueline P. Ball, eds., *Bluegrass Craftsman: Being the Reminiscences of Ebenezer Hiram Stedman, Papermaker 1808–1885* (Lexington: University of Kentucky Press, 1959), 71.
74. Sebastian Derr, "Friends and Countrymen (Bourbon Co.)," 28 July 1814, Political Broadsides Collection, KHS; Gary J. Kornblith, "Self-Made Men: The Development of Middling-Class Consciousness in New England," *Massachusetts Review* 26 (1985): 467–72.
75. Marcus Cunliffe, *Soldiers & Citizens: The Martial Spirit in America, 1775–1865* (Boston: Little, Brown & Co., 1968), 236; Lexington Trustees Minute Book, 7 July 1800, 25 September 1801, 13 February 1812, 17 July 1813, UKSC, microfilm; Robert Breckinridge McAfee, *History of the Late War in the Western Country* (Lexington, Ky.: Worsley & Smith, 1816), 103–4.
76. *Kentucky Gazette,* 16 July 1819; Kierner, "Genteel Balls and Republican Parades," 190; Laver, "Muskets and Plowshares," 23, Appendix A.
77. *Kentucky Gazette,* 9 July 1819; Charles Sellers, *The Market Revolution: Jacksonian America, 1815–1846* (New York: Oxford University Press, 1991), 80.
78. Don Higginbotham, "The Federalized Militia Debate: A Neglected Aspect of Second Amendment Scholarship," *WMQ* 55 (1998): 57; Stone, *A Brittle Sword,* 54.
79. *Kentucky Gazette,* 19 May 1825.

80. G. Glenn Clift, *A History of Maysville and Mason County*, 2 vols. (Lexington, Ky.: Transylvania Printing Co., 1936), 1: 161, 295.
81. Carole Shammas, "The Space Problem in Early United States Cities," *WMQ* 57 (2000): 505–11; Fortesque Cuming, *Sketches of a Tour to the West of the Alleghany Mountains*, in *Early Western Travels, 1748–1846*, ed. Reuben Gold Thwaites, 32 vols. (Cleveland: Arthur H. Clarke, 1904–7), 4: 185. The concept and term of "social zoning" is borrowed from Peter Clark, ed., *The Transformation of English Provincial Towns, 1600–1800* (London: Hutchinson, 1984).
82. See Appendix: Table 9. Lexington Town Lot Ownership, 1802–1804. Tax List, Fayette County, 1804, KDLA; Clay Lancaster, *Vestiges of the Venerable City: A Chronicle of Lexington, Kentucky* (Lexington, Ky.: Lexington–Fayette County Historic Commission, 1978), 11.
83. John Stilgoe, *Common Landscape of America, 1580 to 1845* (New Haven, Conn.: Yale University Press, 1982), 88–99; E. P. Thompson, "The Moral Economy of the Crowd in Eighteenth-Century England," *Past and Present* 50 (1971): 126; Christopher Clark, "The Consequences of the Market Revolution in the North," in *The Market Revolution in America: Social, Political, and Religious Expressions, 1800–1880*, eds. Melvyn Stokes and Stephen Conway (Charlottesville: University Press of Virginia, 1996), 32; T. H. Breen, "Narrative of Commercial Life: Consumption, Ideology, and Community on the Eve of the American Revolution," *WMQ* 50 (1993): 471–501; Thomas Ashe, *Travels in America Performed in 1806* (London: E. M. Blunt, 1808), 191; Lancaster, *Antebellum Architecture*, 126, 131–35, 139–40, 145–46; William B. Scot Jr., "Greek Revival Architecture in Kentucky," *Southern Quarterly* 26 (1987): 102; C. Julian Oberwarth, *A History of the Profession of Architecture in Kentucky* (Louisville, Ky.: Gateway Press, 1987), 3.
84. Matthew Schoenbachler, "The Origins of Jacksonian Politics: Central Kentucky, 1790–1840" (Ph.D. diss., University of Kentucky, 1996), 53; *Kentucky Gazette*, 20 September, 23 August 1803; Robert E. Mutch, "Yeoman and Merchant in Pre-industrial America: Eighteenth-Century Massachusetts as a Case Study," *Societas* 7 (1977): 285.
85. Jessie Roach Davis, comp., "Memoirs of Micah Taul," *RKHS* 27 (1929): 356; Christopher Columbus Baldwin, in Robert Remini, *Henry Clay: Statesman for the Union* (New York; W. W. Norton, 1991), 19; Jackson Turner Main, *The Social Structure of Revolutionary America* (Princeton, N.J.: Princeton University Press, 1965), 203.
86. François André Michaux, *Travels to the West of the Alleghany Mountains*, in Thwaites, ed., *Early Western Travels, 1748–1846*, 203; Dale M. Royalty, "Banking, Politics, and the Commonwealth, Kentucky, 1800–1825" (Ph.D. diss., University of Kentucky, 1971), 6, 9; William Littell, ed., *The Statute Law of Kentucky*, 4 vols. (Frankfort, Ky.: William Hunter, 1811–19), 3: 25–31; Bray Hammond, *Banks and Politics in America from the Revolution to the Civil War* (Princeton, N.J.: Princeton University Press, 1957).

87. Bank of Kentucky Records, Lexington Branch, Minute Book A, 19 August 1811, KDLA; Basil W. Duke, *History of the Bank of Kentucky, 1792–1895* (Louisville, Ky.: John P. Morton & Co., 1895), 14.
88. Royalty, "Banking, Politics, and the Commonwealth," chap. 1; Cuming, *Sketches of a Tour*, 183–84.
89. Wade, *The Urban Frontier*, 109; Tax List, Fayette County, 1808, KDLA; Bank of Kentucky Records, Lexington Branch, Tell-tale Ledger, KDLA; James A. Ramage, "The Hunts and Morgans: a Study of a Prominent Kentucky Family" (Ph.D. diss., University of Kentucky, 1972).
90. Margaretta Brown to John Brown, 31 January 1811, Brown Family Papers, FHS; Elizabeth A. Perkins, "The Consumer Frontier: Household Consumption in Early Kentucky," *JAH* 78 (1991): 486–510; Cuming, *Sketches of a Tour*, 189; *Kentucky Gazette*, 13 October 1800.
91. Cuming, *Sketches of a Tour*, 189; Richard Fowler, *Letters from Lexington and the Illinois 1819*, in Thwaites, ed., *Early Western Travels 1748–1846*, 10: 95; Eliza Todd to Nancy, 9 February 1804, 11CC119, Draper.
92. *Kentucky Gazette*, 18 January, 27 September 1797; 13 October 1800; Thomas Hart, in Aron, *How the West Was Lost*, 133; Lindsey Apple, "Poetry and Politics: The Kentucky Gazette in Verse," *RKHS* 82 (1984): 115–35.
93. Lexington Trustees Minute Book, 1792–1812, UKSC.
94. Michaux, *Travels to the West*, 200; Lexington Trustees Minutes Book, 7 March, 12 July 1808.
95. Groshart, ed., *The Poems and Prose of Alexander Wilson*, 1: 192; Henry Clay to John W. Hunt, 9 May 1812, quoted in C. Julian Overwarth, *A History of the Profession of Architecture in Kentucky* (Louisville, Ky.: Gateway Press, 1987), 8; Lancaster, *Vestiges of the Venerable City*, 24; William Faux, *Memorable Days in America*, part I, in Thwaites, ed., *Early Western Travels, 1748–1846*, 11: 188.
96. *Kentucky Gazette*, 24 January 1798; Mutch, "Yeoman and Merchant in Pre-Industrial America," 291; Lexington Trustees Minute Book, 21 October 1800, 26 December 1808, 4 December 1809, 15 December 1810.
97. See Appendix: Table 10. Means of Purchases by Gender. Norma Basch, *In the Eyes of the Law: Women, Marriage, and Property in Nineteenth-Century New York* (Ithaca, N.Y.: Cornell University Press, 1982); Suzanne Lebsock, *The Free Women of Petersburg: Status and Culture in a Southern Town, 1784–1860* (New York: Norton Books, 1984); Victoria E. Bynum, *Unruly Women: The Politics of Social and Sexual Control in the Old South* (Chapel Hill: University of North Carolina Press, 1992).
98. John W. Hunt daybook 1796–97, UKSC; *Kentucky Gazette*, 4 April 1798; Leavy, "Memoirs of Lexington and Its Vicinity," *RKHS* 41 (1943): 336; Kulikoff, "The Transition to Capitalism," 135; Sonya Charles, "Guardianship, Family Relations, and Gender Roles in Early Kentucky, 1800–1815" (unpublished manuscript, 1994); Jeanne Boydston, "The Woman Who Wasn't There: Women's Market Labor and the Transition to Capitalism in the United States,"

in *Wages of Independence: Capitalism in the Early American Republic,* ed. Paul A. Gilje (Madison, Wis.: Madison House, 1997), 29–30.
99. Perkins, "The Consumer Frontier," 496–97; Benedick Leonard's store ledger (Washington), December 1790–November 1791; John Moylan's store ledger (Lexington), January 1792–November 1794, UKSC; John W. Hunt's daybook (Lexington), July 1790–September 1796, UKSC, microfilm; Daniel Halstead's account books (Lexington), 1806, FHS; William Tureman's daybook, January–May 1807, MCHM; Edmund Martin's journal (Maysville), vol. B, July 1800–February 1811, MCHM.
100. Lucas, *History of Blacks in Kentucky,* 110; Edward Strutt Abdy, *Journal of a Residence and Tour in the United States of North America from April 1833, to October 1834* (London: John Murray, 1835), 347, 353.
101. See Appendix: Table 11. Free Black Population Growth, 1790–1810. Lucas, *History of Blacks in Kentucky,* 112; William Bruce Strother, "Some Aspects of Negro Culture in Lexington, Kentucky" (M.A. thesis, University of Kentucky, 1939), 72–74; Linda Ramsey Ashley and Elizabeth Tapp Wills, eds., *Funeral Notices: Lexington, Ky., 1806–1887* (Rochester, Mich.: n.p., 1982), i; Cuming, *Sketches of a Tour,* 161; R. H. Stanton and E. C. Phister, arrs., *The Charter of the City of Maysville and the Amendments together with the By-laws and Ordinances adopted by the City Council* (Maysville, Ky.: Samuel Pike, 1849), 66; Ira Berlin, *Many Thousands Gone: The First Two Centuries of Slavery in North America* (Cambridge, Mass.: Belknap Press, 1998), 273–74.
102. Eslinger, "The Beginnings of Afro-American Christianity Among Kentucky Baptists," 208–10; Bishop, *An Outline of the History of the Church,* 234.
103. Charles R. Staples, *The History of Pioneer Lexington, 1779–1806* (1939; reprint, Lexington: University Press of Kentucky, 1996), 286; Bishop, *Outline of the History of the Church,* 233.
104. *Biography of London Ferrell, Pastor of the First Baptist Church of Colored Persons* (Lexington, Ky.: A. W. Elder, 1854), 1, 3, 11; Charles Scott to Henry Dearborn, July 1804, Charles Scott Papers, UKSC.
105. Eslinger, "The Beginnings of Afro-American Christianity among Kentucky Baptists," 208–10; William Bruce Strother, "Negro Culture in Lexington, Kentucky" (M.A. thesis, University of Kentucky, 1939), 8; J. Winston Coleman Jr., ed., *Lexington's First City Directory 1806* (Lexington, Ky.: Winburn Press, 1953); *Federal Census, 1810: Kentucky,* Fayette County; Staples, *The History of Pioneer Lexington,* 253–60.
106. *Biography of London Ferrell,* 7; Strother, "Negro Culture in Lexington, Kentucky," 54; Hatch, *The Democratization of American Christianity,* 106–7; Robert Peter, *History of Fayette County, Kentucky,* ed. William Henry Perrin (1882; reprint, Easley, S.C.: Southern Historical Press, 1979), 469–71.
107. Gary B. Nash, "The Social Evolution of Preindustrial American Cities, 1700–1820," in *The Making of Urban America,* ed. Raymond A. Mohl (Wilmington, Del.: Scholarly Resources, 1988), 35.

108. Leavy, "Memoirs of Lexington and Its Vicinity," *RKHS* 41 (1943): 337.
109. Ibid., 323–24.
110. Lancaster, *Vestiges of the Venerable City*, 36–38; Leavy, "A Memoir of Lexington and Its Vicinity," *RKHS* 42 (1944): 36–38; Peter, *History of Fayette County*, 142, 785.
111. Wright, *Transylvania*, 57–58; Sonne, *Liberal Kentucky*, 145–50; "An Act to further Regulate the Transylvania University," 3 February 1818, *Acts Passed at the First Session of the Thirty-Fifth General Assembly for the Commonwealth of Kentucky* (Frankfort: Gerard and Kendall, 1817).
112. Horace Holley to Mary Holley, in Wright, *Transylvania*, 62.
113. Jim Combs, "A Brief History of the Second Presbyterian Church," http://www.2preslex.org/HISTORY.HTM.
114. Tax List, Fayette County, 1804, KDLA.
115. Timothy Flint, *Recollections of the last ten years, passed in occasional residences and journeyings in the valley of the Mississippi, from Pittsburgh to the Gulf of Mexico and from Florida to the Spanish frontier; in a series of letters to the Rev. James Flint, of Salem, Massachusetts* (Boston: Cummings Hilliard & Company, 1826), 67; Patrick Lee Lucas, "It's All Greek to Me: Re-examining the 'Athens of the West' Claim of Lexington, Kentucky, 1820–1829" (M.A. thesis, University of Kentucky, 1998).
116. Alexander Wilson to Alexander Larson, 28 April 1810, in Groshart, ed., *The Poems and Prose of Alexander Wilson*, 1: 195; Cuming, *Sketches of a Tour*, 4: 200; *The Farmers' Almanac* (Lexington, Ky.: William Essex & Son, 1816), UKSC.
117. Aron, "The Poor Men to Starve," 175–93; John Robert Shaw, *A Narrative of the Life and Travels of John Robert Shaw, the Well-Digger* (Lexington, Ky.: Daniel Bradford, 1807), 197–99.

5. CHANGING LANDSCAPES

1. James Flint, *Letters from America*, in *Early Western Travels 1748–1846*, ed. Reuben Gold Thwaites, 32 vols. (Cleveland: Arthur H. Clarke Co., 1904–7), 9: 129.
2. Ibid.; J. Winston Coleman Jr., *Stage-Coach Days in the Bluegrass* (1935; reprint, Lexington: University Press of Kentucky, 1995), 35, 42.
3. Clay Lancaster, *Vestiges of the Venerable City: A Chronicle of Lexington, Kentucky—Its Architectural Development and Survey of Its Early Streets and Antiquities* (Lexington, Ky.: Lexington–Fayette County Historic Commission, 1978), 23; Coleman, *Stage-Coach Days in the Bluegrass*, 48.
4. Richard L. Bushman, *The Refinement of America: Persons, Houses, Cities* (New York: Alfred A. Knopf, 1992), 369.
5. Coleman, *Stage-Coach Days in the Bluegrass*, 163–65.
6. Ibid., 35; Christopher Colles, *A Survey of the Roads of the United States of America 1789*, ed. Walter W. Ristow (Cambridge, Mass.: Belknap Press, 1961),

101; Daniel Drake, *Pioneer Life in Kentucky: A Series of Reminisential Letters*, ed. Charles Drake (Cincinnati: Robert Clarke & Co., 1870), 35; Mary Ann Corlis to Susan Corlis, 15 June 1816, Corlis-Respess Family Papers 1698–1984, FHS.

7. Ellen Eslinger, *Citizens of Zion: The Social Origins of Camp Meeting Revivalism* (Knoxville: University of Tennessee Press, 1999), chap. 3; Daniel Blake Smith, "'This Heaven in Idea': Image and Reality on the Kentucky Frontier," in *The Buzzel about Kentuck: Settling the Promised Land*, ed. Craig Thompson Friend (Lexington: University Press of Kentucky, 1999), 77–98; Edward Pessen, "The Lifestyle of the Antebellum Urban Elite," *Mid-America* 55 (1973): 183; Thomas Ashe, *Travels in America performed in 1806* (London: E. M. Blunt, 1808), 192–93. *Gens comme il faut* literally means "the people that must be"; and *canaille* translates as "the scoundrels."

8. *Kentucky Gazette*, 17 April, 27 April 1793; Jackson Turner Main, *The Social Structure of Revolutionary America* (Princeton, N.J.: Princeton University Press, 1965), 194–95.

9. Ashe, *Travels in America*, 192.

10. Alexander Groshart, ed., *The Poems and Literary Prose of Alexander Wilson, the American Orinthologist*, 2 vols. (Paisley, Scotland: Alexander Gardner, 1876), 1: 193–94. Markethouses served similar symbolic purposes in colonial America; Darrett Rutman and Anita H. Rutman, *A Place in Time: Middlesex County, Virginia 1650–1750* (New York: W. W. Norton, 1984); John J. McCusker and Russel R. Menard, *The Economy of British America, 1607–1789* (Chapel Hill: University of North Carolina Press, 1980), 277–94.

11. Henry Clay to John W. Hunt, 12 September 1810, Hunt-Morgan Papers, UKSC; Bushman, *The Refinement of America*, 357–58; "Travels in Kentucky," in *Travels in the Old South*, ed. Eugene Schwaab, 2 vols. (Lexington: University of Kentucky Press, 1973), 1: 269–70; Thomas Hulme, *A Journal Made During a Tour in the Western Countries of America: September 30, 1818–August 7, 1819*, in Reuben Gold Thwaites, ed., *Early Western Travels 1748–1846*, 10: 67; J. J. Polk, *Autobiography of Dr. J. J. Polk* (Louisville, Ky.: John P. Morton & Co., 1867), 59.

12. *Kentucky Gazette*, 9 April 1802, 16 August 1803, 12 December 1809; François André Michaux, *Travels to the West of the Alleghany Mountains*, in *Early Western Travels, 1748–1846*, ed. Reuben Gold Thwaites, 32 vols. (Cleveland: Arthur H. Clarke, Co., 1904–7), 3: 203–6; Thomas Chapman, "A Journey through the United States," in Schwaab, ed., *Travels in the Old South*, 1: 31; Groshart, ed., *Poems and Literary Prose of Alexander Wilson*, 1: 192; Thomas Hulme, *A Journal Made During a Tour in the Western Countries of America*, in Thwaites, ed., *Early Western Travels, 1748–1846*, 10: 67; Louis B. Wright, *Culture on the Moving Frontier* (Bloomington: Indiana University Press, 1955), esp. chap. 2.

13. Jacob M. Price, "Reflections on the Economy of Revolutionary America," in Ronald Hoffman et al., eds., *The Economy of Early America: The Revolutionary Period, 1763–1790* (Charlottesville: University of Virginia Press, 1988), 309; J. Winston Coleman Jr., ed., *Lexington's First City Directory, 1806* (Lexington,

Ky.: Winburn Press, 1953); "Kentucky Manufactures," in Schwaab, ed., *Travels in the Old South*, 1: 66; John Melish, *Travels through the United States of America in the Years 1806 and 1807, 1809, 1810, and 1811*, 2 vols. (Philadelphia: Thomas and George Palmer, 1812), 402.

14. William Faux, *Memorable Days in America*, part 1, in Thwaites, ed., *Early Western Travels, 1748–1846*, 11: 190; Elizabeth A. Perkins, "The Consumer Frontier: Household Consumption in Early Kentucky," *JAH* 78 (1991): 509.
15. *Kentucky Gazette*, 12 March 1796.
16. William Taylor Barry to John Barry, 26 July 1804, William Taylor Barry Letters 1798–1835, FHS; Bushman, *Refinement of America*, 226; Richard Lee Mason, *Narrative of Richard Lee Mason in the Pioneer West 1819* (New York: Charles Frederick Heartman, 1819), 29; C. Julian Oberwarth, *History of the Profession of Architecture in Kentucky* (Louisville, Ky.: Gateway Press, 1987), 69–70; W. H. Whitley, "A Glimpse of Paris in 1809," *RKHS* 20 (1922): 50, 51, 53, 57.
17. Tench Coxe, *A Statement of the Arts and Manufactures of the United States of America for the Year 1810* (Philadelphia: A. Cornman, 1814), 121–28.
18. Tax Lists for Bourbon County, 1792, 1812, KDLA; Chapman, "A Journey through the United States," 25; Robert Peter, *History of Bourbon, Scott, Harrison, and Nicholas Counties, Kentucky*, ed. William Henry Perrin (1882; reprint, Cincinnati: Art Guild Reprints, 1968), 98.
19. Nancy O'Malley, *A New Village Called Washington* (Washington, Ky.: Old Washington Inc., 1987), 46, 57, 106, 142–46; Michaux, *Travels to the West*, 172; *Federal Census, 1810: Kentucky*, Mason County; Valentine Peers to Eleanor Peers, 3 May 1797, Valentine Peers Correspondence, UKSC; Drake, *Pioneer Life in Kentucky*, 184; Bushman, *Refinement of America*, 110–17.
20. Drake, *Pioneer Life in Kentucky*, 180–81, 186; Tax Lists for Nicholas and Mason Counties, 1807, KDLA.
21. J. Winston Coleman Jr., ed., *Lexington's First City Directory, 1806* (Lexington, Ky.: Winburn Press, 1953), 4–13; *Kentucky Gazette*, 12 December 1809; *Federal Census, 1810: Kentucky*, Mason County.
22. Flint, *Letters from America*, 9: 142.
23. J. Winston Coleman Jr., *Slavery Times in Kentucky* (Chapel Hill: University of North Carolina Press, 1940), 118, 218; Marion B. Lucas, *History of Blacks in Kentucky*, Volume 1: *From Slavery to Segregation, 1760–1891* (Frankfort: Kentucky Historical Society, 1992), 23–24, 42–43; Edward Strutt Abdy, *Journal of a Residence and Tour in the United States*, 2 vols. (London: John Murray, 1835), 2: 348; Drake, *Pioneer Life in Kentucky*, 178–79.
24. (Paris) *Western Citizen*, 24 July 1816; Coleman, *Slavery Times in Kentucky*, 144–45.
25. *Kentucky Gazette*, 29 September 1800; Lexington Trustees Minute Book, 7 August 1800, 25 September 1801, 22 June 1802, UKSC; William Littell, *The Statute Law of Kentucky*, 4 vols. (Frankfort, Ky.: William Hunter, 1811–19), 3: 35–36; Lucas, *History of Blacks in Kentucky*, 59; James Sidbury, *Ploughshares into*

Swords: Race, Rebellion, and Identity in Gabriel's Virginia, 1730–1810 (New York: Cambridge University Press, 1997), chap. 2. Richard C. Wade argued that towns established watches less for the "security of the inhabitants and their property than the control of slaves"; Wade, *The Urban Frontier,* 88. It is difficult to discern how whites differentiated between those goals.

26. Philip D. Morgan, "British Encounters with Africans and African-Americans, circa 1600–1780," in Bernard Bailyn and Philip D. Morgan, eds., *Strangers in the Realm: Cultural Margins of the First British Empire* (Chapel Hill: University of North Carolina Press, 1991), 163–64, 193; Harry Innes, *Executor of Edmund Lyne v. Edmund Lyne's Devisees,* 2 Speed 299, *Kentucky Reports* (1803); *Francis Ratcliff v. Fayette County Court,* 2 Speed 248, *Kentucky Reports* (1803); *Beall v. Joseph (a negro),* 3 Hardin 56, *Kentucky Reports* (1806).

27. David Rice to James Pemberton, 16 January 1790, in Ira Berlin, *Slaves without Masters: The Free Negro in the Antebellum South* (New York: Oxford University Press, 1974), 85; Morgan, "British Encounters with Africans and African-Americans," 164; Lucas, *History of Blacks in Kentucky,* 34, 110; Donald R. Wright, *African Americans in the Early Republic* (Arlington Heights, Ill.: Harlan Davidson, 1993), 81; Stephen Aron, "'The Poor Men to Starve': The Lives and Times of Workingmen in Early Lexington," in Friend, ed., *The Buzzel about Kentuck,* 184; Todd H. Barnett, "Virginians Moving West: The Early Evolution of Slavery in the Bluegrass," *FCHQ* 73 (1999): 233; Leonard P. Curry, *The Free Black in Urban America, 1800–1850: The Shadow of a Dream* (Chicago: University of Chicago Press, 1981), 37–38; *Federal Census 1810: Kentucky,* Fayette, Bourbon, Nicholas, and Mason Counties; Keith Barton, "'Good Cooks and Washer': Slave Hiring, Domestic Labor, and the Market in Bourbon County, Kentucky," *JAH* 84 (1997): 436–60.

    When compared to other cities of the Early American Republic, black communal development in the village West appears anomalous. Gary B. Nash found that in American cities before the 1820s, free blacks lived among enslaved blacks, "formed neighborhoods, established families, and organized churches, schools, and benevolent societies"; Nash, "The Social Evolution of Preindustrial American Cities, 1700–1820," in *The Making of Urban America,* ed. Raymond A. Mohl (Wilmington, Del.: Scholarly Resources, 1988), 29. The only evidence of true African American community along the Maysville Road, however, was Old Captain's church in Lexington.

28. David Hackett Fischer, *Albion's Seed: Four British Folkways in America* (New York: Oxford University Press, 1989), 388–89; John Brand to John W. Hunt, 2 September 1810, Hunt-Morgan Papers, UKSC; Joseph Hornsby diary, 1 January 1803, FHS; Marion Tinling and Godfrey Davis, eds., *The Western Country in 1793: Reports on Kentucky and Virginia* (San Marino, Calif.: The Castle Press, 1948), 132–33; *Kentucky Gazette,* 25 April 1798.

29. Tinling and Davis, eds., *The Western Country in 1793,* 79; Alexis de Tocqueville, *Democracy in America,* ed. J. P. Mayer, trans. George Lawrence (New York: Harper

and Row, 1969), 345–46; Lewis Perry, *Boats against the Current: American Culture between Revolution and Modernity, 1820–1860* (New York: Oxford University Press, 1993), 97–98.
30. Coleman, *Slavery Times in Kentucky,* 101–3.
31. James Darnaby and William Ellis Jr., "A Plat and Survey of the ROAD from LEXINGTON to MAYSVILLE made under the direction of the COMMISSIONERS appointed by An Act of the General Assembly of the Commonwealth of Kentucky," UKSC; Thomas H. Appleton Jr., "An Englishman's Perception of Antebellum Kentucky: The Journal of Thomas Smith, Jr., of Lincolnshire," *RKHS* 79 (1981): 61; Stephen Aron, *How the West Was Lost: The Transformation of Kentucky from Daniel Boone to Henry Clay* (Baltimore: Johns Hopkins University Press, 1996), 141–42.
32. William Leavy, "A Memoir of Lexington and Its Vicinity," *RKHS* 41 (1943): 317–18.
33. Flint, *Letters from America,* 9: 136; Thomas Barton, "Politics and Banking in Republican Kentucky, 1805–1824" (Ph.D. diss., University of Wisconsin, 1968), 234; *The Kentucky Almanac* (Lexington, Ky.: John Bradford, 1819), UKSC; *Kentucky Reporter,* 28 February 1816; Charles Sellers, *The Market Revolution: Jacksonian America, 1815–1846* (New York: Oxford University Press, 1991), chap. 4; Matthew G. Schoenbachler, "The Origins of Jacksonian Politics: Central Kentucky, 1790–1840" (Ph.D. diss., University of Kentucky, 1996), chap. 4; Sandra Van Burkleo, 'The Paws of Banks': The Origins and Significance of Kentucky's Decision to Tax Federal Banks, 1818–1820," *JER* 9 (1989): 457–87.
34. Flint, *Letters from America,* 130–31, 134.
35. William T. Barry to Catherine Barry, 19 November 1819, William Taylor Barry Papers, FHS.
36. *Kentucky Gazette,* 26 April, 9 May 1828; *Louisville Correspondent,* 26 August 1868; Flint, *Letters from America,* 9: 135.
37. George Shannon, in Schoenbachler, "Origins of Jacksonian Politics," 131, 133–37; Flint, *Letters from America,* 9: 238; *Argus of Western America,* 5 November 1819; Dudley Woodbridge Jr. to Samuel and George Trotter, 11 September 1816, in Kim Gruenwald, *River of Enterprise: The Commercial Origins of Regional Identity in the Ohio Valley, 1790–1850* (Bloomington: Indiana University Press, 2002), chap. 4; Patrick Lee Lucas, "It's All Greek to Me: Re-examining the 'Athens of the West' Claim of Lexington, Kentucky, 1820–1829" (M.A. thesis, University of Kentucky, 1998), chap. 4; Aron, "The Poor Men to Starve," 181–83; Robert Ireland, *The County Courts in Antebellum Kentucky* (Lexington: University Press of Kentucky, 1972), 24–26.
38. William Faux, *Memorable Days in America,* part II, in Thwaites, ed., *Early Western Travels, 1748–1846,* 12: 13; *Kentucky Gazette,* 8 July 1824; "To the People of Fayette County," 191.
39. "Soldiers to the Field!!!" Broadside, August 1823, in Schoenbachler, "Origins of Jacksonian Politics," 193.

40. Arndt M. Stickles, *The Critical Court Struggle in Kentucky, 1819–1829* (Bloomington: Indiana University Press, 1929); Sandra Van Burkleo, "'That Our Pure Republican Principles Might Not Wither': Kentucky's Relief Crisis and the Pursuit of 'Moral Justice,' 1818–1826" (Ph.D. diss., University of Minnesota, 1988); Samuel Rezneck, "The Depression of 1819–1822, A Social History," *AHR* 39 (1933): 33, 43–45.
41. Frank F. Mathias, "The Relief and Court Struggle: Half-way House to Populism," *RKHS* 71 (1973): 175; *Kentucky Gazette*, 17 September 1821.
42. Schoenbachler, "Origins of Jacksonian Politics," xx; Sellers, *The Market Revolution*, 169–70; Gordon S. Wood, *The Radicalism of the American Revolution* (New York: Alfred A. Knopf, 1992), 257–58, 342–43; "A Lover of Truth and Justice," 29 January 1825, Old Court–New Court Broadside Collection, KHS; Ronald P. Formisano, *The Transformation of Political Culture: Massachusetts Parties, 1790s–1840s* (New York: Oxford University Press, 1983), 130.
43. Polk, *Autobiography of Dr. J. J. Polk*, 61–63.
44. Francesco Arese, *A Trip to the Prairies and in the Interior of North America*, trans. Andrew Evans (New York: Harbor Press, 1934), 44; Susan J. Yandell to Dr. Wilson Yandell, 22 November 1825, Yandell Family Papers, 1823–1887, FHS; *Western Gazetteer; or Emigrants' Directory* (Auburn, N.Y.: H. C. Southwick, 1817), 94; Horace Holley, in Clara Goode, "Chamiere des Praires," *Journal of American History* 16 (1922): 126; *Niles' Weekly Register*, 11 June 1814; "Travels in Kentucky," in Schwaab, ed., *Travels in the Old South*, 1: 266–67; Joy Carden, *Music in Lexington before 1840* (Lexington, Ky.: Lexington–Fayette County Historic Commission, 1980), chap. 4 and appendix K; Wade, *The Urban Frontier*, 90; Melish, *Travels through the United States*, 1: 404.
45. Frances L. S. Dugan and Jacqueline P. Ball, eds., *Bluegrass Craftsman: Being the Reminiscences of Ebenezer Stedman, Papermaker 1808–1885* (Lexington: University of Kentucky Press, 1959), 75; *Kentucky Gazette*, 11 May 1827; unknown, in Samuel Rezneck, "The Depression of 1819–1822, A Social History," *AHR* 39 (1933): 31, 47.
46. Mary Ann Corlis to John Corlis, 11 April 1816, Corlis-Respess Family Papers; Richard L. Bushman, "Opening the American Countryside," in *The Transformation of Early American History: Society, Authority, and Ideology*, ed., James A. Henretta, Michael Kammen, and Stanley N. Katz (New York: Alfred A. Knopf, 1992), 255; David Meade to Ann Randolph, 1 September 1796, in Bayard Still, ed., "The Westward Migration of a Planter Pioneer in 1796," *WMQ* 21 (1941): 340.
47. Coleman, *Stage-Coach Days in the Bluegrass*, 27; Washington Board of Trustees Act of June 1803, in ibid., 15; Fayette County Court, List of Taverns and Roads 1806, UKSC; *Laws of Kentucky*, 2 vols. (Lexington, Ky.: John Bradford, 1799), 1: 371–77.
48. Donald C. Jackson, "Roads Most Traveled: Turnpikes in Southeastern Pennsylvania in the Early Republic," in *Early American Technology: Making & Doing Things from the Colonial Era to 1850*, ed. Judith A. McGaw (Chapel Hill:

University of North Carolina Press, 1994), 203; Horace Holley to Luther Holley, 22 October 1822, Holley Family Papers, Connecticut Historical Society, Hartford; Dugan and Bull, eds., *Bluegrass Craftsman,* 74; Horace Holley to Mary Holley, 20 July 1824, TU.

49. Samuel M. Wilson, "The Old Maysville Road," in *Proceedings of the Second Annual Meeting of the Ohio Valley Historical Association* (Columbus: Ohio State Archeological and Historical Society, 1909), 101; Kentucky General Assembly, *Journal of the House of Representatives of the Commonwealth of Kentucky, 1809* (Frankfort, Ky.: William Gerard, 1810), 164; "An Act Authorizing a Lottery to improve the Maysville Road, from Maysville to the South end of Washington, in Mason County," *Acts Passed at the First Session of the Nineteenth General Assembly for the Commonwealth of Kentucky* (Frankfort, Ky.: William Garrard, 1811), 145–46.

50. *Kentucky Gazette,* 11 December 1817; "An Act to incorporate the Lexington and Louisville Turnpike Road Company, and to incorporate the Lexington and Maysville Turnpike Company," 4 February 1817, *Acts Passed at the First Session of the Twenty-Fifth General Assembly for the Commonwealth of Kentucky* (Frankfort, Ky.: Gerard & Kendall, 1817), 179; Carol Sheriff, *The Artificial River: The Erie Canal and the Paradox of Progress, 1817–1862* (New York: Hill and Wang, 1996), 5.

51. Robert Peter, *History of Fayette County, Kentucky,* ed. William Henry Perrin (1882; reprint, Easley, S.C.: Southern Historical Press, 1979), 73; "An Act to incorporate the Lexington and Louisville Turnpike Road Company, and to incorporate the Lexington and Maysville Turnpike Company," 179; John R. Stilgoe, *Common Landscape of America, 1580–1845* (New Haven, Conn.: Yale University Press, 1982), 112.

52. Wilson, "The Old Maysville Road," 102; Richard White, *Land Use, Environment, and Social Change: The Shaping of Island County, Washington* (Seattle: University of Washington Press, 1982), 40; *Kentucky Gazette,* 17 November 1826; Adlard Welby, *A visit to North America and the English settlements in Illinois, with a winter residence at Philadelphia: soley to ascertain the actual prospects of the emigrating agriculturist, mechanic, and commercial speculator,* 2 vols. (London: J. Drury, 1821), 1: 213.

53. Cecil Harp and J. Winston Coleman Jr., "The Old Lexington and Maysville Turnpike," *Kentucky Engineer* 1 (1939): 15; Mary Ann Corlis to John Corlis, 21 May 1827, Corlis-Respess Family Papers; "An Act to Incorporate the Maysville and Lexington Turnpike Road Company," 22 January 1827, in *Acts Passed at the First Session of the Thirty-Fifth General Assembly for the Commonwealth of Kentucky* (Frankfort, Ky.: Jacob H. Holeman, 1827), 81–95; Peter, *History of Fayette County,* 73.

54. Appleton, "An Englishman's Perception of Antebellum Kentucky," 62.

55. Susan Yandell to Wilson Yandell, 24 October 1831, Corlis-Respess Family Papers; "Travels in Kentucky," 268.

56. Mann Butler, in Thomas D. Clark, *A History of Kentucky* (Lexington, Ky.: John Bradford Press, 1960), 279; Coleman, *Stage-Coach Days in the Bluegrass,* 132–34; *Laws of Kentucky,* 1: 174–75.

57. Fayette County Court Order, 9 July 1827, in Wilson, "The Old Maysville Road," 104–5.
58. *Western Citizen,* 12 February, 4 April, 19 November 1831; Abdy, *Journal of a Residence and Tour,* 349; Paul C. Heinlein, "Cattle Driving from the Ohio Country," *Agricultural History* 28 (1954): 83–95; *Western Citizen,* 19 November 1831; Charles F. Hoffman, *A Winter in the West: By a New Yorker,* 2 vols. (New York: Harper & Bros., 1835), 2: 158; Schoenbachler, "The Origins of Jacksonian Politics," 228–33. For elaboration on the term "bluestocking," see David S. Shields, *Civil Tongues and Polite Letters in British America* (Chapel Hill: University of North Carolina Press, 1997), 120.
59. Henry Clay, "Speech on Domestic Manufactures," 26 March 1810, in *The Papers of Henry Clay,* eds. James F. Hopkins et al., 10 vols. (Lexington: University of Kentucky Press, 1959–1991), 4: 460; Blair A. Pogue, "'I Cannot Believe the Gospel That Is So Much Preached': Gender, Belief, and Discipline in Baptist Religious Culture," in *The Buzzel about Kentuck: Settling the Promised Land,* ed. Craig Thompson Friend (Lexington: University Press of Kentucky, 1999), 217–42; *Voyage au Kentoukey et sur les bords du Genesée, précédé de conseils aux libéraux, et à tous ceux qui se proosent de passer aux États-Unis* (Paris, France: M. Sollier, 1821), 156.
60. Leonard Curry, "Election Year—Kentucky, 1828," *RKHS* 55 (1957): 196–213; Coleman, *Stage-Coach Days in the Bluegrass,* 184.
61. Peter, *History of Bourbon, Scott, Harrison, and Nicholas Counties, Kentucky,* 429.
62. Joan W. Conley, ed., *History of Nicholas County* (Carlisle, Ky.: Nicholas County Historical Society, 1976), 15–16; William Taylor Barry to Catherine Barry, 23 August 1817, William Taylor Barry Papers.
63. Peter, *History of Bourbon, Scott, Harrison, and Nicholas Counties,* 330; Perry, *Boats against the Current,* 66.
64. Act of 24 January 1829, in Wilson, "The Old Maysville Road," 107.
65. Appleton, ed., "An Englishman's Perception of Antebellum Kentucky," 62.
66. Flint, *Letters from America,* 127; William Faux, *Memorable Days in America,* part I, in Thwaites, ed., *Early Western Travels, 1748–1846,* 11: 185; Zodak Cramer, *The Navigator,* in *Who's Who on the Ohio River,* ed. Ethel C. Leahy (Cincinnati: E. C. Leahy Publishing Co., 1931), 130; G. Glenn Clift, *History of Maysville and Mason County,* 2 vols. (Lexington, Ky.: Transylvania Printing Co., 1936), 1: 170–71.
67. Dugan and Bull, eds., *Bluegrass Craftsman,* 12; Clift, *History of Maysville and Mason County,* 173–74.
68. "An Act to Incorporate the Maysville and Lexington Turnpike Company," 88; Harp and Coleman, "The Old Lexington and Maysville Turnpike," 15–16; Jackson, "Roads Most Traveled," 208; Thomas Metcalfe to Valentine Peers, 17 September 1826, Valentine Peers Collection, KHS; Lewis and Richard Collins, *History of Kentucky,* 2 vols. (Louisville, Ky.: John P. Morton and Co., 1924), 1: 36.

69. *Kentucky Gazette,* 8 August 1826, 18 June, 16 July 1830; *Western Citizen,* 1 July 1831; Aron, *How the West Was Lost,* 133–39; John Lauritz Larson, *Internal Improvements: National Public Works and the Promise of Popular Government in the Early United States* (Chapel Hill: University of North Carolina Press, 2001), 172; Maurice G. Baxter, *Henry Clay and the American System* (Lexington: University Press of Kentucky, 1995), 111–14; Pamela L. Baker, "The Washington Road Bill and the Struggle to Adopt a Federal System of Internal Improvement," *JER* 22 (2002): 456–61.
70. Henry Clay, "Speech at Fowler's Garden," 16 May 1829, in Hopkins et al., eds., *The Papers of Henry Clay,* 2: 53; Journal of John Roche, 22 April 1829, 17CC90, Draper; Henry Clay letter, n.d., KHS; Henry Clay, "Speech to Troops in Georgetown, Kentucky," in *Kentucky Gazette,* 18 August 1812.
71. Peter, *History of Fayette County, Kentucky,* 73; Wilson, "The Old Maysville Road," 104–5; Thomas Metcalfe to Valentine Peers, 17 September 1826.
72. *Western Citizen,* 5 March 1831; Thomas Metcalfe to Valentine Peers, 17 September 1826; Adam Beatty to Valentine Peers, 21 August 1826; An Act to Incorporate the Maysville and Lexington Turnpike Road Company," 81; Matthew Schoenbachler, "'Ill Fares the Land': The Kentucky Bluegrass Region and the Market Economy" (unpublished manuscript, 1994), 10.
73. Nathaniel Cheairs Hughes, ed., *Kentucky Memoirs of Uncle Sam Williams* (Chattanooga, Tenn.: private printing, 1978), 8, 41, 66–67, 70, 71, 77–78, 82; "An Act to Establish a State Bank," 27 December 1806, in Littell, *The Statue Law of Kentucky,* 1: 390–99; Schoenbachler, "The Origins of Jacksonian Politics," 261.
74. Van Burkleo, "'That Our Pure Republican Principles Might Not Wither,'" 238, 284; Hughes, ed., *Kentucky Memoirs of Uncle Sam Williams,* 85–86, 90, 95.
75. Peter, *History of Fayette County,* 73; *Journal of the House of Representatives of the Commonwealth of Kentucky, 1809* (Frankfort, Ky.: William Gerard, 1810), 164; David Waldstreicher, *In the Midst of Perpetual Fetes: The Making of American Nationalism, 1776–1820* (Chapel Hill: University of North Carolina Press, 1997), 292.
76. Robert Wickliffe, in Van Burkleo, "'That Our Pure Republican Principles Might Not Wither,'" 352; Thomas Metcalfe to Valentine Peers, 17 September 1826; Thomas Moore, *To the Citizens of Lincoln, Jessamine, Mercer, and Washington Counties* (Washington: n.p., 1829); Thomas P. Moore to Stephen F. Austin, 28 February 1829, in "The Stephen F. Austin Papers," *Annual Report of the American Historical Association* 3 (1919): 1517.
77. Van Burkleo, "'That Our Pure Republican Principles Might Not Wither,'" 354–57; John Lauritz Larson, "Jefferson's Union and the Problem of Internal Improvements," in *Jeffersonian Legacies,* ed. Peter S. Onuf (Charlottesville: University of Virginia Press, 1993), 360; "Circular to the Citizens of Mason county, Fleming, Nicholas, Bourbon, and Fayette; and the several towns and

counties, binding on or contiguous to the great road leading from Maysville to Lexington," 24 August 1826, Valentine Peers Correspondence, 1789–1848, UKSC; *Kentucky Gazette,* 10 November 1826; Richard Higgins to Valentine Peers, 30 November 1826, Valentine Peers Collection; Michael Zuckerman, "The Fabrication of Identity in Early America," *WMQ* 34 (1977): 111.

78. Peter, *History of Fayette County,* 280; William Hayley petition, in Nancy O'Malley, *A Cultural Evaluation of Archaeological Resources along Paris Pike, Bourbon and Fayette Counties, Kentucky* (Lexington, Ky.: Program for Cultural Resource Assessment, 1987), 32–33; Henry Clay to Nicholas Biddle, 28 November 1829, in Hopkins et al., eds., *The Papers of Henry Clay,* 8: 129–30; Thomas Bodley to William S. Bodley, 2 February, 15 May 1803, Bodley Family Papers, 1776–1939, FHS.
79. Thomas Bodley to William Bodley, 15 May 1830, Bodley Family Papers; Collins, *History of Kentucky,* 540.
80. Hugh S. Bodley to William S. Bodley, 10 June 1830, Bodley Family Papers; Wilson, "The Old Maysville Road," 108; Carter Goodrich, "National Planning of Internal Improvements," *Political Science Quarterly* 63 (1948): 30–31; Daniel M. Jansen, "Andrew Jackson's Maysville Road Veto: A Reappraisal" (M.A. thesis, University of Tennessee, 1992).
81. Henry Adams, *The Life of Albert Gallatin* (New York: Peter Smith, 1943), 350–51; Goodrich, "National Planning of Internal Improvements," 54–56; John Lauritz Larson, "'Bind the Republic Together': The National Union and the Struggle for a System of Internal Improvements," *JAH* 74 (1987): 363–87.
82. Aron, *How the West Was Lost,* 133–39.
83. "Speech on the Roads and Canals Bill," 14 January 1824, *Debates and Proceedings in the Congress of the United States,* 18th Cong., 1st Sess., 999–1000; Robert V. Remini, *Henry Clay: Statesman for the Union* (New York: W. W. Norton, 1991), chap. 13; Drew R. McCoy, "An Unfinished Revolution: The Quest for Economic Independence in the Early Republic," in *The American Revolution: Its Character and Limits,* ed. Jack P. Greene (New York: New York University Press, 1987), 131–48.
84. Daniel Feller, "Politics and Society: Toward a Jacksonian Synthesis," *JER* 10 (1990): 135–61; idem, *The Jacksonian Promise: America, 1815–1840* (Baltimore: Johns Hopkins University Press, 1995), 164; Larson, *Internal Improvements,* 182–83; Edward Pessen, *Jacksonian America: Society, Personality, and Politics* (Urbana: University of Illinois Press, 1985), 124; *Kentucky Gazette,* 28 November 1828; "Jackson's Veto Message," 27 May 1830, UKSC.
85. *Kentucky Gazette,* 4 June 1830.
86. *National Intelligencer,* 18 June 1830; Carlton Jackson, "The Internal Improvement Vetoes of Andrew Jackson," *Tennessee Historical Quarterly* 25 (1966): 261–79.
87. *Kentucky Gazette,* 18 June 1830.
88. Henry Clay to Adam Beatty, 8 June 1830, in Hopkins et al., *The Papers of Henry Clay,* 8: 220–21; Wilson, "The Old Maysville Road," 110; Maurice G. Baxter,

Henry Clay and the American System (Lexington: University Press of Kentucky, 1995), 111–14; Thomas Bodley to Wm. S. Bodley, 20 June 1830, Bodley Family Papers; Leslie Combs to Andrew M. January, 16 February 1831, Leslie Combs Papers, 1793–1881, FHS. Armstrong's response to Combs was written on the back of Combs's letter.

89. Thomas Bodley to Wm. Bodley, 24 Oct. 1831, Bodley Family Papers; Susan Yandell to her father, 24 October 1831, Yandell Family Papers, 1823–1887, FHS.

90. "An Act to Amend an Act to Incorporate Certain Turnpike Road Companies," 22 January 1830, in *Acts of the General Assembly of the Commonwealth of Kentucky Incorporating the Maysville, Washington, Paris, and Lexington Turnpike Road Company* (Maysville: Collins & Co., 1831), sec. 12; Peter, *History of Bourbon, Scott, Harrison, and Nicholas Counties*, 340; Abdy, *Journal of a Residence and Tour*, 2: 340; O'Malley, *A Cultural Evaluation of Archaeological Resources along Paris Pike*, 63.

91. "Travels in Kentucky," 1: 267.

92. Ibid., 1: 285.

93. "An Act to Incorporate the Maysville, Washington, Paris, and Lexington Turnpike Road Company," 29 January 1829, in *Acts of the General Assembly of the Commonwealth of Kentucky Incorporating the Maysville, Washington, Paris, and Lexington Turnpike Road Company*, sec. 17; Coleman, *Stage-Coach Days in the Bluegrass*, 233, 236, 239 n 18; Jackson, "Roads Most Traveled," 230.

94. Coleman, *Stage-Coach Days in the Bluegrass*, 239; Wilson, "The Old Maysville Road," 111.

95. Wade, *The Urban Frontier*, 169; Stuart Seely Sprague, "Town Making in the Era of Good Feelings: Kentucky, 1814–1820," *RKHS* 72 (1974): 337–41; Hector Green to Ellen Green, 4 June 1833, Green Family Papers, 1822–1900, FHS.

96. Francis M. Goddard diary, UKSC; Susan Yandell to David Wendel, 11 June 1833, Yandell Family Papers; Nancy Baird, "Asiatic Cholera's First Visit to Kentucky: A Study in Panic and Fear," *FCHQ* 48 (1974): 230.

97. The 1833 cholera epidemic epitomized the final episode in, as historian William Cronon phrased it, a "complex response of an entire ecosystem—its soil, its vegetation, its animals, its climate—to human actions"; "A Place for Stories: Nature, History, and Narrative," *JAH* 78 (1992): 1373.

98. John H. Ellis, *Medicine in Kentucky* (Lexington: University Press of Kentucky, 1977), 1; *Lexington Observer and Kentucky Reporter*, 11 July 1833; Dr. C. W. Short to the Frankfort *Commonwealth*, 16 June 1833, FHS.

99. *Lexington Observer and Kentucky Reporter*, 15 November 1832; Cholera Broadside, 12 November 1832, UKSC; Dr. Charles H. Caldwell to Dr. George Hayward, 25 November 1832, UKSC.

100. *Lexington Observer and Kentucky Reporter*, 29 May 1833, 2; *Maysville Eagle*, 11 April 1833; Baird, "Asiatic Cholera's First Visit to Kentucky," 230.

101. John A. English, "An Inaugural Dissertation on the Epidemic Cholera, as it appeared in the City of Lexington in June and July 1833" (M.D. thesis,

Transylvania Medical School, 1834), 5; *Lexington Observer and Kentucky Reporter,* 16 May 1833, 8 June 1833; Peter, *History of Fayette County,* 410.

102. Lunsford P. Yandell, "An Account of Spasmodic Cholera, as It Appeared in the City of Lexington," *Transylvania Journal of Medicine* 6 (1833): 198; Susan Yandell to David Wendel, 11 June 1833; Gary O'Dell, "Water Supply and the Early Development of Lexington, Kentucky," *FCHQ* 67 (1993): 432; J. S. Chambers, "The Cholera Epidemic of 1833" (unpublished manuscript), TU, 1; Clay Lancaster, *Vestiges of the Venerable City: A Chronicle of Lexington, Kentucky—Its Architectural Development and Survey of Its Early Streets and Antiquities* (Lexington, Ky.: Lexington–Fayette County Historic Commission, 1978), 48.

103. Samuel McAdow, "On the Epidemic of 1824" (M.D. thesis, Transylvania Medical School, 1825), 1; *Lexington Observer and Kentucky Reporter,* 12 July 1832; R. H. Stanton and E. C. Phister, arr., *The Charter of the City of Maysville and the Amendments together with the By-laws and Ordinances adopted by the City Council* (Maysville, Ky.: Samuel Pike, 1849), 24.

104. C. W. Short to Frankfort *Commonwealth,* 16 June 1833; *Lexington Observer and Kentucky Reporter,* 11 July 1833, 1; "Epidemic Cholera: An Eclectic, Miscellaneous, and Clinical Review," *Western Journal of Medicine* 7 (1833): 91; Rhoda Anderson to George Anderson, 18 June 1833, Anderson Family Papers, UKSC.

105. Abdy, *Journal of a Residence and Tour,* 2: 349; Richard Harrison Shryock, *Medicine and Society in America 1660–1860* (Ithaca, N.Y.: Great Seal Books, 1960), 93–94; Charles W. Short to Frankfort *Commonwealth,* 16 June 1833.

106. *Kentucky Gazette,* 10 August 1833, 1; Robert Davidson, in John B. Boles, *Religion in Antebellum Kentucky* (Lexington: University Press of Kentucky, 1976), 128; Henry Clay to Peter Porter, 2 July 1833, in *The Papers of Henry Clay,* ed. Robert Seager II, 11 vols. (Lexington: University Press of Kentucky, 1984), 8: 650–51; *National Gazette and Literary Register* (Philadelphia), 22 August 1833; J. Winston Coleman Jr., *The Springs of Kentucky* (Lexington, Ky.: Winburn Press, 1955), 33.

107. Micajah and Polly Harrison to Jilson Harrison, 12 July 1833, Harrison Family Papers, KHS; *Maysville Eagle and Monitor Extra,* 31 May 1833; Coleman, *The Springs of Kentucky,* 32–34; idem, *Stage-Coach Days in the Bluegrass,* 76.

108. *Maysville Eagle and Monitor Extra,* 31 May 1833; *O. B.'s Reminiscences: Memories of Old Maysville Between the Years 1832–1848* (Maysville, Ky.: New Republican Print, 1883), 5; A. P. Thompson to Thornton K. Thompson, 24 June 1833, UKSC; *Lexington Observer and Reporter,* 6 June 1833; Thomas Bodley to Wm. S. Bodley, 3 June 1833, Bodley Family Papers, 1776–1939.

109. A. P. Thompson to Thornton K. Thompson, 24 June 1833, UKSC; Peter, *History of Bourbon, Scott, Harrison, and Nicholas Counties,* 100; *Lexington Observer and Kentucky Reporter,* 11 July 1833; Baird, "Asiatic Cholera's First Visit to Kentucky," 230; C. W. Short to William Short, 16 June 1833, Charles Wilkins Short Papers, FHS.

110. Susan Yandell to David Wendel, 7 June 1833; John E. Cooke, "Remarks on Cholera as It Appeared in Lexington, June 1833," *Transylvania Journal of Medicine* 6 (1833): 313; *Lexington Observer and Kentucky Reporter*, 8 June 1833.
111. See Appendix: Table 12. Demographics of Cholera Epidemic in Lexington, 1833. J. S. Chambers, *The Conquest of Cholera* (New York: Macmillan, 1938), 169.
112. Marion B. Lucas, *A History of Blacks in Kentucky*, Vol. 1: *From Slavery to Segregation, 1760–1891* (Frankfort: Kentucky Historical Society, 1992), 13; Leavy, "A Memoir of Lexington and Vicinity," *RKHS* 42 (1944): 28; Todd A. Savitt, "Black Health on the Plantation: Master, Slaves, and Physicians," in *Sickness and Health in America*, ed. Judith Walzer Leavitt and Ronald L. Numbers (Madison: University of Wisconsin Press, 1985), 313–30; Benjamin M. Darnaby, "On Cholera" (M.D. thesis, Transylvania Medical School, 1838), 16; Charles E. Rosenberg, "The Therapeutic Revolution: Medicine, Meaning, and Social Change in Nineteenth-Century America," *Perspectives in Biology and Medicine* 20 (1977): 485–506.
113. Yandell, "An Account of Spasmodic Cholera," 202.
114. Dawson, "The 1833 Cholera Epidemic in Lexington," 41–46.
115. Ibid., 67–69.
116. C. W. Short to Frankfort *Commonwealth*, 18 June 1833; C. W. Short to William Short, 16 June 1833, Charles Wilkins Short Papers, UKSC; *Lexington Observer and Kentucky Reporter*, 11 July 1833; *National Gazette and Literary Register*, 23 July 1833; William Bruce Strother, "Negro Culture in Lexington, Kentucky" (M.A. thesis, University of Kentucky, 1939), 55; *Kentucky Gazette*, 19 October, 22 June 1833.
117. Gayle Kimball, *The Religious Ideas of Harriet Beecher Stowe: Her Gospel of Womanhood* (New York: Edwin Mellon Press, 1982), 205; Harriet Beecher Stowe, *Uncle Tom's Cabin* (1852; reprint, New York: Dodd, Mead, and Co., 1952), 115; Jefferson Davis to George Gohen, 14 June, 30 June 1833, in *The Papers of Jefferson Davis*, ed. Haskell M. Moore Jr. and James T. McIntosh, 10 vols. (Baton Rouge: Louisiana State University Press, 1952–1999), 1: 269–70; James C. Klotter, *The Breckinridges of Kentucky, 1760–1981* (Lexington: University Press of Kentucky, 1986), 96–97; Jean H. Baker, *Mary Todd Lincoln: A Biography* (New York: W. W. Norton, 1987), 40–41.
118. Funeral Notices, 1806–1887, Lexington Public Library, Lexington, Ky.
119. Charles E. Rosenberg, *The Cholera Years: The United States in 1832, 1849, and 1866* (Chicago: University of Chicago Press, 1962), chaps. 1–5; Yandell, "An Account of Spasmodic Cholera," 201; *Kentucky Gazette*, 6 July 1833; "Epidemic Cholera," 91; Horace Hawley to J. Finklin, Charleton Hunt, Dr. Dudley, and Dr. Cooke, 14 June 1833, UKSC; Martin S. Pernick, "Politics, Parties and Pestilence: Epidemic Yellow Fever in Philadelphia and the Rise of the First Party System," in Leavitt and Numbers, eds., *Sickness and Health in America*, 356–71.

120. *Lexington Observer and Kentucky Reporter,* 11 July 1833; *Kentucky Gazette,* 10 August 1833.
121. Goddard diary; Circular "From the Pastor of Christ Church, Lexington to his Parishioners," 2 June 1833, Benjamin Bosworth Smith Papers, UKSC.
122. Rhoda Anderson to George Anderson, 6 June 1833, Anderson Family Papers, UKSC; Susan Yandell to Mrs. David Wendel, 28 July 1833, Yandell Family Papers.
123. Circular, Benjamin Bosworth Smith Papers, UKSC.
124. Peter, *History of Bourbon,* 371–72; Charles G. Talbert, ed., "Looking Backward through One Hundred Years: Personal Recollections of James B. Ireland," *RKHS* 57 (1959): 123.
125. Robert Davidson, *History of the Presbyterian Church in the State of Kentucky* (New York: Robert Carter, 1847), 335; *Lexington Observer and Reporter,* 10 April 1834; Second Presbyterian Church Records 1818–1956, UKSC; George W. Ranck, *History of Lexington, Kentucky* (Cincinnati: Robert Clarke, 1872), 328.
126. Peter, *History of Fayette County, Kentucky,* 411; Samuel W. Price, *The Old Masters of the Bluegrass* (Louisville, Ky.: John P. Morton & Co., 1902), x.

INDEX

Numbers in **bold** refer to illustrations and tables

Abdy, Edward, 205, 228, 265
account-book credit: evidenced in store ledgers, 137, 198, 204, **286** table 7, **288**; merchants' use of, 135–36
"Act Authorizing a Lottery to improve the Maysville Road, from Maysville to the South end of Washington, in Mason County, An" (1811), 240–41
"Act Concerning Public Roads, An" (1797), 71
"Act more effectually to provide for the National Defence, by establishing an Uniform Militia throughout the United States, An" (1792), 179, 180
"Act to Amend the Act Concerning Public Roads" (1801), 71
"Act to Incorporate the Lexington and Louisville Turnpike Road Company, and to incorporate the Lexington and Maysville Turnpike Company, An" (1817), 241
"Act to Incorporate the Maysville and Lexington Turnpike Road Company, An" (1827), 242
"Act to Prevent dealing with Slaves" (1792), 84
"Acts Concerning Towns in this Commonwealth" (1800), 79
Adams, James, 136
Adams, John, 84
African Americans: and cholera epidemic, 273, 274–75, **289**; community among, 35, 36–37, 231–32, 352n27; and market participation, 204–6; and pioneer masculinity, 35; populations of, 230, **288** table 11; and revivalism, 174–75; and slave code, 229–30; and social compact, 84–85; and supernaturalism, 40; and view of West, 35–36, 84–85, 230; westward migration of, 36
Age of Reason, The (Paine), 88
agriculture: corn production, 115–16; cycle of, 116–17; decline of, 248; and domestic productions, 119–20; and environmental changes, 148–49; families and, 35; folk advise about, 154–55; frolics and, 117; and garden ideal, 104; livestock and, 118; and moral economy, 108, 118; and republicanism, 105–7; and slave ownership, 117–18
Alanantowamiowee: cultural significance of, 51; geologic origins of, 10; as old buffalo trace, 2, **6**; as route for white migration, 15; as warpath, 15–16; and Woodland culture, 10. *See also* Limestone Road, Maysville Road, Maysville Turnpike
Alexandria, Va., 136
alligator-horse mythology, 189–90
American Episcopal Church, 86, 169, 208, 211, 279
American Philosophical Society, 149–50
American System, 2, 253, 261
Anderson, Rhoda, 279
Anti–Christian Republican, ideal of, 174
apprenticeship, 128–29, 130, 323n18
architecture: brick genteel home, **227**; evolution of, 71–75, 149; farmhouse, **2**, **113**; frame house, **163**, **226**; gentry view of pioneer, 72, 74–75; in Lexington, 210; log cabin, **75**; in Mayslick, 75; meetinghouse, **160**; middle-class use of, **195**, **212**; in Millersburg, **225**; in Nicholas and Mason Counties, 226–27; in Paris, 223, **224**; station, **62**; taverns, **224**, **250**; at Transylvania University, **213**; in Washington, **226**, **227**
Arminianism: criticized, 86, 89; and purists, 167–68; and revivalism, 160, 162, 167
Armstrong, John, 131, 193, 251, 264
Arnold, Thomas, 223
artisans: and apprenticeship, 128–29; and cloth production, 326n37; conspicuous consumption of, 129–30; ethnic compositions of, 33; land ownership among, **287**; on Lexington board of trustees, 200–202, **286** table 8; Lexington neighborhoods of, 209; moral economy of, 127–29; and Panic of 1819,

363

artisans (cont.)
  233–34; and refinement, 218, 221–22; and rock wall construction, 149; and village West, 124, 127, 129; in Washington, 225. *See also* men of commerce
Asbury, Francis, 44
Ashe, Thomas, 67, 195, 196, 220, 222
Atheism, 89
Aundinaria gigantea, 50, 146
Austin, Moses, 45, 142
Ayres, Godfree, 35
Ayres, Richard, 16
Ayres, Samuel, 35

Backcountry culture. *See* pioneers
Bailey's Station, 74
Baker, William F., 224
Baltimore, Md., 127, 131, 136
Bank of Kentucky, 197–98, 199, 232
Bank of Limestone, 233
Bank of the United States, 232–33
Baptists, 171, 162
Barker, Elihu, "Map of the State of Kentucky, from actual survey," 55, **57**, 58
Barr, Robert, 130
Barr, Thomas, 88
Barrow, David, 88, 148
Barry, William T.: on republicanism, 83–84; on superstition, 87, 167; on environmental changes, 150; as member of middle class, 183, 223, 250; at Transylvania University, 213
barter: evidenced in store ledgers, 137, 204, **286** table 7, **288** table 10; merchants' use of, 135–36
Bartlett & Cox, 136
Bartlett, William, 250
Barton, Hunt & Co., 136
Batson, Robert, 224
Battle of Blue Licks: cultural relevance of, 51, 81, 185; effects on Indians, 15–16; military conflict, 9–10; and Revolutionary imagery, 98, 251, 268
Battle of New Orleans, 188
Battle of the Thames, 80
Battle of Tippecanoe, 186
Baxter, George, 165, 167
Beaden, Jacob, 29
Beargrass Creek settlements, attack on, 15
Bedford, Thomas, 186

Bell, Rebecca, 168
Betts, Raymond, 56
Bibb, George M., 194, 262
Bickham, Gellig & Co., 136
Big Man: civic roles of, 93–94; ideal of, 23, 25–26. *See also* masculinity
Bird, Frank, 36
Bishop, Robert Hamilton, *An Apology for Calvinism*, 176–77
Bison americanus: consumption of, 147–48; densities of, 12; demise of, 147; misidentification of, 294n4; and origins of Alanatowamiowee, 10
Blair v. Williams (1823), 257
Bledsoe, Jesse, 185, 194, 213
Bledsoe, William, 88
Blue Licks, Ky.: depicted as civic disorder, 67; and Maysville Turnpike, 265; origin of name, 51; pioneers in, 62; as resort, 250–51; as sacred space, 21; soil around, 51, 143; springs at, 147, 250–51; use of waters, 154, 271
Blue Licks, Battle of. *See* Battle of Blue Licks
Blue Licks State Battlefield Park, 8
Blue, Rolla, 206, 274
bluegrass, origins of, 147
Bluegrass region, 52–55
Blythe, James, 90, 176, 184, 212
Bodley, Breckinridge, 276
Bodley, Thomas: and cholera epidemic, 276; and Market Street Presbyterian Church, 213; as member of middle class, 182, 195, 210; as proponent of Maysville Turnpike, 259–60, 263; as proponent of railroad, 264; and Transylvania University, 212
Bolingbroke, 88
Book of Common Prayer, 86, 168, 315–16n51
Boone, Daniel: at Battle of Blue Licks, 10; as Big Man, 14, 23, 55, 94; death of, 158; and Indians, 13; as republican Stoic, 18–19; and Wilderness Road, 43
Boonesborough, Ky., 55, 94
Boston, Mass., 138, 151–52
Bourbon County, Ky.: African Americans in, 230, **288** table 11; agriculture in, 229, 282; Cane Ridge revival in, 159; church growth in, 171; division of, 51, 95, 145; grazer system in, 248; landownership in, 107, **284** table 4; manufacturing in, 223–24; populations of, 144, 223, 230; refinement of,

247–48; roadwork in, 259; settlement patterns of, 27; slavery in, 117–18, **285** table 5; social distinctions in, 248; tenancy rates in, 109
Bourbon County Court, and milling, 114
Bourbontown, Ky. See Paris, Ky.
Bowed, A., 279
Brackenridge, Hugh Henry, "The Rising Glory of America," 67
Bradford, Daniel, 201
Bradford, Fielding, 236
Bradford, James, 173
Bradford, John: and Cane Ridge revival, 162; on superstition, 167; as editor of *The Kentucky Almanac*, 20, 97; as editor of *The Kentucky Gazette*, 83, 97, 151; "Fairfield" home of, 195; and garden ideal, 20, 22, 104; as gentry, 157–58; and Lexington markethouse, 55; as republican leader, 97–98; and Transylvania University, 90, 91, 176
Bradford, John Jr., 201
Bradley, Robert, 221
Brand, James, 275
Brand, John: attitude toward slaves, 230; as immigrant, 33; "Rose Hill" home of, 195
Breckinridge, John: "Cabell's Dale" home of, 72, 195; death of, 158; on environmental changes, 142; as gentry, 64, 70, 72, **73**, 83, 119, 181; migration of, 46, 196; in ruling class, 100; and tenancy, 109
Breckinridge, John C., 277
Breckinridge, Joseph C., 93, 213, 214, 277
Breckinridge, Mary Hopkins, 72
Breckinridge, Robert, 109
Brent, Hugh, 256
Brown, James, 91
Brown, John, 92, 158, 197
Brown, Margaretta, 92, 199
Brown, Samuel, 91, 237
Bryan's Station, Ky.: and Bryan family, 29–30; Indian siege of, 9; origins of settlers, 27; as symbolic of pioneers, 53; women in, 24
Bryan's Station Baptist Church, 175
Buckner, William Thomas, "Xalapa" home of, 223
buffalo. See *Bison americanus*
Bullock, Edmund, 212
Bunyan, John, *The Pilgrim's Progress*, 86
Burkley, William, 247

Burr, Aaron, 93
Butler, Mann, 17, 113, 120, 247

Cabin Creek Revival, 162
Caldwell, Charles, 269
Caldwell, William, 223
Calhoun, John C., 255
Calvinism: and Arminianism, 162; defined, 302n68; and Deists, 89; among purists, 38, 40
Campbell, James, 33
Campbell, John, and Big Man ideal, 23
Cane Ridge Presbyterian Meetinghouse, 159–60, **160**
Cane Ridge Revival: attendance at, 159; compared to Green River Revival, 171–72; cosmology of, 268; cultural significance of, 158; as frolic, 168–69; revivalist ethos of, 166–68; as sacred space, 21; and slaveholdings, 207; spatial patterns at, 164; and supernaturalism, 163, 165–66
canebrakes. See *Arundinaria gigantea*
capitalism: and agriculture, 120–23; theorized, 68–69. *See also* market economy
Carey, Matthew, 177
Carlisle, Ky., 249, 265
carriages, 71, 218, 223, 267
cash: and Bank of Kentucky, 197–98; centrality of, 141; devaluation of, 233; evidenced in store ledgers, 136–40, 204, **286** table 7, **288** table 10
Chambers, John: "Cedar Hill" home of, 225; as immigrant, 33
Chapman, Thomas, 55, 130, 151, 223
Charless, Joseph: as bookstore owner, 88, 177, 178, 222; as immigrant, 33
Cherokees, and Kentucky hunting grounds, 13
Chillicothe, Ohio, 14, 131
cholera epidemic of 1833, 267–78, 282, **289**
Christian Mother, ideal of, 174, 177. *See also* femininity
Christian Republican, ideal of, 174, 187. *See also* masculinity
Christmas: contrasting pioneer versions of, 32; gentry version of, 86–87
churches: membership of, 37; purists in, 38–43. *See also* religion; *and specific denominations and churches*
Cincinnati, Ohio: and cholera, 268; education in, 153, 276; and national urban development,

Cincinnati (cont.)
  57; and railroad, 264; and steamboat economy, 241, 252, 267; as village West, 55
Clark, George Rogers, 16
Clark's Mill, 50
Clarke, William, 33
class: as analytical category, 76; defined, 312–13n31; distinctions in, 219–28, 238–39, 248
Clay, Henry: and agriculture, 55; and American System, 2, 249, 253, 261; "Ashland" home of, 157, 195, 196, 214; celebrity of, 280; and cholera epidemic, 272; and family, 129, 180, 248; law office of, 194, **195**; on Lexington courthouse, 202; and Maysville Turnpike, 253–55, 256; and Maysville Turnpike bill veto, 2, 260, 263; as member of middle class, 182, 186, 197, 199, 221, 248, 253–55, **255**; and slavery, 231; and Transylvania University, 182, 212
Clay, Lucretia, 129, 157
Clinkenbeard, William, 150
cloth production, 119–20, 222, 223–24, **285** table 6, 326n37
coaches: described, 218; difficulty on Limestone Hill, 48; travel along Maysville Road, 217, 218
Colburn, John, 130
Coleman, J. Winston, "The Old Lexington and Maysville Turnpike," 2
Collot, Georges-Henri-Victor: career of, 60; describes stations, 60–62; on Lexington, 55; on Limestone, 46; on Limestone Hill, 48; on Mayslick, 50; on Nicholas County, 52; "Road from LIMESTONE to FRANKFORT in the state of KENTUCKY," 60–62, **61**, 65, 242; on Stoner's Fork, 53; view of social evolution, 62–64, 65, 66, 68
Combs, Leslie, 157, 192, 213, 264
commodity money, 115, 119
Concord Presbyterian Meetinghouse: as sister congregation to Cane Ridge Presbyterian Meetinghouse, 160; revival at, 162
Condict, Lewis, 57
conjuring, 40. *See also* supernaturalism
Conover, Betsy, 280
Conover, Billy, 280
Constitution of 1792, 76, 77–78, 84, 258
Constitution of 1799, 72, 84–85, 171, 205

Cooke, John E., 274
Cooper, William, 107
Corlis, Elizabeth, 169, 170
Corlis, George: describes flatboats, 46; and farming, 103–4, 112, 149; home of, **113**; as New Englander, 29
Corlis, John, 103–4
Corlis, Joseph: and farming, 103–4, 112, 149; home of, **113**; and revivalism, 170
Corlis, Mary Ann: and femininity, 83, 248; home of, **113**; on Maysville Road, 242; as member of younger generation, 179; and revivalism, 170; and social distinctions, 238–39
Corlis, Susan, 103–4
corn production, 115–16
Cotterill, R. S.: interpretation of village West, 56; "The Old Maysville Road," 2
countryside: blurred with urban landscape, 58; cholera in, 270, 271; ethnic diversity of, 26
Coyle, Cornelius, 33
Coyle, John, 33
Craig, Joseph, 36
Cramer, Zodak, 135
Crawford, Hugh, 33
Crèvecoeur, J. Hector St. John, view of social evolution, 63–64, 67
Cumberland Gap, 43, 55
Cumberland Road, 254
Cuming, Fortesque: on African Americans, 206; on artisans, 128; on springs at Blue Licks, 154; on frontier mythology, 98; on refinement, 130, 199, 214
Cumming, Kitty, 168

Danville, Ky., 55
Darnaby, James: "A Plat and Survey of the Road from Lexington to Maysville," 242, **243–46**, 267; surveys Maysville Turnpike, 242–52
Darnaby, Ned, 49
Davidson, Billy, 39
Davidson, Robert, 272
Davis, Jefferson, 277
Davis, Thomas T., 132
de Lafayette, Marquis: Fayette County named for, 95; cultural significance of, 238; tours Kentucky, 157–58, 192–93, **194**
de Toqueville, Alexis, 231

debt relief, 235–36
Declaration of Independence, 81, 93, 95, 96
Deism, 88, 89, 174
Delisle, John, 34, 127
democracy: defined, 76, 313n34; and frontier mythology, 76–80
Democrats: and Maysville Turnpike, 258–59; political economy of, 261–62; reaction to Maysville Turnpike bill veto, 262
Democratic-Republican societies, 93. *See also* Lexington Democratic Society
Derr, Sebastian, 190
Dickinson, John, 76
Discovery, Settlement, and Present State of Kentucky, The (Filson), 18
domestic productions, 119–20, 135, 248–49, **285** table 6, 332–33n81
Drake, Abraham, home of, 1, **2**
Drake, Daniel: on environmental changes, 145; on ethnic differences, 28, 30, 31, 226; and family farm, 98, 110, 118; and frontier mythology, 17; on living alongside road, 68; and masculinity, 32; and medical practice, 152–53; migration of, 22; and purists, 41, 42; on religion, 37, 41, 42, 86; on slavery, 228; "Some Accounts of the Epidemic Disease which prevail at Mays-Lick, in Kentucky," 153; on supernaturalism, 39–40, 166; *A Systematic Treatise, Historical, Etiological, and Practical, on the Principal Diseases of the Interior Valley of North America,* 153; and traditional view of nature, 21; at Transylvania University, 153, 212–13
Drake, Elizabeth Shotwell, 31, 42
Drake, Isaac, 16, 31, 117, 145
Drake, John, 39
drovers, 218, 267
dressways, affect of French on, 34, 238
Duncan, Henry, 23, 94, 273
Duncan, Joseph, 223
Duncan Tavern, 223, **224**
Dunkards, 33
Durham, Josiah, 157
Durrett, Peter: death of, 158; as preacher, 36–37, 175, 206, 207–8
Dutch Station, 17

Eagle Tavern, 249
Edmondson, John, 180

Edwards, Jonathan, *Some Thoughts Concerning the Present Revivalism on Religion in New England,* 177
Ellis, James, 17, 249
Ellis, William: "A Plat and Survey of the Road from Lexington to Maysville," 242, **243–46**, 267; surveys Maysville Turnpike, 242–52
Ellis's Station. *See* Ellisville
Ellis's Tavern, 249, **250**
Ellisville, Ky.: as county seat, 60, 249; origin of name, 17, 249
Elkhorn Baptist Association, 207–8
Elkhorn Creek, 53, 207
Elkins, Stanley, 76
English, drawn to village West, 33
English, John, "An Inaugural Dissertation on the Epidemic Cholera, as it appeared in the City of Lexington in June and July 1833," 153
Episcopal Burying Ground, 280
Episcopalians: and African Americans, 208; American Episcopal Church, 86, 169, 208, 211, 279; as Arminians, 86, 89; and Christmas, 32, 86–87; ethos of, 315–16n51; as gentry, 86–87
Eskippakithiki, 12
Essex, William: *The Farmers Almanac,* 155; as bookstore owner, 222
ethnic diversity, among pioneers, 26–37
Ewing, Ephram P., 180, 185–86

families: centrality of farmhouse, 112–13; and domestic productions, 248–49; and farming, 35
Farmers Almanac, The (Essex), 155, 214
Farmers and Mechanics Bank, 233
farming. *See* agriculture, livestock
Farrar, Asa, 48
Faux, William, 202, 222, 234
Fayette County, Ky.: African Americans in, 230, **288** table 11; agricultural fair in, 229; church growth in, 171; landownership in, 107, 112, 214, **284** table 4; origin of name, 95, 96; populations of, 144, 230; refinement of, 247; Revolutionary War veterans in, **283** table 1; settlement patterns of, 27; slavery in, 117–18, 207, **285** table 5; social distinctions in, 180
Fayette County Court: and apprenticeship regulation, 128–29; and inoculation, 151;

Fayette County Court (cont.)
  and licensing of ministers, 87–88; and
  roadwork, 239–40, 247
femininity: pioneer expressions of, 24–25;
  gentry expressions of, 82–83; middle-class
  expressions of, 183; and revivalism, 173–74,
  248; and War of 1812, 187, 190. *See also*
  Christian Mother, Spartan Mother, women
Ferrell, London, 206–8, 276
Ficklin, John, 162
Ficklin, Joseph, 30, 33, 92
Filson, John: *Discovery, Settlement, and Present
  State of Kentucky, The,* 18; "Map of Kentucke,"
  50, 104; and republican Stoicism, 18, 67
Finley, James, 165, 166, 168, 173
Finley, John, 167
Finley, Robert, 159, 160
First Baptist Church, 208, 276
First Presbyterian Church: founded, 41; and
  gentry, 86, 90, 169; and middle class, 208;
  and Washington memorial procession, 81
Fishback, James, 174, 216
Fishette, Michael, 25
flatboats, 46, **47**
Fleming, John, 96
Fleming County, Ky., 95
Fleming's Station, 96
Flint, James: on agriculture, 116; on Christmas,
  32; on cloth production, 120; on Maysville,
  251; on Panic of 1819, 233; on slavery, 228;
  on traveling along Maysville Road, 217
Flint, Timothy, 104, 214
flooding: responses to, 151–52; of Town Creek,
  149, 270
Florence, Ala., 255
Floyd, John, 146
foodways: changing patterns of, 147–48;
  contrasting pioneer versions of, 31–32
Fort Ancients culture, 12
Fort Harrod, 94
Fort Lexington, 9, 53, 75, 94
Fort Washington. *See* Cincinnati, Ohio
Fourth of July. *See* Independence Day
Fowler, Richard, 199
Fox, Arthur, 48
Frankfort, Ky., 55, 205
Franklin, Benjamin, 68, 83
Fraser, Alexander, 33
Fraser, Robert, 33, 127
Free and Easy Club, 89

French, drawn to village West, 34
French Revolution, 34, 59, 60, 97
Freneau, Philip, "The Rising Glory of
  America," 67
frolics: and agriculture, 117; and Backcountry
  pioneers, 22; contrasting pioneer purposes
  for, 30–31; and purists, 37, 39, 42; and
  revivalism, 168–69; and witchcraft, 39

Gallatin, Albert, 260
garden, as ideal, 20, 22
gentry: and African Americans, 84; appropriation of Revolutionary imagery, 94; architecture among, 74–75; characteristics of, 4, 70;
  conflict with middle class, 181, 183–86,
  199–200, 236; differences among, 83–84;
  gender roles among, 80–84; homes of,
  71–75; literacy among, 29; and Maysville
  Road, 71; and Old Light theology, 86; and
  revivalism, 167, 168, 169–71; as ruling class,
  100, 124; and rural elegance, 72–73; suspicions of pioneers, 74–75, 77, 92; view of
  slavery, 84–85
Georgetown, Ky., 55, 94
Georgia, settlement patterns in, **283** table 2
Germans: cultural integrity of, 33; drawn to
  village West, 33; Lexington neighborhood
  of, 209
German Lutheran Church, 33
Germany Station, 17
Giron, Mathurin, 34
God: as punisher, 40, 42, 161–62; and purists, 41;
  reinterpretation of, 166–67; and revivalism,
  165–67; and traditional view of nature, 20
Goddard, Francis, 268, 273, 279
Grand Masonic Lodge, 157
Gratz, Benjamin, 136, 274, 275
Gratz, Michael, 136
grazer system, 248
Great Revival. *See* Cane Ridge Revival
Green, Joseph, 127
Green, Rebecca, 203–4
Green River Revival, 171–72, 341n38
Guadaloupe, 60

Hallack's Inn, 247
Halstead, Daniel, store ledger of, 116, 137, 198,
  204, **285** table 6, **286** table 7
Hamilton, John, 180
Harriman, Herekiah, 162

Harrodsburg, Ky., 55
Harp, Cecil, "The Old Lexington and Maysville Turnpike," 2
Harris, Edward, 26, 48, 109
Harrison, Ann, 35
Harrison, Robert, 35
Harrison, William Henry, 80, 83, 179, 249
Hart, Bartlett, & Cox, 133
Hart, Nathaniel, 113, 114, 129, 133
Hart, Nathaniel G. S., 180
Hart, Thomas Jr., 110, 200, 210
Heckwelder, John, 48
hemp production, 223–24
Herndon's Inn, 247
Higsbee, John, 137
Hildreth, John, "Mt. Airy" home of, 223
Hill, John, 28, 110
Hinkston Creek, 52, 150
Hinkston Exporting Company, 233
History of Kentucky, The (Marshall), 183–84
Hodge, John, 28
Holley, Horace, 212–13, 240
Hopewell, Ky. *See* Paris, Ky.
Hornsby, Joseph, 117, 230
Hostetter, Joseph, 33
housing. *See* architecture
Houston, Peter, 21
Howe, Abram, 33
Hulme, Thomas, 129
Hume, David, 88
Hunt, Abijah, 131
Hunt, John W.: "Hopemont" home of, 195, 210; as merchant, 130, 131, **133**, 135–36; store ledger of, 137, 139–40, 198–99, 203, 204, **285** table 6, **286** table7; view of slaves, 230
Hunt, William, 125
Hunt, Wilson, 134
Hunter of Kentucky, ideal of, 188–90. *See also* masculinity
"Hunters of Kentucky; or the Battle of New Orleans, The" (Woodsworth), 188–89
hunting, 19

Imlay, Gilbert, 9, 18–19, 51
Importance of Family Religion (unknown), 177
"Inaugural Dissertation on the Epidemic Cholera, as it appeared in the City of Lexington in June and July 1833, An" (English), 153

Independence Day, 92–93, 94, 97–98, 99, 121, 184–85
Indians: attack on Beargrass Creek settlements, 15; attack on Bryan's Station, 11, 15–16; attack on Martin's Station, 15; attack on Ruddle's Station, 15; attack on Strode's Station, 15; depicted as savages, 19, 81; Fort Ancients culture of, 12; Mississippian culture of, 12; multinational villages of, 13; Woodland culture of, 10. *See also specific nations*
Indian Creek Revival, 162
Indiana, settlement patterns in, **283** table 2
Innes, Harry, 77
Inquiry into the Nature and Causes of the Wealth of Nations (Smith), 69
internal improvements: Henry Clay and, 253–55; and political economy, 253–59
Ireland, James, 38, 40
Irish: Lexington neighborhood of, 209; drawn to village West, 33
Irish Station, 17, 75
Iroquois, and Ohio River valley, 12
Irwin, Thomas, 125, 126, 130

Jackson, Andrew: and Battle of New Orleans, 188; as president, 83, 180; tours Kentucky, 191–92; vetoes Maysville Turnpike bill, 2, 260–62
January, Andrew, 264, 273
Jefferson, Thomas: and Declaration of Independence, 81; as Deist, 92; describes sectional differences, 27; as president, 93; and internal improvements, 260; and Kentucky Resolutions, 72; literature of, 88; as republican theorist, 66, 67, 76, 96; and western development, 64, 77, 91, 133
Jefferson County, Ky., 110
Jeffersonian-Jacobin literature, sale of, 88
Johnson, Richard, 186
Johnston, John: home of, **226**; as New Englander, 32, 225
Johnston's Fork, 43, 50
Jones, Cyrus Parker, 206, 277
Jordon, John, 130
Julian, Charles: and agricultural ideal, 105, **106**, 108, 118, 122; and cultivation of corn, 115

Keen's Tavern, 192
Keene, John, 158

Kendall, Amos, 180–81
Kennedy, James, 221
Kennedy, Matthew, 196, 213
Kenton, Simon, 23, **24**, 46, 48–49
Kenton's Station, 17, 46
Kentucke, as Native American place-name, 18
Kentucky: landownership in, 284; settlement patterns in, 283; slave ownership in, 285
Kentucky Almanac, The: and agroecological practices, 154–55; and commercial ethos, 232; denounces superstition, 87; and garden ideal, 20; and natural medicines, 154
Kentucky Court of Appeals, 145, 257
Kentucky Gazette, The: and Cane Ridge Revival, 162; and cholera epidemic, 276; denounces superstition, 87; land advertised in, 107; and President Monroe's tour, 191; as republican newspaper, 83, 97–98, 104, 111, 184
Kentucky general assembly: charters Bank of Kentucky, 197–98; charters Kentucky Insurance Company, 197; and Old Court–New Court struggle, 235–36; and place-naming, 95–96; and poverty relief, 111–12; and roadwork, 25, 71, 240–42, 257; and tavern rates, 247; and toll rates, 266–67
Kentucky Hotel, 221
Kentucky Insurance Company, 136, 196, 197
Kentucky land office, 19
Kentucky Resolutions, 72
Kentucky Supreme Court, 323n18
Key, Marshall, 277
King, Nancy, 39, 40, 42, 168
Krinkle, Francis, 33

La Chaumière des Prairies, 73
LaFayette Female Academy, 157
land: distribution of, 107–8, 112, **284** table 4; values of, 108, 194, 198, 209, **287**
Lane Academy, 276
Langhorne, Maurice, 193
Latrobe, Benjamin Henry, 196
lawyers, 196–97
Leavy, William, 130, 185, 208–9, 232
Lee's Creek, 43, 50
Lee's Mill, 50
Leonard, Benedict, store ledger of, 137, 204, **286** table 7, 328n46
Levi, Julius, 193

Lexington, Ky.: African Americans in, 205, 230, **288** table 11; artisans in, 127, 149, 218, 233–34, 238; Big Men in, 94; cholera in, 268–69, 270, 271, 272, 274–76, 277–78, 289; churches in, 178; as county seat, 53; courthouse in, 201–2; Democratic-Republican society in, 93; depicted as civic order, 67; diversity of population, 33–34, 208; economic decline of, 232, 252; flooding in, 149, 151–52, 270; and Lafayette's tour, 192; manufacturing in, 222, 223; markethouse in, 55, 137–40, 202–3, 206, 220–21, 221–22, 275, 277; as marketplace, 56, 131, 271; and middle class, 193–216; and national urban development, 57–58, 158–59, 193, **284** table 3; neighborhoods in, 208–9, **211**, 193–95, 213–14; origin of name, 53, 94, 96; origins, 94; and Panic of 1819, 233–34; population of, 53–55, 151–52, 193, 230; and promotion of Kentucky, 115; property values in, 108, 194, 198, 209, **287**; and President Monroe's tour, 191–92; and railroad, 259, 264; refinement of, 55, 193–96, 237–38; as resort, 221, 269–70; revivals in, 162; and state capital, 55; settlement patterns of, 27; social zoning in, 208–14; spatial patterns of, 53, **54**; station at, **62**; suffrage in, 79
Lexington and Maysville Turnpike Road Company, 240–42
Lexington board of trustees: and African Americans, 207–8, 229; and cholera, 276, 278; and markethouse, 55, 137–40, 202–3, 237; members of, 200–201, **286** table 8; in memorial procession, 81; and roadwork, 259
Lexington Democratic Society, 72, 132
Lexington Female Academy. *See* LaFayette Female Academy
Lexington Medical Society, 183
Licking River, 51, 265
Licking River Revival, 162
Limestone, Ky., as settlers' point of arrival, 46. *See also* Maysville
Limestone Creek, 143
Limestone Hill: roadwork on, 240–41, 257; soils atop, 48, 143; travel up, 48
Limestone Road: becomes known as, 17; condition of, 49; and pioneer masculinity, 25–26; as urban corridor, 55, 57. *See also*

Alanantowamiowee, Maysville Road, *See* Maysville Turnpike
livestock, 118–19, 148–49
log cabins: architecture of, **75**, 311n24; contrasted to log houses, 72; contrasted to genteel homes, 75
longhunters, and Kentucke hunting grounds, 13
Lord Dunmore's War, 80
Louisiana Purchase, 133
Louisville, Ky., 55, 94, 110, 241, 267, 278
lower Shawnee Town, 14
Luckey, John, 159
Lyle, John, 111, 162
Lyle, John T.: at Cane Ridge Revival, 165, 168, 172; as preacher, 90, 170; relationship to father, 111, 162; and War of 1812, 187

MacAdam, John, 253
macadamization, 253
Maccoun, David, 232
Maccoun, James, 232
Madison, James: and internal improvements, 260–61; as president, 186; as republican theorist, 66, 67, 99; and western development, 64
"Map of Kentucke" (Filson), 50, 104
"Map of the State of Kentucky, from actual survey" (Barker), 55, **57**, 58
market economy: and agriculture, 116, 120–23; and markethouses, 137–42; and Mississippi River, 131–34; and slavery, 118; and social evolution, 68–69; and tenancy, 109–10. *See also* capitalism
Market Street Episcopal Church, 196
Market Street Presbyterian Church, 208, 211, 213, 280
markethouses: in Lexington, 55, 137–42, 200, 202–3, 220–21, 237; in Maysville, 252; moral economy of, 139–42; and republicanism, 138–39; in Washington, 140
Marshall, Humphrey, *The History of Kentucky*, 183–84
Marshall, Robert, 172
Marshall, Thomas, "The Hill" home of, 225, **227**
Martin, Edmund: store ledger of, 137, 198, 204, **286** table 7; tavern of, 249
Martin, Henry, 269
Martin, John, 256

Martin's Station, attack on, 15
Martin's Tavern, 249
masculinity: Big Man ideal of, 22–26; contrasting pioneer versions of, 32; gentry ideal of, 80–82; middle-class ideals of, 180–81, 183; and militias, 179–80; and parades, 81–82; and revivalism, 172–73, 174; and roadwork, 25–26; and War of 1812, 186–87, 188–90. *See also* Big Man, Christian Republican, Hunter of Kentucky
Mason, George, 95
Mason, Peter, 129
Mason, Richard Lee, 223
Mason County, Ky.: African Americans in, 230, **288** table 11; Backcountry culture in, 30, 251; church growth in, 171; county seat of, 46, 49; division of, 145; as draw for New Englanders, 32; Dunkards in, 33; landownership in, 107, **284** table 4; merchant-millers in, 116; origins of, 51, 95; populations of, 144–45, 226, 230; refinement of, 224–27; revival in, 162; Revolutionary War veterans in, **283** table 1; roadwork in, 240–41, 257; settlement patterns of, 27; slavery in, 117–18, 227, **285** table 5
Mason County Court, and roadwork, 25
Maxwell, John, 33, 36, 41, 207
May, John, 17, 45, 46
May, William, 17, 50
Mayslick, Ky.: architecture of, 75; artisans in, 227; cholera in, 273; contrasting pioneer cultures in, 30–32; depicted as civic disorder, 67; and natural environment, 152–53; neighborliness in, 112; origin of name, 17; origins of settlers, 27, 30, 50; population of, 50
Maysville, Ky.: African Americans in, 205, **288** table 11; artisans in, 227; Big Men in, 93–94; cholera in, 268, 269, 272–73; commercial appeal of, 46–47; as county seat, 46; depicted as civic order, 67; environmental changes in, 148, 150; Lafayette's visit to, 193, **194**; fire in, 269; flooding in, 149; as jerkwater, 48; markethouse in, 252; as marketplace, 56, 136; and Maysville Turnpike, 259, 262; origin of name, 17, 46; population of, 47; refinement of, 251–52; suffrage in, 79; yellow fever in, 270–71
Maysville and Washington Turnpike Road Company, 257, 259

Maysville Road: conditions of, 142, 217; gentry purpose for, 71; improvement of, 3; populations along, 144–45; and socioeconomic distinctions, 218–19; as symbol of nature, 241–42. *See also* Alanatowamiowee, Limestone Road, Maysville Turnpike

Maysville board of trustees: and African Americans, 84; and Committee of Public Health, 271

Maysville Turnpike: Henry Clay and, 253–55; incorporated, 241–42; marker, 1, **3**; and political economy, 253–59; roadwork on, 252–53, 254, 264–65; surveyed, 242–52, **243–46**; tolls along, 265–67; veto of bill, 260–62. *See also* Alanatowaniowee, Limestone Road, Maysville Road

McAdow, Samuel, "On the Epidemic of 1824," 153, 270–71

McAfee, John Breckinridge, 191

McAfee, Robert, 88

McCalla, Andrew, 130

McCalla, John: as member of middle class, 210; "Mount Hope" home of, 196, **212**

McChord, James, 213

McConnell, William, 94–95, 96, 147

McConnell's Station: attack on, 15; and environmental changes, 147–48; origin of name, 17

McGee's Station, 75

McGready, James, 174

McGregor, Alexander, 33

McIlvain, Hugh, 239

McKee, James, home of, 225

McKitrick, Eric, 76

McNair, John, 130

McNemar, Richard, 172

Meade, David: and architectural distinctions, 72, 195; and environmental changes, 142, 143, 146, 149; as gentry, 237–38; "La Chaumière des Prairies" home of, 73; and land purchases, 107; migration of, 45, 48, 71; and rural elegance, 73–74; and village West, 67, 239

Medical College of Ohio, 153

Megowan, Robert, 41

Megowan, Stewart, 180

Megowan's Tavern, 180

Melish, John, 108, 238

men of commerce: as identifiable community, 4; on Lexington's board of trustees, 137–38, **286** table 8; and flooding, 151–52; and hiring of architects, 195–96; and markethouses, 137–42; and Maysville Turnpike, 253–59, 264; and moral economy, 126–27; and Panic of 1819, 233–35; and poverty, 111–12; as ruling class, 124, 130; and village West, 124. *See also* artisans, merchants

Mentelle, Charlotte, 34

Mentelle, Waldemarde, 34

Mentelle's for Young Ladies, 34

merchants: and account-book credit, 135, 137, 198, 204, **286** table 7; as Backcountry elite, 130–31; and Bank of Kentucky, 199; and barter, 135–36, 137, 204, **286** table 7; and cash, 136–40, 204, **286** table 7; and domestic productions, 120, 135; and importation and exportation, 131, **132**, 132–34, 136, 141; and Kentucky Insurance Company, 197; land ownership among, **287**; on Lexington board of trustees, 137–38, 151–52, **286** table 8; and markethouses, 137–42, 200; and moral economy, 120–23, 125–26, 131, 134–35, 139–42; and Panic of 1819, 235; and village West, 124, 130–31. *See also* men of commerce

merchant-manufacturers, 222

Meredith, Elizabeth Preston, 71

Metcalfe, Charles, 256

Metcalfe, Thomas: "Forest Retreat" home of, 249; as inn keeper, 265; as proponent of Maysville Turnpike, 253, 255, 258

Methodists, church growth among, 171

Michaux, François André, 28, 35, 49, 115, 118, 134, 145, 197, 225

middle class: conflict with gentry, 181, 183–86, 199–200, 236; generational nature of, 178–80; homes in Lexington, 195–96; as lawyers, 196–97; Lexington neighborhood of, 210–13, **211**; and Market Street Presbyterian Church, 213; and masculinity, 180–81, 182–83, 185; and militias, 180; and social distinctions, 219–28; temperance among, 271–72; and Transylvania University, 212–13; and War of 1812, 183–89, 191–93

militias: and gentry, 92–93, 179–80; and middle class, 180; and pioneers, 25, 92

Miller, John, 17, 52

Miller, Joseph, 224

Miller's Station. *See* Millersburg, Ky.

Millersburg, Ky.: depicted as civic disorder, 67; and environmental changes, 150; land prices in, 108; as marketplace, 56; manufacturing in, 224; and Maysville Turnpike, 259, 262, 265; origin of name, 17, 52; origins of settlers, 27, 52; population of, 52; refinement of, 249; and War of 1812, 186
milling, as moral activity, 114–15
Mississippi River: and regional trade, 135; opening of, 3, 99–100, 121, 122, 132–34
Mississippian culture, 12
Moffitt, Peggy, 168
Monroe, James, 191–92
Montesquieu, 88
Moore, James, 90–91, 176
Moore, Patrick, 125
Moore, Thomas, 258
Mooreland, William, 264
Mooreland Inn, 247
moral economy: and artisans, 127–29; and barter, 135; and community, 4; and domestic productions, 119–20; and farming, 108, 113–14, 118; among gentry, 77–80; and merchants, 120–23, 126–27, 134–35, 139–42; and milling, 114–15
Morris, Deacon, 31
Morrow, Josiah, 89
Morton, William, 130, 176, 182, 195
Mount Zion Presbyterian Church, 41–42, 90
Moylan, John: store ledger of, 136, 198, **286** table 7; store stock of, 139–40
Myer, Melchoir, 33

Narrative of the Life and Travels of John Robert Shaw, the Well-Digger, A (Shaw), 219
Nashville, Tenn., 264
Natchez, Miss., 134
National Republicans: and internal improvements, 249; political economy of, 261; as proponents of Maysville Turnpike, 262, 263
National Road, 2, 254
nationalism. *See* patriotism
nature: destruction of, 142–52, 241–42; and garden ideal, 20, 22; and medicinal practices, 154; meteor shower, 280; and Native Americans, 294n6; as ripe for cholera, 268; supernatural manipulation of, 40; and traditional view, 20–21; as wilderness, 18–20
Neave, Jeremiah, 33

New Chillicothe, Ohio, attack on, 16
New Lights: congregational courts among, 173; and femininity, 173–74; and gentry, 168; and masculinity, 172–73, 174; radicalism among, 169–70; worldview of, 164–65, 167–69, 172. *See also* revivalism
New Orleans, La., 99, 125, 131, 132, 133, 134, 282
New York City, N.Y., 56, 127, 136, 138
Nicholas, George: and constitution of 1792, 76, 77–80, 84; death of, 158; as gentry, **78**, 83, 96, 181; migration of, 196; and suspicion of pioneers, 78; at Transylvania University, 182
Nicholas, Mary, 173
Nicholas County, Ky.: African Americans in, 230, **288** table 11; artisans in, 227; church growth in, 171; landownership in, **284** table 4; and Maysville Turnpike, 259; and meteor shower, 280; origins, 96, 145; populations of, 145, 226, 230; seat of, 249; settlement patterns of, 27; slavery in, 228, **285** table 5
Noel, Loftus, 34
North Licking River, 50, 150

Ohio River: and environmental changes, 148, 150–51; as migration route, 45; as Native American place-name, 18; as trade route, 135, 136; as western route, 45–46
Old Captain. *See* Peter Durrett
Old Court–New Court struggle, 235–36
"Old Lexington and Maysville Turnpike, The" (Harp and Coleman), 2
Old Lights: and purists, 86–89; and supernaturalism, 87
"Old Maysville Road, The" (Cotterill), 2
"Old Maysville Road, The" (Wilson), 2
Olympian Springs, Ky., 221, 272
"On the Epidemic Fever of 1823" (Shackleford), 153
"On the Epidemic of 1824" (McAdow), 153
Overton, James, 206
Overton, Samuel, 207

Paine, Thomas: *The Age of Reason,* 88; denounced, 89; and revivalism, 169
Panic of 1819, 233–34, 261
parades, 81–82
Paris, Ky.: African Americans in, 205, 230, **288** table 11; artisans in, 127, 248; Big Men in,

Paris, Ky.: African Americans in (cont.) 93–94; cholera in, 273–74, 278–79; depicted as civic order, 67; manufacturing in, 223, 256; as marketplace, 56; and Mayville Turnpike, 259; origin of name, 53, 94, 96; origins of, 53; pioneer culture in, 23, 55, 159; population of, 53, 230; refinement of, 53, 223, 247–48; social distinctions in, 248; suffrage in, 79; and village West, 58
Paris Bank, 233
parlors, 74
Parry, Needham, 48
patriotism: and Revolutionary imagery, 94–98; and veterans, 98–99
Patterson, Robert: as Big Man, 23, 94; and Cane Ridge Revival, 162, 165; home of, **75**; and Mount Zion schism, 41; and settlement of Lexington, 51, 53
Peers, Valentine, 53, 119, 225
Pelham, Charles, 193
Perrin, William, 280
Philadelphia, Penn., 56, 59, 130, 131, 136, 138
Phoenix Hotel, 221, 275
Pilgrim's Progress, The (Bunyan), 86
Pindell, Thomas, 210
pioneers: and Biblical allusion, 21; characteristics of, 4, 20–26, 30; contrasting cultures among, 28–37; foodways among, 31–32; gender roles among, 24–26; homestead patterns of, 20; merchants among, 130–31; origin of term, 62–63; purist theology in, 39–43; religious expression among, 31, 168–71, 178; sense of community, 22, 30; view of nature, 20–21; view of society, 22–24
Pittsburgh, Penn., 45, 131
"Plat and Survey of the Road from Lexington to Maysville, A" (Darnaby and Ellis), 242, **243–46**, 267
Point, the, **47**. *See also* Limestone, Ky.
Point Pleasant Revival, 162
Polk, J. J., 99
Pope, John: home of, 195, 196; and Market Street Presbyterian Church, 213; as member of middle class, 182, 194
Pope, William, 182, 214
Postlethwaite, John, 276
Postlethwaite's Hotel. *See* Phoenix Hotel
Postlethwaite's Tavern, 157. *See also* Phoenix Hotel

poverty: and garden ideal, 111; and land prices, 108; in Mason County, 110; relief of, 111–12; and republicanism, 79–80
Presbyterians: church growth among, 171; revivals among, 158, 159, 162
preemption, 19
Prentiss, James, 83
Priestley, Joseph, 90
Price, Samuel, 110, 280
Price, William, 87, 99
Proclamation Line of 1763, 95, 318–19n74
Prosser, Gabriel, 229
Protzman, Lawrence, 33, 53, 96
Providence, R.I., 103
purists: and Arminians, 167–68; and baptism, 43; as critics of gentry, 88; as critics of infidelity, 89–90; and frolics, 37, 39, 42; response to revivalism, 175–78; as scriptural literalists, 38–43, 159; and supernaturalism, 39–41; and *Watt's Psalms*, 41–42
Purviance, David, 159

Randolph, John, 70
Rankin, Adam: and Cane Ridge Revival, 163, 165–66; death of, 158; on ethnic diversity, 27; home of, **163**; as purist, 41–42, 121
Ray's Fork Baptist Church, 38–39, 40, 168
refinement: among artisans, 129–30; and social evolution, 66–67; vernacularization of, 124–25, 222
Relief War, 235–36, 253, 256–57
religion: among African Americans, 36, 37; and cholera, 278–80; contrasting pioneer expressions of, 31; among gentry, 86–89; and pioneer culture, 37–43; polarization of, 161, 164–65; purist version of, 38–43; revivalist version of, 164–65, 166–67, 169, 172. *See also specific denominations and churches*
religious literature, sale of, 177
republicanism: and agriculture, 119, 120–23; and elections of 1820s, 236; and hunting, 19; and localism, 99–100; and markethouses, 138–39; and poverty, 79–80; and preemption, 19; and social evolution, 65–68; and subsistence, 19; at Transylvania University, 91–92; and wilderness, 18–19, 65–66
revivalism: and African Americans, 174–75; attraction of, 158, 162–63; and cholera,

279–80; cultural significance of, 158–59; as disorder, 169–71; and femininity, 248; as frolic, 168–69; and masculinity, 172–73, 174; and purists, 175–78; and reinterpretation of God, 166; and Revolutionary ideals, 167. *See also* New Lights

revivalists. *See* New Lights.

Revolutionary War: as democratic impulse, 40–41, 77–79, 130, 261; enabling capitalism, 69; enabling republicanism, 69; imagery of, 238, 251; and nascent national structure, 96–97; as revivalist impulse, 167, 168; in trans-Appalachia, 94–95, 98; veterans of, 97–99, 157–58, **283** table 1

Rice, David, 37–38, 89, 161, 169, 172, 175, 230

Ridgley, Frederick, 91

"Rising Glory of America, The" (Freneau and Brackenridge), 67

"Road from LIMESTONE to FRANKFORT in the state of KENTUCKY" (Collot), 60–62, **61**, 65

roadwork: in Fayette County, 239–40; legislation of, 25–26; on Limestone Hill, 240–41; on Maysville Turnpike, 252–53; and middle class, 241–42; origins of, 298–99n38; and pioneer masculinity, 25–26

Roche, John, 254

Rogers, James, 239

Ross, James, 127

Ruddle's Station, attack on, 15

rural elegance, ideal of, 66, 68, 72

Rush, Benjamin, 66

Russell, James, 182

Russell, Mary Owen, 187

Ryan's Inn, 247

Salem Presbyterian Church, 90

Samuel, William, 249

Sanders, Lewis: and economic troubles, 232; "Placentia" home of, 195; and Transylvania University, 212

Sanders' Manufacturing Company, 233

Satan, 20, 41, 89, 165–67

Savary, John, 34

Sayre, David, 129

Schatzell, John Peter, 33

Scotch Station, 17

Scots: drawn to village West, 33; Lexington neighborhood of, 209

Scott, Charles, 179, 207

Scott, Jefferson, "New Forest" home of, 223

Scott, Joseph, 49

Scott, Patrick, 115, 159, 161

Seitz, John A., 125, 130

settlement patterns: group migrations, **27**; of pioneers, 20

Shackleford, John, "On the Epidemic Fever of 1823," 153

Shaw, John Robert: injuries to, **215**; *A Narrative of the Life and Travels of John Robert Shaw, the Well-Digger*, 219; success of 215–16

Shawnees: and alcohol, 14–15; alliance with British, 15; at Battle of Blue Licks, 10; Chillicothe, 14; and Christianity, 14–15; Eskippakithiki, 12; forced migration of, 12; as imagined threat, 16–17; and Kentucke hunting grounds, 13; lower Shawnee Town, 14; multinational villages, of, 13; New Chillicothe, 16; recolonization of Ohio River valley by, 12–13; settlement map of, **16**

Shedil, John, 33

Shelby, Isaac, 80

Shelby, Susan Hart, 80

Shipp, Laban, 114–15, 119

Short, Charles W., 52, 268, 272, 274, 276

Shryock, Mathias, 196

slavery: and cholera epidemic, 274–75; and farming, 117–18; gentry form of, 84–85; middle-class form of, 228–32; pioneer form of, 35, 85; and religion, 159–60, 161–62

Sloo, Thomas, 135–36

smallpox, 151–52

Smelzer, Peter, 33

Smith, Adam, *An Inquiry into the Nature and Causes of the Wealth of Nations*, 69

Smith, James, 159, 161

Smith, John, 33

Smith, William, 247

social compact: ideal of, 68; contrasted to pioneer ideal of community, 76; implemented in constitution of 1792, 79; revised in constitution of 1799, 84–85

social evolution: Enlightenment views of, 62–69; and urbanization, 67–68

social zoning, 208–14, **209**, 274

soils: around Blue Licks, 51, 143; of Bluegrass region, 52; fertility markers of, 143–44,

soils (cont.)
    146–47; hardening of, 150; around Johnston's Creek, 50; south of Licking River, 52; atop Limestone Hill, 48, 143; around Washington, 49
Solomon, William, 280–81, **281**
"Some Accounts of the Epidemic Diseases which prevail at Mays-Lick, in Kentucky" (Drake), 153
Some Thoughts on the Present Revivalism on Religion in New England (Edwards), 177
South Kentucky Baptist Association, 36
Spangter, John, 139
Spartan Mother, ideal of, 187
Spears, Jacob, "Stonecastle" home of, 223
Springle, Jacob, 33, 201
Springle, John, 33
St. Andrew's Society, 33
St. Genevieve, Ill., 131
St. John's River, 189
St. Louis, Mo., 134
stagewagons, 217
Standeford, Nathan, 223
stations: described, 60–62; as origins of village West, 56; and pioneer identity, 17–18, 60; map of, **16**; as symbols of white culture, 15. *See also specific stations*
steamboats, 252
Stedman, Ebenezer: on artisans, 127, 128; on condition of roads, 240; on de LaFayette's visit, 238; as New Englander, 29, 32; on steamboats, 252
Steele, Robert, 41
Stilfield, Simon, 275
Stone, Barton: criticized, 176; describes Cane Ridge Revival, 164, 170, 172; organizes camp meeting, 162; as preacher, 160
Stone, Edward, 229
Stoner's Fork, 52, 114
Stoney Creek, 249
Strode's Station, attack on, 15
stores, purchases in, **286** table 7, **288** table 10
Stout, Benjamin, 137
Stowe, Harriet Beecher: *Uncle Tom's Cabin*, 277; in Washington, 276–77
Straws, Hannibal, 275
suffrage: and constitution of 1792, 79–80; in village West, 79

supernaturalism: at Cane Ridge revival, 163, 165–67; correlation to economic growth, 303n71; as traditionalism, 39–41
Sutton, David, 196
Systematic Treatise, Historical, Etiological, and Practical, on the Principal Diseases of the Interior Valley of North America, An (Drake), 153

Taylor, Samuel, 112
Taylor, Zachary, 277
Templin, Tereh, 167
tenancy, 108–11, 323n13
Terasse, Henri, 34
theology: purist, 38–43; in First Presbyterian Church, 41
Thompson, A. P., 273
Todd, Andrew, 160–61
Todd, Eliza, 183, 199
Todd, Levi: death of, 158; describes canebrakes, 52; "Ellerslie" home of, 74, 195; and Fayette County land titles, 112; and rural elegance, 74; and Transylvania University, 176
Todd (Lincoln), Mary, 277
Todd, Robert S., 213, 277
Tombigbee District, Miss., 64
Toqueville, Alexis de. *See* de Toqueville, Alexis
Toulmin, Harry: on artisans, 127; *A Description of Kentucky*, 64; and European view of West, 34; as immigrant, 33; on poverty, 110; on property values, 108; on slavery, 230, 231; as traditionalist, 121; as a Unitarian, 87–88, 90; view of social evolution, 64, 66; and yeoman agricultural ideal, 105, 112, 118, 122, 123, 138, 141, 321n8
Town Creek: and cholera, 274; flooding of, 151–52, 201, 207; and markethouse, 237
traditionalism: and nature, 20; among New Lights, 165–69; origins of, 20; and social construction, 5; and supernaturalism, 39–41
Transylvania University: 1806 seal of, **177**; as arena of religious conflict, 90–92, 175–78; board of trustees of, 90; as center of gentry ideology, 91; Main Building, 196, **213**; and middle-class ideology, 182–83; in middle-class neighborhood, 210–13; presidency of, 64, 90–91; professorships at, 90, 153, 182; redefinition of purpose for, 176–77
Transportation Revolution, 3, 292n1

Traveller's Hall. *See* Kentucky Hotel
Treaty of Greenville (1795), 137
Treaty of Paris (1783), 94
trees: as markers of soil fertility, 143–44; and property boundaries, 145–46; utility of, 73, 145
Trenton, N.J., 130
Trotter, George: death of, 158; as editor of the *Kentucky Gazette*, 276; as Democrat, 235, 236, 262, 263; on Lexington board of trustees, 201; as member of middle class, 180, **202**; as merchant, 131, 136, 193, 234
Trotter, James: death of, 158; as merchant, 130; and Mount Zion schism, 41; "Woodlands" home of, 195, 196
Trotter, Samuel: on Lexington board of trustees, 201; as merchant, 88, 131, 136, 193, 234
Trotter & Sons Co., **132**
Tureman, William, store ledger of, 137, 140, 141, 198, 204, 205, **286** table 7

U.S. Constitution, 100
U.S. Engineering Department, 255
U.S. Highway 68, 1, 8
Unitarians, 87, 89, 174
urban development. *See* village West
Usher, Luke, 127

Versailles, Ky., 55, 94
veto of Maysville Turnpike bill, 2, 260–63
Vibrio cholerae. *See* cholera epidemic of 1833
village West: and artisans, 127, 129; as best poor man's country, 215–16; and cholera, 278; as compared to countryside, 58, 214–15; as compared to other Wests, **283** table 2; defined, 55–57, 307n109; as draw for Europeans, 33–34; ethnic diversity of, 26–27; flooding in, 149, 270; and market economy, 69, 123–24; as market for domestic productions, 120; markethouses in, 137–42; and Maysville Turnpike, 256; and merchants, 130–31; and middle class, 193–216, 282; and national urban development, 57–58, **284** table 3; and natural medicines, 154; and opportunity for African Americans, 35–36; and suffrage, 79; as symbol of civilized order, 66–67. *See also specific villages and towns*

Vimont, Louis, 34
Virginia legislature: and Kentucky land distribution, 19, 70; and place-naming, 95, 96; and preemption, 19; and roadwork, 25

W. W. Ater Company, 275
Wallace, John, 108
War of 1812: cultural significance of, 158–59; as distraction from improvements, 260; enthusiasm for, 185–86; and femininity, 187, 190; lessons of, 261; and masculinity, 186–87, 188–90
Warfield, Elisha Sr., 240
Washington: African Americans in, 205, **288** table 11; architecture in, 225; artisans in, 149, 225, 227; Big Men in, 93–94; as county seat, 49; depicted as civic order, 67; and environmental changes, 150; improvements to, 49; and Indian threat, 49; markethouse in, 140; as marketplace, 56; origin of name, 94, 96; origins, 48–49; population of, 49; and promotion of Kentucky, 115; refinement of, 224–27; soils around, 49; suffrage in, 79; as village West, 55, 58
Washington board of trustees, and markethouse, 140
Washington, George, 36, 81
Washington, D.C., 157
waterways: flooding of, 149; inconsistent levels of, 50, 149–50; and pioneer identity, 18. *See also specific creeks and rivers*
Watt's Psalms, and purists, 41–42
weeds, 146–47
Weir, James, 33
Weir, Joshua, 275
Welby, Adlard, 242
Weld, Isaac, 45
Welsh, James: animosity toward James Blythe, 90; as Federalist, 91; and Mount Zion schism, 41; as target of student complaints, 91–92
Wernwag, Lewis V., 253
West, Edward, 201
West, William, 33, 127, 201
Western Conference (Methodist), 171
Western Monitor, The, 174
Whiskey Rebellion, 93
Wickliffe, Robert, 212

widows, 276
Wigglesworth, John, 33
wilderness: destruction of, 142–52; and republicanism, 18–19, 65–66; and savagery, 65; and traditionalism, 20
Wilderness Road, 43–44, 55
Wilkinson, James, 93
Williams, Samuel, 256–57, 267
Williams v. Blair (1822), 257
Wilson, Alexander, 110, 123, 178, 201, 214, 220, 221
Wilson, Joshua, 89, 176
Wilson, Robert, 33
Wilson, Samuel M., "The Old Maysville Road," 2
Winchester, James, 186
Winchester, Va., 256
Winthrop, John, 42
witchcraft: folklore about, 302n69, 39; and superstition, 39; trial, 40–41. *See also* supernaturalism
Wolf, Charles, gravemarker of, **273**

Wollstonecraft, Mary, 88
women: and cholera epidemic, **289**; and market participation, 203–4, **288** table 10; and revivalism, 173–74; and War of 1812, 187, 190. *See also* femininity
Wood, Benjamin, 129
Wood, William, 43, 48, 50
Woodland culture, 10
Woodruff, Ezra, 129
Woodsworth, Samuel, "The Hunters of Kentucky; or, the Battle of New Orleans," 188–89
Wright's Pond, 247
Wyandots, at Battle of Blue Licks, 10

Yandell, Lunsford, 270, 275
Yandell, Susan, 32, 237, 247, 264, 274, 279
yellow fever, 270–71

Zane's Trace, 45
Zanesville, Ohio, 255

www.ingramcontent.com/pod-product-compliance
Lightning Source LLC
Chambersburg PA
CBHW022210090526
44584CB00012BA/379